Mathematical Methods of Physics

Mathematical Methods of Physics

Problems with Solutions

Igor V. Kolokolov | Evgeny A. Kuznetsov
Alexander I. Milstein | Evgeny V. Podivilov |
Alexander I. Chernykh | David A. Shapiro
Elena G. Shapiro

JENNY STANFORD
PUBLISHING

Published by

Jenny Stanford Publishing Pte. Ltd.
101 Thomson Road
#06-01, United Square
Singapore 307591

Email: editorial@jennystanford.com
Web: www.jennystanford.com

British Library Cataloguing-in-Publication Data
A catalogue record for this book is available from the British Library.

Mathematical Methods of Physics: Problems with Solutions

ISBN 978-981-5129-21-2 (Hardcover)
ISBN 978-1-003-51360-5 (eBook)

Contents

Preface to the English Edition

This book is the English translation of a very specific collection of problems on mathematical methods of physics, which has already gone through many editions in Russian. The authors of this book are professional theoretical physicists and have taught a course in methods of mathematical physics at the Physics Department of Novosibirsk State University for many years. Methods of mathematical physics are traditionally considered as a branch of mathematics and are taught by mathematicians. In teaching physics, it is very important to focus on the qualitative analysis of problems and the ability to use a variety of ways to solve them. In addition, from our point of view, specific mathematical methods should be tied to the solution of physical problems, but physical problems should not be tied to the mathematical methods used. Therefore, our book contains a discussion of a very wide range of mathematical methods and their application to a wide variety of physical problems. Each section contains a theoretical introduction and a large number of problems and examples of various levels. Almost all problems contain solutions and instructions for solving them. This makes the book very useful for self-study. Numerous discussions with young researchers, who are currently successfully working in the fields of theoretical physics, plasma physics, hydrodynamics, and others and studied the mathematical methods of physics using our textbook, fully confirmed the correctness of the chosen teaching methodology. We are sure that mastering modern methods of mathematical physics with the help of our book will also be of great benefit to the readers of the English edition.

Preface to the First Edition

In many respects, the teaching style in mathematical physics at the physics departments of many Universities has not changed for a long time. The traditional courses are still focused on the theoretical base and remain outside the applications, such as quantum mechanics, electrodynamics, optics, plasma physics, etc. It is often difficult for students to understand physics because of the absence of sufficient knowledge in mathematical methods. Feeling dissatisfied with the traditional program, we decided to experiment in the Physics Department of the University of Novosibirsk (Russia). Physicists lecture this course and include the additional mathematical techniques required for a better understanding of physics.

This course combines classical mathematical physics, with its applications in quantum mechanics, hydrodynamics, and electrodynamics. For more than 15 years, we have improved and developed the course structure to collect the most exciting problems illustrating the main mathematical approaches of theoretical physics. This book is the result of our efforts in this direction. It reflects our ideas, strategies, and experience. It is not a traditional textbook such as, for instance, the well-known books by Courant & Hilbert or Morse & Feshbach. Our work aims to train readers in solutions to various applied problems in physics, mechanics, engineering, etc. exploiting primary mathematical education. We assume that readers are familiar with the basics of linear algebra, advanced calculus, ordinary differential equations, and the theory of analytic functions.

This book begins with the elements of the theory of linear operators (Chapter 1). It serves as the introduction to the book. It gives a student the mathematical preliminaries from linear algebra. It is devoted mainly to such topics as the completeness condition of the space basis and decomposition of a linear operator over projectors. From the very beginning, we use

Dirac's notations accepted in quantum mechanics. First, we consider the linear operators in finite-dimensional spaces (matrices). Next, we give some prescriptions on how to extend formulas to the infinite-dimensional (Hilbert) space.

Chapter 2 covers the conventional method of characteristics for both homogeneous and inhomogeneous linear partial differential equations of the first order. The discussion accompanies the quasilinear equations on the uniqueness of their solutions. In particular, we study solutions of the Hopf equations for different initial conditions. We also consider linear and quasilinear systems of the first-order equations. The problem of Riemann invariants and the so-called simple Riemann waves is a crucial topic of this chapter that is of importance for many applications in hydrodynamics, kinetics, etc.

Chapter 3 is devoted to the linear equations of the second order. Here, their classifications and canonical forms, the method of separation of variables (the Fourier method) for the elliptic, hyperbolic, and parabolic equations are discussed in detail and illustrated by many examples, both formal and physical. In this chapter, one can find many usual exercises that exploit the Fourier method in numerous physical problems.

We analyze self-similar substitutions and solutions for linear and nonlinear equations in Chapter 4. Though self-similar solutions are partial, they describe many unusual and nontrivial physical processes such as the diffusion, spreading wave packages in quantum mechanics, formation and propagation of weak discontinuities in gas. The self-similar solutions serve very often as asymptotic solutions. In this chapter, we treat a particular class of self-similar solutions, the traveling waves, for the Burgers, Korteweg–de Vries (KdV) equation, and hydrodynamics of polytropic gas. They include a fundamental class of solutions in the form of a soliton or kink. Collapse as a process of a singularity formation in a finite time for smooth initial conditions is also analyzed in this chapter for gas dynamics and nonlinear optics. Another essential question considered in this chapter is solutions of first-order nonlinear equations of the type of Hamilton–Jacobi, well known from classical mechanics.

Chapter 5 deals with special functions as eigenfunctions of the Sturm–Liouville differential operator. This operator, in particular, occurs after the separation of variables for the linear PDEs of the second order. In this chapter, we discuss in detail the properties of the Bessel functions with both integer and half-integer indices, orthogonal polynomials (Hermite,

Legendre, Laguerre) and spherical harmonics, etc. Generating functions, series expansion, recurrence relations, and integral representations are found and examined. The functions selected in this chapter are associated with typical symmetries of physical problems. For example, Bessel functions originate when we separate variables in a system with axial symmetry, the Legendre functions arise from the equations with spherical symmetry. This chapter also includes some aspects of the theory of hypergeometric and confluent hypergeometric functions.

Chapter 6 covers the asymptotic methods, such as the Laplace technique of integration and the stationary phase. We illustrate the saddle point method for integrals along a path in the complex plane by asymptotics of special functions considered in the previous chapter. This chapter also touches upon the averaging process (or multi-scale expansion) for ordinary differential equations by introducing "fast" and "slow" variables. Nowadays, all these methods are traditional tools in many fields of physics, e.g., in plasma or accelerator physics.

Chapter 7 is devoted to Green functions. First, we give a fundamental solution of linear ordinary differential equations for homogeneous boundary conditions. When a linear operator (together with zero boundary conditions) has zero eigenvalues, we introduce the generalized Green function. In the next sections, we build up the Green functions for elliptic, parabolic, and hyperbolic PDE. We also show how the Green functions help to determine the spectrum of linear differential operators.

Chapter 8 covers the integral equations. We analyze the Fredholm and Volterra equations (mainly of the second kind). We also consider the inverse scattering problem for a one-dimensional Schrödinger operator. We choose the most straightforward modification of the inverse scattering problem based on the integral Gelfand–Levitan–Marchenko integral equation. The main exercises here consist of finding the reflectionless potentials and their eigenfunctions.

In the last three chapters, we treat the groups and their representations. First, we introduce the main concepts of finite group theory. Many exercises in Chapter 9 include subgroups, cosets, indices, conjugate classes, group representations, characters, etc. Chapter 10 deals with continuous groups. We emphasize here the Lie groups and their algebras. Since the irreducible representations of the rotation group **SO(3)** are most important for physics, we devote a separate section to them. Chapter 11 includes the applications of group theory in physics. Many examples demonstrate how to use the representations in physical systems, particularly the spectra of a molecule with some symmetry, the degeneracy order of levels, the selection rules in quantum mechanics, the invariant tensors. We discuss the Clebsch–Gordan decomposition, used in solid-state physics or spectroscopy.

Each chapter contains a brief theoretical introduction illustrated by solving typical problems. We outline a mathematical background without proofs of theorems. The bibliography, in which one can find a more in-depth explanation with the proofs, is placed at the end of the book.

More than half of the problems, called the examples, are presented together with detailed solutions. We reserve other problems (exercises) at the end of each chapter for readers on his/her own. We supply them by answers and sometimes with hints or remarks. Ordering problems should help a reader acquire complicated mathematical concepts and new skills in solving physical problems. Therefore we hope this textbook could also be handy for self-education. In our opinion, this way of learning mathematical physics methods is more productive and fruitful than the traditional one. We believe that careful learning of this textbook will substantially improve the readers experience in their future professional activities. If our book helps them to solve physical problems, we consider our mission to be accomplished.

Igor Kolokolov
Evgeny Kuznetsov
Alexander Milstein
Evgeny Podivilov
Alexander Chernykh
David Shapiro
Elena Shapiro
Novosibirsk, 2004

Chapter 1

Linear Operators

1.1 Finite Dimensional Space

Operator $\hat{A}$ acting in the linear space X is called linear when for any two vectors f and g from X and for two arbitrary complex numbers α and β the following property is satisfied[1]:

$$\hat{A}(\alpha f + \beta g) = \alpha \hat{A} f + \beta \hat{A} g.$$

If X is a n-dimensional space, then linear operators can be written as $n \times n$ matrices.

Let A be a square matrix of order n. The matrix $A^\dagger$ obtained by transposing and complex conjugation of A is called *Hermitian conjugate* to A. If $A = A^\dagger$, then such a matrix is called Hermitian. The matrix A is called *unitary* if $AA^\dagger = A^\dagger A = E$ where E is the identity matrix.

Trace $\operatorname{tr} A$ of the matrix A is the sum of its diagonal elements. The number λ is called *eigenvalue* of A if there is a vector $v \neq 0$ such that $Av = \lambda v$, and v is called *eigen vector* of the matrix A corresponding to the eigenvalue λ. Polynomial $\det |\lambda E - A|$ is called the *characteristic polynomial* of matrix A.

A sequence of square matrices of the same order

$$A_1, \ldots, A_m, A_{m+1}, \ldots$$

[1] The operator $\hat{A}$ is called *antilinear* when $\hat{A}\,(\alpha f + \beta g) = \alpha^* \hat{A} f + \beta^* \hat{A} g$, where the asterisk denotes the complex conjugation.

Mathematical Methods of Physics: Problems with Solutions
Igor V. Kolokolov, Evgeny A. Kuznetsov, Alexander I. Milstein, Evgeny V. Podivilov, Alexander I. Chernykh, David A. Shapiro, and Elena G. Shapiro

ISBN 978-981-5129-21-2 (Hardcover), 978-1-003-51360-5 (eBook)
www.jennystanford.com

is called *convergent* to the matrix A if the sequence of the corresponding matrix elements converges.

Obviously, if the matrices A_m and B_m tend to A and B as m increases, then $A_m + B_m$ and $A_m B_m$ tend respectively to $A + B$ and AB. In particular, if T is non-degenerate matrix of the same order n and matrix A_m with increasing m tends to A, then $T^{-1}A_mT$ will tend to $T^{-1}AT$.

Direct sum of matrices is called a block-diagonal matrix

$$A = A^{(1)} \oplus A^{(2)} \oplus \cdots \oplus A^{(s)} = \begin{pmatrix} \boxed{A^{(1)}} & 0 & \ldots & 0 \\ 0 & \boxed{A^{(2)}} & \ldots & 0 \\ \ldots & \ldots & \ldots & \ldots \\ 0 & 0 & \ldots & \boxed{A^{(s)}} \end{pmatrix},$$

where $A^{(j)}$ are square matrices of order $n^{(j)}$, standing along the diagonal of A, and all other elements of A are equal to zero.

Let $a_0 + a_1\lambda + a_2\lambda^2 + \ldots$ be formal power series of the complex variable λ. Consider the corresponding power series in terms of the matrix A:

$$a_0E + a_1A + a_2A^2 + \ldots \tag{1.1}$$

Denote $f_n(A) = a_0E + a_1A + a_2A^2 + \cdots + a_nA^n$. Set (1.1) is called convergent if the sequence of partial sums $f_1(A), \ldots f_n(A), \ldots$ has a limit. This limit is called the sum of the series.

If λ is not equal to any eigenvalue λ_i of matrix A, then the matrix $A - \lambda E$ is invertible. Matrix inverse to it $R_\lambda = (A - \lambda E)^{-1}$ is called the *resolvent*. An integral from matrix means a matrix composed of integrals of matrix elements. Let a be an arbitrary constant that is greater than all $|\lambda_i|$. Then the following equalities are valid:

$$A^m = -\frac{1}{2\pi i} \oint_{|\lambda|=a} \lambda^m R_\lambda \, d\lambda, \quad m = 0, 1, 2, \ldots, \tag{1.2}$$

which follow from the expansion of the resolvent in a series for $|\lambda| \geqslant a$:

$$R_\lambda = -\frac{1}{\lambda}\left(E + \frac{A}{\lambda} + \frac{A^2}{\lambda^2} + \ldots\right).$$

Figure 1.1 shows the contour of integration.

As it follows from (1.2), for any function $f(\lambda)$ defined by a power series for $|\lambda| \leqslant a$ one can write down equality

$$f(A) = -\frac{1}{2\pi i} \oint_{|\lambda|=a} f(\lambda) R_\lambda \, d\lambda.$$

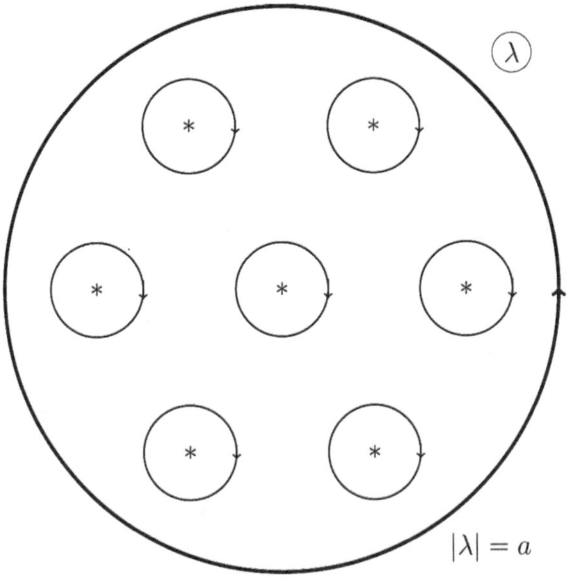

Figure 1.1 Integration loop $|\lambda| = a$ in complex plane of spectral parameter λ. The circle with λ inside denotes this plane, asterisks are poles on the plane.

This formula is a generalization of the Cauchy theorem to the case of functions of matrices.

Now we tighten the integration loop so that it turns into a set of small contours, each of which surrounds only one eigenvalue (Fig. 1.1). Then the formula (1.2) when $m = 0$ gives

$$\sum_j P_j = E, \quad P_j = \frac{1}{2\pi \mathrm{i}} \oint^{(\lambda_{j-})} R_\lambda \, d\lambda,$$

where the symbol (λ_{j-}) denotes the integration loop in a negative direction (clockwise), containing the eigenvalue of λ_j but spanning no other singular points of the integrand. The matrix P_j is the *projection operator (projector)* to a subspace corresponding to an eigenvalue λ_j, i.e.,

$$P_j P_i = P_i P_j = 0 \text{ if } \lambda_i \neq \lambda_j, \quad P_j^2 = P_j.$$

Any linear Hermitian operator A can be decomposed into projectors: $A = \sum_j \lambda_j P_j$ where $\operatorname{Im} \lambda_j = 0$ and $P_j^\dagger = P_j$.

1.2 Functionals and Generalized Functions

Functional acting on the given space $\mathcal{M}$ of functions depending on n variables is the mapping of this space into complex plane

$$\Phi := \{f(x), x \in \mathbb{R}^n\} \to \mathbb{C}.$$

Denote $\Phi[f]$ the value of functional Φ for function $f \in \mathcal{M}$.

The functional is called *linear* if the mapping is linear:

$$\Phi[\alpha f + \beta g] = \alpha\Phi[f] + \beta\Phi[g].$$

Here α, β are complex numbers, f and g are functions.

By means of a given function $F(x)$ one can always construct a linear functional acting on some subset of the functional space

$$\Phi_F[g] = \int_{\mathbb{R}^n} dx\, F(x)g(x). \tag{1.3}$$

However, not every linear functional can be written as (1.3) using only smooth functions $F(x)$. In the general case, it is possible to define a *generalized function* $F(x)$ concerning this functional Φ, so that $\Phi[F]$ expressed by (1.3). If Φ is the limit of the sequence of functionals Φ_n, each of which has the form (1.3) with a smooth function $F_n(x)$, then the generalized function $F(x)$ can be considered as a limit sequences $\{F_n(x)\}$ and some pointwise properties can be attributed to it (as a function of the argument x).

An important example of a generalized function for applications is Dirac δ-function corresponding to a linear functional Φ_δ acting on smooth functions $g(x)$. By definition

$$\Phi_\delta[g] = \int_{\mathbb{R}^n} dx\, \delta(x)g(x) = g(0)$$

for any smooth function g. This equality is equivalent the following definition of a δ function:

$$\delta(x) = 0 \quad \text{at} \quad x \neq 0, \quad \int_{\Omega_0} dx\, \delta(x) = 1,$$

where Ω_0 is arbitrary region of the n-dimensional space $\mathbb{R}^n$ containing the point $x = 0$. Derivatives of one-dimensional δ-functions are defined through functionals

$$\int_{\Omega_0} dx\, \frac{d\delta(x)}{dx} g(x) = -g'(0), \quad \int_{\Omega_0} dx\, \frac{d^n\delta(x)}{dx^n} g(x) = (-1)^n g^{(n)}(0),$$

where function $g(x)$ is assumed sufficiently times differentiable.

One should always remember that equalities containing δ-function and its derivatives mean only equal values of functionals in the case of smooth functions.

1.3 Hilbert Space and Completeness

A linear space $\mathcal{L}$ is called *the Hilbert space* if:

(1) For each pair f and g of its elements, there is defined the *scalar product* (f, g) with values in $\mathbb{C}$, satisfying the following properties:

(a) Linearity concerning the second argument

$$(f, \alpha g_1 + \beta g_2) = \alpha(f, g_1) + \beta(f, g_2)$$

for any $f, g_{1,2} \in \mathcal{L}, \ \alpha, \beta \in \mathbb{C}$;

(b) Hermitian property

$$(f, g) = (g, f)^*;$$

(c) non-negativeness of the norm

$$\|f\|^2 \equiv (f, f) \geqslant 0, \quad \text{and from} \quad (f, f) = 0 \quad \text{follows} \quad f = 0;$$

(2) The space $\mathcal{L}$ has a countable infinite-dimensional basis

$$\{f_n, \ n = 1, \ldots, \infty\},$$

such that any element of $g \in \mathcal{L}$ can be represented as a linear superposition $f_n : \ g = \sum c_n f_n$.

A Hilbert space is the linear (vector) space, therefore, its elements can be called vectors.

Any basis in $\mathcal{L}$ can be turned into *orthonormal* with respect to given scalar product using the Gram–Schmidt procedure. The basic example of a Hilbert space is the space of $L^2(\Omega)$-functions $f(x)$ defined in the region Ω of space $\mathbb{R}^n$ and integrable squared modules. The scalar product of functions f and g

is defined as follows:

$$(f, g) = \int_{\Omega} dx f^*(x) g(x)$$

and satisfies all the conditions. The existence of a countable basis in $L^2(\Omega)$ is proved in the standard mathematical analysis course.

Consider a fixed *orthonormal* in $L^2(\Omega)$ infinite sequence of functions

$$\{f_n, \ n = 1, \ldots, \infty, \quad (f_n, f_m) = \delta_{nm}\}.$$

They form a basis in $L^2(\Omega)$ if and only if the following *completeness relation* holds:

$$\sum_{n=1}^{\infty} f_n^*(x) f_n(y) = \delta(x - y). \tag{1.4}$$

This formula means the equality of generalized functions (of the integrals of the products of both parts with smooth function from $L^2(\Omega)$ or as the limit of a sequence equalities with both sides belonging to $L^2(\Omega)$. The latter is possible because, although the δ-function itself is $\notin L^2(\Omega)$, but in this space, there is a sequence of elements with a limit to the δ-function (see, for example, problems to this chapter).

1.4 Self-Adjoint Operators

Every linear operator acting in space of functions can be written as *integral* operator, i.e.,

$$\hat{A} f(x) = \int K(x, y) f(y)\, dy,$$

where $K(x, y)$ is called *kernel* of the integral operator $\hat{A}$. Kernel $K(x, y)$ can be understood as a generalized function of two variables. In particular, if $K(x, y)$ is a linear combination δ-function and its derivatives $K(x, y) = \sum_{k=0}^{n} c_k \delta^{(k)}(x - y)$, then $\hat{A}$ is nothing more than the *differential* operator of order n.

Consider the linear operator $\hat{A}$ acting in Hilbert space. This means that there is a linear correspondence rule $f \to \hat{A} f$ and *area of definitions* $\mathcal{D} : f \in \mathcal{D}$. The latter area may be smaller than the total space L^2. For example, in $L^2(0, 1)$ (i.e., the space with the integrable square of functions on the closed interval [0,1]), the symbol d/dx can act only on those functions which have the first derivative. A similar statement can be made about d^2/dx^2. Besides, for symbols of differential operators, their definition should include some boundary conditions. Subsequently, when specifying the area of the definitions of $\mathcal{D}$ we will explicitly give only the boundary conditions or conditions equivalent to them. For example, in $L^2(\mathbb{R})$ operator d^2/dx^2 includes the boundary conditions:

$$\frac{d^2}{dx^2}, \quad \mathcal{D} = \{f \in L^2(\mathbb{R})\}. \tag{1.5}$$

In other words, for functions from L^2, defined on the real axis, the boundary condition follows from the quadratic integrability of the function, in particular, this means $f(x) \to 0$ at $x \to \pm\infty$.

The operator $\hat{A}^\dagger$, *conjugate* to the given operator $\hat{A}$, is determined by the equality

$$(v, \hat{A}u) = (\hat{A}^\dagger v, u). \tag{1.6}$$

This formula also fixes the area of definition $\mathcal{D}^*$ for the operator $\hat{A}^\dagger$. The area consists of such vectors v, that for any u from the area $\mathcal{D}$ of the operator $\hat{A}$ the scalar product $(v, \hat{A}u)$ can be rewritten as (w, u). (Therefore, if such w exists, then w is $\hat{A}^\dagger v$.) The area $\mathcal{D}^*$, generally speaking, does not coincide with the area $\mathcal{D}$.

As an example, consider in $L^2(0, 1)$ linear operator id/dx with a domain of definition consisting of functions turning to zero at the ends of the interval:

$$\mathcal{D} = \{u(0) = u(1) = 0\}. \tag{1.7}$$

Scalar product

$$(v, \hat{A}u) = \int_0^1 v^* \, \mathrm{i}\frac{d}{dx} u \, dx = \int_0^1 (\mathrm{i}\frac{d}{dx} v)^* \, u \, dx + \mathrm{i}v^* u|_0^1 = (\hat{A}v, u) \tag{1.8}$$

has the form $(w, u) = (\hat{A}v, u)$ for any v, i.e.,

$$\hat{A}^\dagger = \mathrm{i}d/dx, \quad \mathcal{D}^* = \text{boundary conditions are arbitrary}.$$

Therefore, in this case, the adjoint operator $\hat{A}^\dagger$ has the same correspondence rule as $\hat{A}$ but has a broader domain of definition.

The operator $\hat{A}$ is called *self-adjoint (Hermitian)* if $\hat{A}^\dagger$ coincides with $\hat{A}$ together with the scope. Examples of self-adjoint operators are

(1)

$$\mathrm{i}\frac{d}{dx}, \quad \mathcal{D} = \{f(0) = f(1)\}.$$

(2)

$$\frac{d^2}{dx^2}, \quad \mathcal{D} = \{f(0) = f(1) = 0\}.$$

(3)

$$-\Delta + V(\mathbf{r}), \quad \mathcal{D} = \{f \in L^2(\mathbb{R}^3)\}.$$

In the last example, we assume that for considered functions f the integral

$$\int V(\mathbf{r})|f(\mathbf{r})|^2\, d\mathbf{r}$$

converges.

An operator U is called *unitary*

$$UU^\dagger = U^\dagger U = I, \tag{1.9}$$

where I is the identity operator. Note that Hermitianity and unitarity are defined concerning this scalar product and do not necessarily remain under another definition of the scalar product.

1.5 Ket- and Bra- Vectors

In an abstract Hilbert space consider a mapping of vectors from the Hilbert space into $\mathbb{C}$ satisfying the usual linearity relations. To any vector v of the Hilbert space $\mathcal{H}$, we can associate linear functional as follows:

$$\Phi_v[u] = (v, u).$$

From the definition of $\Phi_v[u]$ follows that functionals Φ_v themselves form a linear space:

$$\Phi_{\alpha v + \beta w} = \alpha^* \Phi_v + \beta^* \Phi_w.$$

In other words, this space is isomorphic to $\mathcal{H}$. This space $\mathcal{H}^*$ is called the space of elements conjugate to the elements $\mathcal{H}$. In the physical literature the elements are called bra-vectors and denote Φ_v as $\langle v|$. The elements $\mathcal{H}$ are called ket-vectors $|u\rangle$, and the scalar product v and u acting as the functional Φ_v on the vector u is written as $\langle v|u\rangle$.

In terms of bra- and ket- vectors

(1) The projector on the vector v can be written as $P_v = |v\rangle\langle v|$.

(2) The completeness condition reads as

$$\sum_n |u_n\rangle \langle u_n| = I.$$

(3) Decomposition of the vector $|v\rangle$ over the complete orthonormal system $|u_n\rangle$ looks like

$$|v\rangle = \sum_n |u_n\rangle \langle u_n|v\rangle .$$

One could find a detailed narration of the Hilbert spaces in the textbooks (Courant and Hilbert, 1989; Kolmogorov and Fomin, 1963; Morse and Feshbach, 1953). Generalized functions and self-adjoint operators are discussed, for instance, in books (Sobolev, 1964; Vladimirov, 1971). Dirac's bra- and ket- notation is introduced in monograph (Dirac, 1967).

Worked Examples

1.1. Prove that the linear transformation U of the vector space V is unitary if and only if U does not change the lengths of the vectors.

Solution. Let $a, b \in V$. Put $U a = a'$, $U b = b'$. Then, for any number α we have

$$(a + \alpha b, a + \alpha b) = (U(a + \alpha b), U(a + \alpha b)) = (a' + \alpha b', a' + \alpha b').$$

Performing the multiplication and using $(a, a) = (a', a')$, $(b, b) = (b', b')$, we get

$$\alpha^*(b, a) + \alpha(a, b) = \alpha^*(b', a') + \alpha(a', b').$$

When $\alpha = 1$

$$(b, a) + (a, b) = (b', a') + (a', b'), \tag{1.10}$$

putting $\alpha = \mathrm{i}$ and cancelling out by i, we get

$$-(b, a) + (a, b) = -(b', a') + (a', b').$$

The last equality together with (1.10) gives $(a, b) = (a', b')$, $(U a, U b) = (U U^\dagger a, b)$. Check out that if U is Unitary matrix, then the scalar product is safe.

1.2. Let A, B, C, D be square matrices $n \times n$, besides $AC = CA$ with A, C being nondegenerate matrices. Prove that

$$\begin{vmatrix} A & B \\ C & D \end{vmatrix} = \begin{vmatrix} D & C \\ B & A \end{vmatrix} = |AD - CB|,$$

$$\begin{vmatrix} A & C \\ B & D \end{vmatrix} = \begin{vmatrix} D & B \\ C & A \end{vmatrix} = |DA - BC|.$$

Solution.

$$\begin{vmatrix} A & B \\ C & D \end{vmatrix} = |AC|^{-1} \begin{vmatrix} C & 0 \\ 0 & A \end{vmatrix} \begin{vmatrix} A & B \\ C & D \end{vmatrix} = |AC|^{-1} \begin{vmatrix} CA & CB \\ AC & AD \end{vmatrix} =$$

$$= |AC|^{-1} \begin{vmatrix} CA & CB \\ 0 & AD - CB \end{vmatrix} = |AD - CB|.$$

Similar equalities can be established analogously.

1.3. Prove the equality:

$$\det A = \begin{vmatrix} a + x_1 & a & \dots & a \\ a & a + x_2 & \dots & a \\ \dots & \dots & \dots & \dots \\ a & a & \dots & a + x_n \end{vmatrix} =$$

$$= x_1 \cdot x_2 \cdot \dots \cdot x_n \left(1 + \frac{a}{x_1} + \dots + \frac{a}{x_n}\right).$$

Solution. We introduce the diagonal matrix

$$X = \mathrm{diag}(x_1, \dots, x_n)$$

and rewrite the determinant of the desired matrix $\det A = \det X \det (I + J X^{-1})$, where all the elements of matrix J are equal to a, and I is the identity matrix. The rank of matrix $J X^{-1}$ is equal to unity, therefore, the only non-zero eigenvalue of this matrix is equal to its trace (all other eigenvalues occur zero):

$$\lambda_1 = \mathrm{tr}\, J X^{-1} = \frac{a}{x_1} + \frac{a}{x_2} + \dots + \frac{a}{x_n}.$$

Hence $\det (I + J X^{-1}) = 1 + \lambda_1$, and determinant

$$\det A = x_1 \cdot x_2 \cdot \dots \cdot x_n \left(1 + \frac{a}{x_1} + \dots + \frac{a}{x_n}\right).$$

1.4. Let H be a Hermitian matrix. Prove that $U = \exp(\mathrm{i}H)$ is unitary.

Solution. We need to show that $UU^\dagger = E$. It is easy to show (see the problem **1.20**) that series

$$U = E + (\mathrm{i}H) + \frac{1}{2!}(\mathrm{i}H)^2 + \ldots$$

and

$$U^\dagger = E + (-\mathrm{i}H) + \frac{1}{2!}(-\mathrm{i}H)^2 + \cdots = \exp(-iH)$$

converge. Substituting these series in the product $UU^\dagger$ we have

$$UU^\dagger = \left[E + \mathrm{i}H + \frac{1}{2!}(\mathrm{i}H)^2 + \ldots\right] \cdot \left[E + (-\mathrm{i}H) + \frac{1}{2!}(-\mathrm{i}H)^2 + \ldots\right].$$

Changing the order in summation, we obtain

$$UU^\dagger = \sum_{j=0}^{\infty} \frac{(\mathrm{i}H)^j}{j!} \sum_{m=0}^{j} (-1)^m C_j^m = E, \quad C_j^m = \frac{j!}{m!(j-m)!},$$

because of

$$\sum_{m=0}^{j} (-1)^m C_j^m = (1-1)^j = \delta_{j0}.$$

1.5. Prove that for an arbitrary matrix A

$$\det \mathrm{e}^A = \exp(\mathrm{tr}\, A). \tag{1.11}$$

Solution. First, note that the equality to be proved is invariant relative to similarity transformations: if (1.11) is valid then

$$\begin{aligned}\det\left(\exp(QAQ^{-1})\right) &= \det\left(E + QAQ^{-1} + \frac{1}{2!}(QAQ^{-1})^2 + \ldots\right)\\ &= \det\left(Q\left(E + A + \frac{1}{2!}(A)^2 + \ldots\right)Q^{-1}\right)\\ &= \det \mathrm{e}^A = \exp(\mathrm{tr}\, A) = \exp(\mathrm{tr}\,(QAQ^{-1})),\end{aligned}$$

where Q is an arbitrary non-degenerate matrix. Therefore, proving statement in some fixed basis, we prove it immediately for all bases. If A is diagonalizable, then let Q be transition matrix to the diagonal basis for A: $QAQ^{-1} = \mathrm{diag}(\lambda_1, \ldots, \lambda_n)$. Then

$$\det\left(\mathrm{e}^A\right) = \det\left(\exp(QAQ^{-1})\right) = \det \mathrm{diag}(\mathrm{e}^{\lambda_1}, \ldots, \mathrm{e}^{\lambda_n}) =$$
$$= \exp(\lambda_1 + \cdots + \lambda_n) = \exp(\mathrm{tr}\,(QAQ^{-1})) = \exp(\mathrm{tr}\, A).$$

In the general case, Q can be chosen so that QAQ^{-1} is a direct sum of Jordan cells. Arbitrary Jordan cell J_m,

$$J_m = \begin{pmatrix} \lambda & 1 & 0 & \dots\dots & 0 \\ 0 & \lambda & 1 & 0 \dots & 0 \\ \cdot & \cdot & \cdot & \cdot & \cdot \\ 0 & \dots\dots\dots & & 0 & \lambda \end{pmatrix},$$

can be represented as $J_m = \lambda I_m + \tilde{J}_m$, where I_m is the identity matrix, $\tilde{J}_m$ is the upper triangular nilpotent (all eigenvalues are zero) matrix. Since the identity matrix commutes with any other, then

$$\exp J_m = \exp(\lambda I_m) \exp(\tilde{J}_m).$$

We immediately see that $\exp(\tilde{J}_m)$ is the upper triangular matrix with unit elements on the diagonal. Consequently,

$$\det \exp J_m = \det \exp(\lambda I_m) \det \exp(\tilde{J}_m) = \det \exp(\lambda I_m) = \exp(\operatorname{tr} J_m).$$

Due to the property for both trace and determinant,

$$\det \exp \left(J_{m_1} \oplus J_{m_2} \oplus \dots\right) = \det \left(\exp J_{m_1} \oplus \exp J_{m_2} \oplus \dots\right) =$$

$$= \det \exp J_{m_1} \det \exp J_{m_2} \cdots =$$

$$= \exp \left(\operatorname{tr} J_{m_1} + \operatorname{tr} J_{m_2} + \dots\right) = \exp \left(\operatorname{tr} (J_{m_1} \oplus J_{m_2} \oplus \dots)\right),$$

that proves the equality for any matrix A.

1.6. Prove that for any matrix A unitary matrix B exists so that the matrix $A' = B^{-1}AB$ is an upper triangular (Schur theorem).

Solution. Let T be the matrix that transforms A to the Jordan form (namely, $T^{-1}AT$ is the direct sum of Jordan cells (1.13)). Orthogonalize the columns of the matrix T using the Gram–Schmidt procedure. Note that the orthogonalization is equivalent to multiplying the matrix T by some non-degenerate upper triangular matrix S from the right. For example, to j-th column of matrix T add its i-th column multiplied by α, it is enough to multiply the matrix T from the right by the matrix S', for which the matrix elements $S'_{ll} = 1$, $l = 1, \dots, n$, $S'_{ij} = \alpha$, and the rest are zero. If S is such an upper triangular matrix, then the columns of TS are an orthonormal system of n vectors. Therefore, $U = TS$ is unitary. Matrix $S^{-1}(T^{-1}AT)S$ is the product of an upper triangular matrices S^{-1}, $T^{-1}AT$, S. Therefore, $U^{\dagger}AU = (S^{-1}T^{-1})A(TS)$ is an upper triangular matrix.

1.7. Consider the matrix Ω in the form

$$\Omega = \begin{pmatrix} A & B \\ -B^\dagger & C \end{pmatrix},$$

where A, B, C are $n \times n$ matrices with A and C being Hermitian. Since $\Omega \neq \Omega^\dagger$, its eigenvectors are no longer be orthogonal regarding the ordinary scalar product, and its eigenvalues may be complex. Let the matrix Σ have the following form

$$\Sigma = \begin{pmatrix} I_n & 0 \\ 0 & -I_n \end{pmatrix},$$

where I_n is the identity $n \times n$ matrix. Show that

(a) if the eigenvector $\mathbf{e}$ is such that $(\mathbf{e}, \Sigma\mathbf{e}) \neq 0$, then the corresponding eigenvalue being real;

(b) for any two eigenvectors $\mathbf{e}_1$ and $\mathbf{e}_2$ of operator Ω

$$\Omega\mathbf{e}_1 = \lambda_1\mathbf{e}_1, \;\; \Omega\mathbf{e}_2 = \lambda_1\mathbf{e}_2, \;\; \lambda_1 \neq \lambda_2,$$

at $(\mathbf{e}_1, \Sigma\mathbf{e}_1) \neq 0$ and $(\mathbf{e}_2, \Sigma\mathbf{e}_2) \neq 0$ the condition of "generalized" orthogonality is satisfied

$$(\mathbf{e}_1, \Sigma\mathbf{e}_2) = (\mathbf{e}_1)_k^*(\Sigma)_{kj}(\mathbf{e}_2)_j = 0.$$

Solution. 1.7(a). First note that the following relations are explicitly verified:

$$\Sigma^2 = I_{2n}, \;\; \Omega^\dagger = \Sigma\Omega\Sigma, \;\; \Omega^\dagger\Sigma = \Sigma\Omega.$$

From $\Omega\mathbf{e} = \lambda\mathbf{e}$ and the Hermitianity of Σ it follows that

$$(\Sigma\mathbf{e}, \Omega\mathbf{e}) = \lambda(\Sigma\mathbf{e}, \mathbf{e}), \;\; (\Omega\mathbf{e}, \Sigma\mathbf{e}) = \lambda^*(\mathbf{e}, \Sigma\mathbf{e}).$$

Taking into account $\Omega^\dagger\Sigma = \Sigma\Omega$ we get:

$$(\lambda - \lambda^*)(\mathbf{e}, \Sigma\mathbf{e}) = 0,$$

whence at $(\mathbf{e}, \Sigma\mathbf{e}) \neq 0$ we have $\lambda = \lambda^*$.

1.7(b). Let use equalities

$$(\mathbf{e}_1, \Sigma\Omega\mathbf{e}_2) = \lambda_2(\mathbf{e}_1, \Sigma\mathbf{e}_2), \;\; (\mathbf{e}_1, \Omega^\dagger\Sigma\mathbf{e}_2) = \lambda_1^*(\mathbf{e}_1, \Sigma\mathbf{e}_2).$$

Subtracting these equalities from one another and using $\Omega^\dagger\Sigma = \Sigma\Omega$, we obtain:

$$(\lambda_2 - \lambda_1^*)(\mathbf{e}_1, \Sigma\mathbf{e}_2) = 0.$$

By the condition of the problem $(\mathbf{e}_i, \Sigma\mathbf{e}_i) \neq 0$, $i = 1, 2$. Therefore, λ_i is real and from $\lambda_1 \neq \lambda_2$ it follows that $(\mathbf{e}_1, \Sigma\mathbf{e}_2) = 0$.

1.8. Find a projector of matrix

$$A = \begin{pmatrix} 0 & 0 & 1 \\ 1 & 0 & 0 \\ 0 & 1 & 0 \end{pmatrix}$$

onto the subspace corresponding to the eigenvalue $\lambda_1 = 1$.

Solution. First, we find the resolvent:

$$R_\lambda = \frac{1}{1-\lambda^3} \begin{pmatrix} \lambda^2 & 1 & \lambda \\ \lambda & \lambda^2 & 1 \\ 1 & \lambda & \lambda^2 \end{pmatrix}.$$

Then the projector is given by integral:

$$P_1 = \frac{1}{2\pi \mathrm{i}} \oint^{(1-)} R_\lambda \, d\lambda = -\operatorname*{Res}_{\lambda=1} R_\lambda = \frac{1}{3} \begin{pmatrix} 1 & 1 & 1 \\ 1 & 1 & 1 \\ 1 & 1 & 1 \end{pmatrix}.$$

1.9. Prove that in any Hilbert space, the triangle inequality holds:

$$\|f\| + \|g\| \geqslant \|f + g\|.$$

Solution. Note that if one of the norms is zero, then the inequality becomes an identity. Let $\|f\| > 0, \ \ \|g\| > 0$. Since the norm of the vector $h = \|f\|g - \|g\|f$ is non-negative, we get

$$2\|f\| \cdot \|g\| \geqslant (f, g) + (g, f).$$

Hence it follows that

$$\|f+g\|^2 = \|f\|^2 + \|g\|^2 + (f, g) + (g, f) \leqslant \|f\|^2 + \|g\|^2 + 2\|f\| \cdot \|g\| =$$
$$= (\|f\| + \|g\|)^2.$$

Extracting the square root of the latter gives the required inequality.

1.10. Chebyshev polynomials are defined as follows:

$$T_n(x) = \frac{1}{2^{n-1}} \cos\left(n \arccos x\right).$$

Prove that the function family

$$f_n(x) = \frac{2^n}{\sqrt{2\pi}} (1 - x^2)^{-1/4} T_n(x) =$$

$$= \sqrt{\frac{2}{\pi}} (1 - x^2)^{-1/4} \cos\left(n \arccos x\right), \ \ n = 0, 1, \ldots, \infty$$

forms an orthonormal basis in $L^2(-1, 1)$.

Solution. We use the completeness of the trigonometric basis:

$$\frac{2}{\pi}\sum_{n=0}^{\infty}\cos(nx_1)\cos(nx_2)=\delta(x_1-x_2),\qquad 0<x_{1,2}<\pi.$$

Consequently,

$$\sum_{0}^{\infty} f_n(x)f_n(y)=\frac{2}{\pi}(1-x^2)^{-\frac{1}{4}}(1-y^2)^{-\frac{1}{4}}\sum_{n=0}^{\infty}\cos(n\arccos x)\cos(n\arccos y)=$$

$$=(1-x^2)^{-\frac{1}{4}}(1-y^2)^{-\frac{1}{4}}\delta(\arccos x-\arccos y)=$$

$$=\delta(x-y).$$

The last equality follows from equality:

$$\delta\left(f(x)\right)=\frac{1}{|f'(x_0)|}\delta(x-x_0),$$

where x_0 is the only solution to the equation $f(x_0)=0,\ f'(x_0)\neq 0$ in the considered interval (see problem **1.26**). One can verify the orthonormality by direct calculation of the integrals.

1.11. Hermit polynomials are defined as follows:

$$H_n(x)=(-1)^n\mathrm{e}^{x^2}\frac{d^n}{dx^n}\mathrm{e}^{-x^2}.$$

Prove that functions

$$\phi_n(x)=\frac{1}{\pi^{1/4}\sqrt{2^n n!}}\exp\left(-x^2/2\right)H_n(x)$$

form an orthonormal basis in $L^2(-\infty,+\infty)$.

Solution. Note that $H_n(x)$ is a n-th polynomial in x with leading term $2^n x^n$. Scalar product

$$N_{nm}=\int_{-\infty}^{+\infty}dx\phi_n(x)\phi_m(x)=\frac{(-1)^m}{\sqrt{\pi\,2^{n+m}n!m!}}\int_{-\infty}^{+\infty}dx\,H_n(x)\frac{d^m}{dx^m}\mathrm{e}^{-x^2}=$$

$$=\frac{1}{\sqrt{\pi\,2^{n+m}n!m!}}\int_{-\infty}^{+\infty}dx\,\mathrm{e}^{-x^2}\frac{d^m}{dx^m}H_n(x)$$

for $m > n$ is equal to zero because of the m-th derivative of any polynomial with degree less m vanishes.

At $m = n$ given that

$$\frac{d^n}{dx^n} H_n(x) = \frac{d^n}{dx^n} 2^n x^n = 2^n n!$$

and computing the Poisson integral, we get $N_{nn} = 1$. Since m and n enter symmetrically, then

$$N_{mn} = \delta_{mn},$$

i.e., the orthonormality is proved.

When checking the completeness (necessary for the basis), we will use the following equality:

$$\frac{d^n}{dx^n} \mathrm{e}^{-x^2} = \sqrt{\pi} \int_{-\infty}^{+\infty} \frac{dp}{2\pi} (\mathrm{i}p)^n \exp\left(-p^2/4 + ipx\right),$$

which can be obtained by direct Fourier transform of the left side, n-times integration by parts and taking the remaining Gaussian integral. Series

$$\sum_{n=0}^{\infty} \phi_n(x)\phi_n(y) = \frac{1}{\sqrt{\pi}} \exp\left(\frac{x^2+y^2}{2}\right) \sum_{n=0}^{\infty} \frac{1}{2^n n!} \frac{d^n}{dx^n} \mathrm{e}^{-x^2} \frac{d^n}{dy^n} \mathrm{e}^{-y^2} =$$

$$= \exp\left(\frac{x^2+y^2}{2}\right) \int_{-\infty}^{+\infty} \frac{dp}{2\pi} \exp\left(-p^2/4 + ipx\right) \sum_{n=0}^{\infty} \frac{(\mathrm{i}p)^n}{2^n n!} \frac{d^n}{dy^n} \mathrm{e}^{-y^2}$$

sums up as a Taylor series:

$$\sum_{n=0}^{\infty} \frac{a^n}{n!} \frac{d^n}{dy^n} f(y) = f(y+a).$$

Thus,

$$\sum_{n=0}^{\infty} \phi_n(x)\phi_n(y) =$$

$$= \exp\left(\frac{x^2+y^2}{2}\right) \int_{-\infty}^{+\infty} \frac{dp}{2\pi} \exp\left(-p^2/4 + ipx - (y + ip/2)^2\right) =$$

$$= \exp\left(\frac{x^2-y^2}{2}\right) \int_{-\infty}^{+\infty} \frac{dp}{2\pi} \exp(\mathrm{i}p(x-y)) = \delta(x-y),$$

Q.E.D.

1.12. Show that operator

$$\hat{A} = \exp\left(a\frac{d}{dx}\right)$$

is unitary in $L^2(-\infty, +\infty)$. Here a is a real number.

Solution. $\hat{A}$ acts on the function $f(x)$ as follows:

$$\hat{A} f(x) = f(x + a).$$

One could verify this fact by expanding the exponent in the expression $\hat{A} f(x)$ in a series and summing up the resulting Taylor series for $f(x)$. Thus,

$$(\hat{A} f, \hat{A} g) = \int_{-\infty}^{+\infty} dx f^*(x + a)g(x + a) = \int_{-\infty}^{+\infty} dy\, f^*(y)g(y) = (f, g).$$

1.13. In $L^2(0, l)$, there acts the operator

$$\mathcal{H} = -\frac{d^2}{dx^2} + G\delta(x - l/2),$$

$$\mathcal{D} = \{f(0) = f(l) = 0,\ f \in L^2(0, l)\}.$$

Find its spectrum and eigenfunctions for various values of G.

Solution. The eigenfunctions of the operator $-d^2/dx^2$, turning into 0 at the point $x = l/2$ are also eigenfunctions for $\mathcal{H}$:

$$\mathcal{H}\psi_m = \frac{4m^2\pi^2}{l^2}, \quad \psi_m = \sin\frac{2\pi m}{l}x, \quad m = 1, \ldots$$

Another set of eigenfunctions does not vanish at $x = l/2$ and satisfies the matching condition:

$$\psi'(l/2 + 0) - \psi'(l/2 - 0) = G\,\psi(l/2),$$

obtained from the equation $\mathcal{H}\psi = \lambda\psi$ by integration over a small interval including the point $x = l/2$. If one substitutes

$$\psi_n = \sin k_n x,\ x < l/2, \quad \psi_n = \sin k_n(l - x),\ x > l/2, \quad \lambda = k_n^2,$$

then the matching condition yields an equation for k_n:

$$2k_n \operatorname{ctg}\frac{k_n l}{2} = -G.$$

Real λ gives both real and purely imaginary k_n. For $k_n \in \mathbb{R}$, we have an infinite number of solutions defining a second set of eigenvalues. To study the purely imaginary k_n, we make the substitution:

$$k_n = \mathrm{i}\kappa, \quad \kappa \in \mathbb{R},$$

which leads to the equation

$$\operatorname{cth}\frac{\kappa l}{2} = -\frac{G}{2\kappa}.$$

This equation has a solution only for $G < 0$ starting only from some threshold value $|G|$. Indeed, the limit of the left-hand side as $\kappa \to +0$ is $2/\kappa l$ and for $|G| < 4/l$ function $\operatorname{cth}(\kappa l/2)$ goes everywhere higher than $\frac{|G|}{2\kappa}$. Thus, there are no intersection points, and consequently, there are no solutions. If $|G| > 4/l$, then the right-hand side of the equation for small κ is greater the left-hand side; at larger κ, the opposite case takes place. There is one intersection point in this case, and therefore the operator $\mathcal{H}$ has one negative eigenvalue λ.

Note that for $G < 0$ and $|G| > 4/lG < 0$, the minimum real value k_1 is greater $2\pi/l$ (there is no intersection of the hyperbola with the first cotangentoid), therefore the corresponding $\psi_1(x)$ has zeros on the interval $(0, l)$. In accordance with oscillation theorem[2] ψ_1 is not a ground state and, respectively, $\lambda_1 = k_1^2$ is not a smallest eigenvalue. The ground state corresponds to the solution with negative λ.

Exercises

1.14. Let $A_1, \ldots, A_k$ be square matrices of the same size. Prove that

$$\operatorname{tr}(A_1 A_2 \ldots A_k) = \operatorname{tr}(A_2 \ldots A_k A_1).$$

1.15. Let the eigenvalues of the matrix A of size $n \times n$ be equal

$$\lambda_1, \lambda_2, \ldots, \lambda_n,$$

and I is the identity matrix. Prove that

$$\det(I + A) = (1 + \lambda_1) \cdot (1 + \lambda_2) \cdot \cdots \cdot (1 + \lambda_n).$$

[2] See, for example, (Landau and Lifshitz, 1977, §21).

1.16. Calculate the characteristic polynomial and eigenvalues of the matrix

$$\begin{pmatrix} \cos\alpha & \sin\alpha \\ -\sin\alpha & \cos\alpha \end{pmatrix}.$$

1.17. The Pauli (or σ) matrices are defined as

$$\sigma_1 = \begin{pmatrix} 0 & 1 \\ 1 & 0 \end{pmatrix}, \quad \sigma_2 = \begin{pmatrix} 0 & -\mathrm{i} \\ \mathrm{i} & 0 \end{pmatrix}, \quad \sigma_3 = \begin{pmatrix} 1 & 0 \\ 0 & -1 \end{pmatrix}. \tag{1.12}$$

(a) Find the coefficients of decomposition of an arbitrary 2×2 matrix A over Pauli matrices: $A = c_0\sigma_0 + c_1\sigma_1 + c_2\sigma_2 + c_3\sigma_3$, where σ_0 is the identity 2×2 matrix.
(b) Show that for the Hermitian matrix A, all the coefficients are real.
(c) Find the general form of 2×2 projector.
(d) Derive the general formula for a product of two σ-matrices.

1.18. Let

$$A = \begin{pmatrix} \rho & 1 & 0 & \dots & 0 \\ 0 & \rho & 1 & \dots & 0 \\ \dots & \dots & \dots & \dots & \dots \\ 0 & 0 & 0 & \dots & \rho \end{pmatrix} \tag{1.13}$$

be the Jordan cell of order n. Show that for all positive integers m the following formula holds

$$A^m = \begin{pmatrix} \rho^m & C_m^1\rho^{m-1} & C_m^2\rho^{m-2} & \dots & C_m^{n-1}\rho^{m-n+1} \\ 0 & \rho^m & C_m^1\rho^{m-1} & \dots & C_m^{n-2}\rho^{m-n+2} \\ 0 & 0 & \rho^m & \dots & C_m^{n-3}\rho^{m-n+3} \\ \dots & \dots & \dots & \dots & \dots \\ 0 & 0 & 0 & \dots & \rho^m \end{pmatrix},$$

where the binomial coefficients C_m^k, $k \leqslant m$ are determined by the formula

$$C_m^k = \frac{m(m-1)\dots(m-k+1)}{k!},$$

and at $k > m$ $C_m^k = 0$.

1.19. Let $f(\lambda) = a_0 + a_1\lambda + \dots + a_k\lambda^k$ be some polynomial in λ. Prove that if A is Jordan cell (1.13) of order n then

$$f(A) = \begin{pmatrix} f(\rho) & \frac{1}{1!}f'(\rho) & \frac{1}{2!}f''(\rho) & \dots & \frac{1}{(n-1)!}f^{(n-1)}(\rho) \\ 0 & f(\rho) & \frac{1}{1!}f'(\rho) & \dots & \frac{1}{(n-2)!}f^{(n-2)}(\rho) \\ 0 & 0 & f(\rho) & \dots & \frac{1}{(n-3)!}f^{(n-3)}(\rho) \\ \dots & \dots & \dots & \dots & \dots \\ 0 & 0 & 0 & \dots & f(\rho) \end{pmatrix}.$$

1.20. Let $f(\lambda) = a_0 + a_1\lambda + \dots + a_m\lambda^m + \dots$ be a formal expansion in powers of parameter λ. Show that for the power series of matrix A to converge, it is necessary and sufficient that each eigenvalue ρ_i of the matrix A either be inside the circle of convergence of the corresponding power series $f(\lambda)$, or lay on the boundary of the circle of convergence. If ρ_i lies on the boundary of the circle of convergence, then the convergence of the series obtained by $(n_i - 1)$ by multiple differentiation of the series $f(\lambda)$ is required, where n_i is the order of the Jordan cell corresponding to the value of ρ_i.

1.21. Find

$$\ln \begin{pmatrix} 1 & x \\ x & 1 \end{pmatrix}, \quad |x| < 1.$$

1.22. Find

$$\sin \begin{pmatrix} 0 & 0 & x \\ 0 & x & 0 \\ x & 0 & 0 \end{pmatrix}.$$

1.23. Prove the equalities:

(a) $\lim\limits_{\epsilon \to +0} \dfrac{1}{\pi} \dfrac{\epsilon}{x^2 + \epsilon^2} = \delta(x);$

(b) $\lim\limits_{\epsilon \to +0} \dfrac{2}{\pi} \dfrac{x^2 \epsilon}{(x^2 + \epsilon^2)^2} = \delta(x);$

(c) $\lim\limits_{\epsilon \to +0} \dfrac{2}{\pi} \dfrac{x\epsilon}{(x^2 + \epsilon^2)^2} = -\frac{d\delta(x)}{dx};$

(d) $\lim\limits_{\epsilon \to +0} \dfrac{1}{\sqrt{\pi\epsilon}} \exp\left(-x^2/\epsilon\right) = \delta(x);$

(e) $x d\delta(x) dx = -\delta(x);$

(f) $\lim\limits_{n \to \infty} \dfrac{1 - \cos(nx)}{\pi n x^2} = \delta(x).$

1.24. Prove that

$$\frac{1}{2\pi} \int\limits_{-\infty}^{\infty} dp \exp(\mathrm{i} px) = \delta(x).$$

1.25. Prove that for any smooth function the following equality holds:

$$f(x)\frac{d\delta(x-a)}{dx} = f(a)\frac{d\delta(x-a)}{dx} - \delta(x-a)\frac{df(x)}{dx}.$$

1.26. Prove that if $f'(a_n) \neq 0$ where $\{a_n\}$ is the set of simple zeros of the function $f(x)$, $f(a_n) = 0$, $f'(a_n) \neq 0$, then

$$\delta\left(f(x)\right) = \sum_n \frac{1}{|f'(a_n)|}\delta(x - a_n).$$

1.27. Prove that two-dimensional δ- function can be written in the cylindrical coordinates as follows:

$$\delta(\mathbf{r}-\mathbf{r}')=\frac{1}{r'}\delta(r-r')\delta(\phi-\phi').$$

Here (r,ϕ), (r',ϕ') are polar coordinates of the point $\mathbf{r}$, $\mathbf{r}'$, respectively.

1.28. Prove that in the three-dimensional case

$$\left(\frac{\partial^2}{\partial x^2}+\frac{\partial^2}{\partial y^2}+\frac{\partial^2}{\partial z^2}\right)\frac{1}{r}=-4\pi\delta(\mathbf{r}).$$

1.29. Check whether the following sequences of functions form a basis in a Hilbert space:

(a) the orthonormal sequence of functions in the space $L^2(0,\pi)$

$$f_0(x)=\sqrt{\frac{1}{\pi}};\ f_n(x)=\sqrt{\frac{2}{\pi}}\cos(nx),\quad n=1,\ldots,\infty;$$

(b) the orthonormal sequence of functions in the space $L^2(0,2\pi)$

$$f_n(x)=\sqrt{\frac{1}{\pi}}\sin(nx),\quad n=1,\ldots,\infty.$$

1.30. Find the operator $L^\dagger$ adjoint to the operator L given by the boundary value problem with conditions of the Sturm type:

$$Ly=p(x)y''+q(x)y'+r(x)y,\quad x\in[0,1],$$

$$\alpha_0y(0)+\beta_0y'(0)=0,\quad \alpha_1y(1)+\beta_1y'(1)=0.$$

(a) Determine what kind of restrictions should satisfy the functions p, q, r and the coefficients α, β in order the operator L will be self-adjoint;

(b) Show that if the boundary conditions on the closed surface S have the form

$$u|_S=0\quad\text{or}\quad \left.\frac{\partial u}{\partial n}\right|_S=0,$$

where n is the normal to the surface, then the Laplace operator $L=\Delta$ will be self-adjoint.

(b) Find out what kind of restrictions should satisfy a_{ij}, b_i, c in the boundary value problem

$$L=a_{ij}(x)\frac{\partial^2}{\partial x_i\partial x_j}+b_i(x)\frac{\partial}{\partial x_i}+c(x),\ \ u|_S=0,$$

so that the operator L will be self-adjoint.

1.31. Find under which restrictions on the coefficients the Sturm–Liouville operator:

$$L = \frac{d}{dx} q(x) \frac{d}{dx} + r(x)$$

with linearly independent boundary conditions

$$\alpha_\mu y(a) + \beta_\mu y'(a) = \gamma_\mu y(b) + \delta_\mu y'(b), \quad \mu = 1, 2,$$

is self-adjoint on the interval $x \in [a, b]$.

1.32. Is operator $\hat{p} = -id/dx$ Hermitian on interval $x \in [0, 2\pi]$ with boundary conditions:

(a) $u(0) = u(2\pi) = 0$;

(b) $u(0) = u(2\pi)$?

1.33. Show that for unitarity it is sufficient to preserve the norm: if for any y: $(Uy, Uy) = (y, y)$, then U is unitary.

1.34. Show that the product of unitary operators is a unitary operator. Is there a similar property for Hermitian operators?

1.35. Show explicitly that operator

$$\exp\left(a\frac{d}{dx}\right) \mathrm{e}^{\mathrm{i}bx}$$

is unitary in $L^2(\mathbb{R})$ for $a, b \in \mathbb{R}$.

1.36. Prove the equality

$$\exp\left(-a\frac{d}{dx}\right) \mathrm{e}^{-\mathrm{i}bx} \exp\left(a\frac{d}{dx}\right) \mathrm{e}^{\mathrm{i}bx} = \mathrm{e}^{\mathrm{i}ab}.$$

1.37. Show explicitly that the operator

$$U = \exp\left(\mathrm{i}a\frac{d^2}{dx^2}\right)$$

is unitarian in $L^2(\mathbb{R})$ for $a \in \mathbb{R}$.

Answers

1.14. **Hint.** Use the trace definition of matrices.

1.15. **Hint.** Eigen values of matrix $I + A$ are equal to $(1 + \lambda_i)$, $i = 1, \ldots, n$.

1.16. Answer.

$$\lambda^2 - 2\lambda\cos\alpha + 1.$$

1.17. Answer.

$$c_0 = \frac{1}{2}\mathrm{tr}\,A, \quad c_i = \frac{1}{2}\mathrm{tr}\,(\sigma_i A), \quad i = 1, 2, 3.$$

1.17. Answer.

$$P = \frac{1}{2}\left(\sigma_0 + \mathbf{n}\cdot\boldsymbol{\sigma}\right),$$

where $\mathbf{n}$ is a unity vector, $\boldsymbol{\sigma} = (\sigma_1, \sigma_2, \sigma_3)$.

1.17. Answer.

$$\sigma_i\sigma_j = ie_{ijk}\sigma_k + \delta_{ij}, \quad i, j, k = 1, 2, 3,$$

where e_{ijk} is the Levi-Civita symbol: $e_{123} = e_{312} = e_{321} = 1$, $e_{213} = e_{132} = e_{231} = -1$, all the other components are equal zero.

1.18. Hint. Use the method of mathematical induction.

1.19. Hint. Take advantage of the result of the previous problem.

1.20. Hint. It is enough to prove the theorem for the Jordan cells, since $f(TAT^{-1}) = Tf(A)T^{-1}$. To do this, use the result of the problem (**1.18**).

1.21. Hint. Calculate first the resolvent R_λ

$$R_\lambda = \frac{1}{(\lambda-1)^2 - x^2}$$

and then use a generalization of the Cauchy formula for matrices.

Answer.

$$\frac{1}{2}\begin{pmatrix} \ln(1-x^2) & \ln\frac{1+x}{1-x} \\ \ln\frac{1+x}{1-x} & \ln(1-x^2) \end{pmatrix}.$$

1.22. Answer.

$$\begin{pmatrix} 0 & 0 & \sin x \\ 0 & \sin x & 0 \\ \sin x & 0 & 0 \end{pmatrix}.$$

1.23. Hint. Use the definition of δ-function.

1.24. Hint. Consider the limit at $\alpha \to +0$ for the integral as for an absolutely convergent integral

$$\lim_{\alpha \to +0} \frac{1}{2\pi} \int_{-\infty}^{\infty} dp \, \exp(ipx - \alpha|p|) = \delta(x).$$

1.25. Hint. Use the definition of a derivative of a δ-function.

1.26. Hint. Use the definition of δ-function.

1.27. Hint. Consider the integral $\int d\mathbf{r}$ from both sides of the equality in cylindrical coordinates.

1.28. Hint. Write the Laplacian in spherical coordinates and take the integral $\int d\mathbf{r}$ from both sides of the equality.

1.29(a). Hint. Sum

$$\sum_{n=1}^{\infty} \cos(nx) \cos(ny)$$

should be calculated as the limit for $\alpha \to +0$ of absolutely convergent series

$$\sum_{n=1}^{\infty} \exp(-\alpha n) \cos(nx) \cos(ny).$$

Answer. Yes, the sequence forms the basis.

1.29(b). Answer. No, the sequence does not form a basis. Only the complete set

$$\left\{ \sqrt{\frac{1}{\pi}} \sin(nx), \quad \sqrt{\frac{1}{\pi}} \cos(nx) \right\}$$

forms a basis. The expansion within this basis is the Fourier series of any function from $L^2(0, 2\pi)$.

1.30. Answer.

$$a_{ij} = a_{ji}, \quad b_i = \frac{\partial a_{ij}}{\partial x_j}.$$

1.31. Answer.

$$q(a)\left(\gamma_1\delta_2 - \gamma_2\delta_1\right) = q(b)\left(\alpha_1\beta_2 - \alpha_2\beta_1\right).$$

In particular, hence one can get the boundary value problems of the I, II, and III kinds and periodic boundary value problem:

$$1^\circ\ y(a) = y(b) = 0,\ \ 2^\circ\ y'(a) = y'(b) = 0,\ \ 3^\circ \text{ Linear combination } 1^\circ, 2^\circ,$$
$$4^\circ\ y(a) = y(b),\ \ y'(a) = y'(b),\ \ q(a) = q(b).$$

1.32(a). **Answer.** No.

1.32(b). **Answer.** Yes.

1.33. Hint. For any u and v, by the problem condition, the norm of linear combination $\lambda u + \mu v$ is preserved with λ and μ being arbitrary complex numbers (see the problem **1.1**).

1.35. Hint. Act with the operator on an arbitrary function $f(x)$, check the norm conservation and use the result of the previous problem.

1.36. Hint. Consider the effect of the left-hand side on an arbitrary function $f(x)$.

1.37. Hint. Prove the equality $(Uf, Uf) = (f, f)$ using representation of the function $f(x)$ in the form of a Fourier integral.

Chapter 2

Method of Characteristics

2.1 Linear First-Order PDE

Real equation

$$a\frac{\partial u}{\partial x} \equiv a_1\frac{\partial u}{\partial x_1} + a_2\frac{\partial u}{\partial x_2} + \cdots + a_n\frac{\partial u}{\partial x_n} = 0, \tag{2.1}$$

where vector $a = (a_1, \ldots, a_n)$ and the unknown function $u(x)$ depend only on n variables $x = (x_1, \ldots, x_n)$, is called the homogeneous linear *partial differential equation* (PDE) of the first order. Consider the functions $x_j(t)$ and limit their dynamics with respect to t by the following system of ordinary differential equations (ODE) of the first order:

$$\frac{dx_j}{dt} = a_j(x), \quad j = 1, 2, ..., n. \tag{2.2}$$

This system is called *equations for characteristics* and, respectively, its solution (trajectory) is called *characteristic*. The autonomous system (2.2) has $n-1$ first integrals $F_i(x)$ independent of t:

$$\frac{dF_i}{dt} = 0, \quad i = 1, \ldots, n-1.$$

A solution in the form of an arbitrary function of all the first integrals,

$$u(x) = g\left(F_1(x), \ldots, F_{n-1}(x)\right), \tag{2.3}$$

Mathematical Methods of Physics: Problems with Solutions
Igor V. Kolokolov, Evgeny A. Kuznetsov, Alexander I. Milstein, Evgeny V. Podivilov, Alexander I. Chernykh, David A. Shapiro, and Elena G. Shapiro

ISBN 978-981-5129-21-2 (Hardcover), 978-1-003-51360-5 (eBook)
www.jennystanford.com

is called the *general solution (general integral)* of equation (2.1). Here we assume the function $g(F_1, F_2, ..., F_{n-1})$ being two times continuously differentiable. For the sake of brevity, below we shall suppose all necessary conditions for function smoothness are fulfilled. The direct substitution of $u = g$ into (2.1) leads to the following chain of equalities:

$$a\frac{\partial u}{\partial x} = a_i\frac{\partial g}{\partial F_j}\frac{\partial F_j}{\partial x_i} = \frac{dx_i}{dt}\frac{\partial g}{\partial F_j}\frac{\partial F_j}{\partial x_i} = \frac{dF_j}{dt}\frac{\partial g}{\partial F_j} = 0,$$

where repeated subscripts imply summation.

Another approach to the equation (2.1) consists in the consideration of coefficients $a(x)$ as components of some vector in the n-dimensional space. In such representation the equation (2.1) means the equality to zero of the first derivative of function u along the vector a. In this case, the solution of the equation by means of the method of characteristics is reduced to the reconstruction of integral curves Γ from the tangent vectors a given at each point x. Along these curves a solution $u(x)$ is constant. In the vicinity of any non-singular point[1] one can make a transformation to a new system of coordinates $t, F_1, F_2, \ldots, F_{n-1}$, with parameter t varying along integral curves or *characteristics*. In that system, the equation takes the form $du/dt = 0$. Its general solution will be an arbitrary function of $n - 1$ coordinates $F_1, F_2, \ldots, F_{n-1}$, the first integrals of the equations for characteristics (2.2).

The *Cauchy problem* for equation (2.1) is formulated by the following way: find the solution $u(x)$ of the equation satisfying the *initial conditions*

$$u(x)|_{x=x_0\in S} = f(x_0), \tag{2.4}$$

where $f(x_0)$ is a given function of variables x_0 and S is some hyperplane of dimensionality $n - 1$. In S we can introduce a system of coordinates and write any vector $s \in S$ as $s(x_0) = s_1, \ldots, s_{n-1}$. The Cauchy problem allows the solution in some vicinity of the initial hyperplane S, if S does not go along the tangent to any characteristic. In this case, any trajectory leaving S will be marked by the label s. In such formulation, the method of characteristics is nothing more than a certain change of variables $(s, t) \to x$. This mapping, generally speaking, can be ambiguous for all space, but it remains univalent in the vicinity of S where the problem solution is possible. When the general

[1]A point is called *singular* if the transformation Jacobian at this point $\frac{\partial(t, F_1, F_2, \ldots, F_{n-1})}{\partial(x_1, x_2, \ldots, x_n)}$ tends to zero.

solution is known, the Cauchy problem is solved easily. In this case, it is enough to know the initial conditions (2.4) to determine function $g(F)$.

The general solution to the *inhomogeneous* linear equation

$$a\frac{\partial u}{\partial x} = b, \tag{2.5}$$

with a given function $b(x)$ should be found, as for any inhomogeneous linear equation, as a sum of the general solution (2.3) of the homogeneous equation (2.1) and a particular solution of the inhomogeneous one (2.5). To determine the latter, it is reasonable to pass on the characteristic $x(t)$, then, moving along the characteristic the left hand side of (2.5) has a form of the complete derivative with respect to t,

$$a\frac{\partial u}{\partial x} = \dot{x}\frac{\partial u}{\partial x} = \frac{du}{dt} = b\left(x(t)\right).$$

Hence, by integrating, we obtain

$$u_{inh} = \int_{t_0}^{t} b\left(x(\tau)\right)\, d\tau.$$

Unlike the homogeneous solution, this solution is no longer constant along the characteristic.

2.2 Quasilinear Equation

Equation (2.5) is called *quasilinear* if its coefficients a and b depend not only on the coordinates x, but also on the unknown function $u(x)$: $a = a(x, u),\ b = b(x, u)$.

Quasilinear equations also can be solved by the method of characteristics. It is important to remember that the superposition principle does not work in this case; the solution of the quasilinear equation can never be represented as a sum of the "homogeneous" solutions and the "inhomogeneous" one. The only modification of this method for quasilinear equations is an extension of the phase space for the auxiliary dynamic system, where one needs to find the integral curves Γ. Besides n coordinates x_i $(i = 1, \ldots, n)$ let us introduce the $(n+1)$-th coordinate u. In the $(n+1)$-dimensional extended phase space along characteristic $\dot{x} = a(x, u)$ the equation (2.5) transforms into ODE. As a result, for x_i and u as functions of parameter t, we have the following auxiliary

ODE system of the first order:

$$\begin{cases} \dot{x}_i = a_i(x, u), \\ \dot{u} = b(x, u). \end{cases} \tag{2.6}$$

The dot denotes a derivative with respect to parameter τ. The general solution to (2.5) is given by an arbitrary function $g(F_1, F_2, \ldots, F_n) = 0$ of its n first integrals F_i independent of τ.

2.3 System of Equations

The method of characteristics in some cases can be generalized to systems of the PDEs of the first order. In the case of two variables, x and t, a system of linear equations has the form

$$Au_t + Bu_x = f, \tag{2.7}$$

where $A(x, t)$ and $B(x, t)$ are matrices of order m, $u(x, t)$ is an unknown vector-column of dimension m, $f(x, t)$ a given vector function of x and t. A deviation of the vector function u, while the infinitesimal shift is (dt, dx), will be equal to:

$$du = u_t dt + u_x dx. \tag{2.8}$$

For the known differential du, one can find derivatives u_t and u_x by solving $2m$ linear equations (2.7), (2.8). However, it cannot be solved for any right side of this system. A solution is absent if the system determinant is equal to zero:

$$\begin{vmatrix} A & B \\ Edt & Edx \end{vmatrix}, \tag{2.9}$$

where E is the unit matrix of order m. Real solutions $x(t)$ of the ordinary differential equation (2.9) are called *characteristics* for the system (2.7).

If Eq. (2.9) has m different real roots dx/dt, the system (2.7) is referred to systems of the *hyperbolic* type. If some roots are complex, the system belongs to the *elliptic* type. In the intermediate case when Eq. (2.9) has at least one degenerate root, the system relates to the *parabolic* type.

To the hyperbolic systems with two variables, one can apply the method of characteristics. Solution of the system (2.7), (2.8) exists on characteristics if the rank of the system matrix is equal to the rank of the extended matrix,

$$\operatorname{rank}\begin{pmatrix} A & B \\ Edt & Edx \end{pmatrix} = \operatorname{rank}\begin{pmatrix} A & B & f \\ Edt & Edx & du \end{pmatrix}. \tag{2.10}$$

This equality is called the *relations on characteristics*. It must be satisfied along a characteristic.

Formulas (2.9), (2.10) can refer not only to the linear systems but also to the quasilinear ones.

Example. Equations of motion for one-dimensional barotropic gas are of the form

$$\frac{\partial \rho}{\partial t} + \frac{\partial}{\partial x}(\rho v) = 0, \qquad \frac{\partial v}{\partial t} + v\frac{\partial v}{\partial x} = -\frac{c^2}{\rho}\frac{\partial \rho}{\partial x}, \tag{2.11}$$

where $c = \sqrt{dp/d\rho}$ is the sound velocity, p, ρ and v are gas pressure, density, and velocity, respectively. For this quasilinear system

$$u = \begin{pmatrix} \rho \\ v \end{pmatrix}, \quad A = \begin{pmatrix} 1 & 0 \\ 0 & 1 \end{pmatrix}, \quad B = \begin{pmatrix} v & \rho \\ c^2/\rho & v \end{pmatrix}.$$

The formula (2.9) allows equations for characteristics to be found

$$\dot{x} = v \pm c, \tag{2.12}$$

and, respectively, Eq. (2.10) gives the relation $dv \pm c d\rho/\rho = 0$. Integrals of the relations on characteristics

$$J_{\pm} = v \pm \int \frac{c\, d\rho}{\rho}, \tag{2.13}$$

are called the *Riemann invariants*. Equations for the Riemann invariants have the diagonal form

$$\frac{\partial J_+}{\partial t} + (v + c)\frac{\partial J_+}{\partial x} = 0, \qquad \frac{\partial J_-}{\partial t} + (v - c)\frac{\partial J_-}{\partial x} = 0.$$

In this particular case, each Riemann invariant is a constant down a characteristic. It is impossible to find the general solution to this specific system implying the method because v and c are expressed by means of both J_+ and J_-. However, in one special case, when one of the invariants, e.g., J_- is independent of coordinate x (for instance, due to the initial or boundary conditions) and, therefore, is constant also in t, the only quasilinear equation remains for J_+. We find the solution of this equation with the help of the method of characteristics. This solution is called the *simple Riemann wave*, see problems **2.17, 2.18**.

The characteristic of a system of linear PDEs is not a line if the number of variables exceeds two. For such systems, they are surfaces (more exactly hypersurfaces) on which the boundary conditions of the Cauchy type do not allow finding partial derivatives of the desired solution. Consequently, the solution itself does not exist even in the vicinity of these surfaces.

Example. Let us consider the following system of m equations for functions of $n+1$ variables:

$$Au_t + B_1 u_{x_1} + \cdots + B_n u_{x_n} = f,\ u = \begin{pmatrix} u_1 \\ \cdot \\ \cdot \\ u_m \end{pmatrix},\ f = \begin{pmatrix} f_1 \\ \cdot \\ \cdot \\ f_m \end{pmatrix}, \tag{2.14}$$

where A_i, B_i are matrices of order m. Let boundary conditions be given on the hypersurface $\phi(t, x_1, .., x_n) = 0$ in the $(n+1)$-dimensional space. Introduce such functions $\alpha_1(t, x_1, \ldots, x_n), \ldots, \alpha_n(t, x_1, \ldots, x_n)$ for which the Jacobian

$$\frac{\partial(\phi, \alpha_1, \ldots, \alpha_n)}{\partial(t, x_1, \ldots, x_n)} \neq 0,$$

and make transformation from the old variables $\{t, x_i\}$ to $\{\phi, \alpha_i\}$. In terms of the new variables the system (2.14) will be written as

$$\left(A\phi_t + \sum_{i=1}^{n} B_i \phi_{x_i}\right) \frac{\partial u}{\partial \phi} + \cdots = f, \tag{2.15}$$

where dots denote terms not containing the derivative $\partial u/\partial\phi$. The system (2.15) cannot be resolved with respect to $\partial u/\partial\phi$, if the following determinant is equal to zero:

$$\left|A\phi_t + B_1\phi_{x_1} + \ldots + B_n\phi_{x_n}\right| = 0. \tag{2.16}$$

Condition (2.16) defines the characteristics of Eq. (2.14). In the given case these are the level surfaces of the function $\phi(t, x_1, \ldots, x_n)$. The gradient of this function $(\phi_t, \phi_{x_1}, \ldots, \phi_{x_n})$, orthogonal to the characteristic surfaces, is called the *characteristic normal.*

Textbooks on ordinary differential equations usually give the theory of linear and quasilinear PDEs, see (Arnold, 1983).

Worked Examples

2.1. Define characteristics for the following equation

$$\frac{\partial u}{\partial x} + y\frac{\partial u}{\partial y} = 0.$$

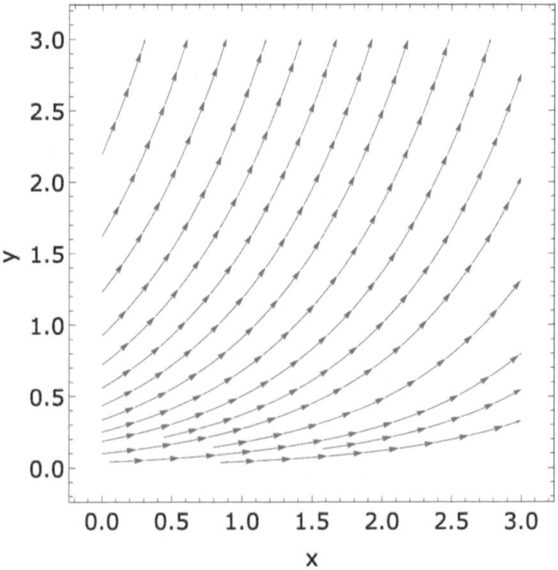

Figure 2.1 Integral curves and tangent to them vectors for equation $\dot{y} = y$.

Solution. The equations for characteristics

$$\dot{x} = 1, \ \dot{y} = y$$

are easily integrated:

$$x(t) = t + C_1, \ y(t) = C_2 e^t.$$

They have only one first integral:

$$F(x, y) = ye^{-x} = \text{const.}$$

Their integral curves for different constants are shown in Fig. 2.1. Therefore $u(x, y) = g\left(ye^{-x}\right)$ with g being an arbitrary function is the general solution. In Fig. 2.1, the characteristics shown coincide with the level lines of the function g. Arrows denote directions of the tangent vectors $a = (1, y)$ to the integral curves. Integral curves $y = \text{const} \cdot e^x$ correspond to solid lines. The solution of the homogeneous equation is a constant along the characteristics.

2.2. Find and depict on the (x, y)-plane characteristics for the following homogeneous equations:

(a)

$$\frac{\partial u}{\partial x} - y^2 \frac{\partial u}{\partial y} = 0;$$

(b)

$$x\frac{\partial u}{\partial x} - y\frac{\partial u}{\partial y} = 0;$$

(c)

$$y\frac{\partial u}{\partial x} - x\frac{\partial u}{\partial y} = 0.$$

Find general solutions and check them by direct substitution.

Solution. 2.2(a). Introduce the parameter t and write down the equations for characteristics:

$$\frac{dx}{dt} = 1, \quad \frac{dy}{dt} = -y^2.$$

Their solution

$$x(t) = t - t_1, \quad y(t) = \frac{1}{t - t_2}$$

gives the parametric family in the (x, y)-plane (see Fig. 2.2). The equations for characteristics have the only first integral

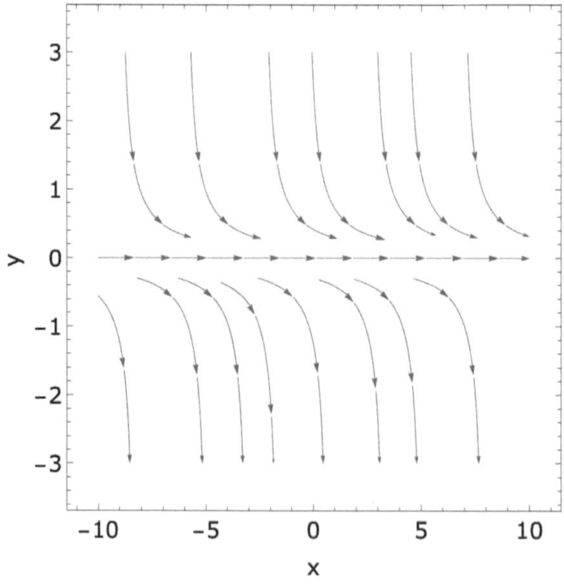

Figure 2.2 The characteristics family to the problem **2.2**(a).

$$F(x, y) = x - \frac{1}{y} = \text{const.}$$

Thus, $u(x, y) = g(x - 1/y)$ is the general solution, where g is an arbitrary function. Indeed, computing first derivatives

$$\frac{\partial u(x, y)}{\partial x} = g', \quad \frac{\partial u(x, y)}{\partial y} = \frac{1}{y^2}g',$$

one can see that the equations transform into the identity for any smooth function g.

Solution. 2.2(b). The solution of the equations for characteristics

$$x = C_1 e^t, \quad y = C_2 e^{-t}$$

gives the parametric family in the (x, y)-plane as shown in Fig. 2.3. The general solution has the form

$$u(x, y) = g(xy).$$

Solution. 2.2(c). Introduce parameter t and write down the equations for characteristics:

$$\frac{dx}{dt} = y, \quad \frac{dy}{dt} = -x.$$

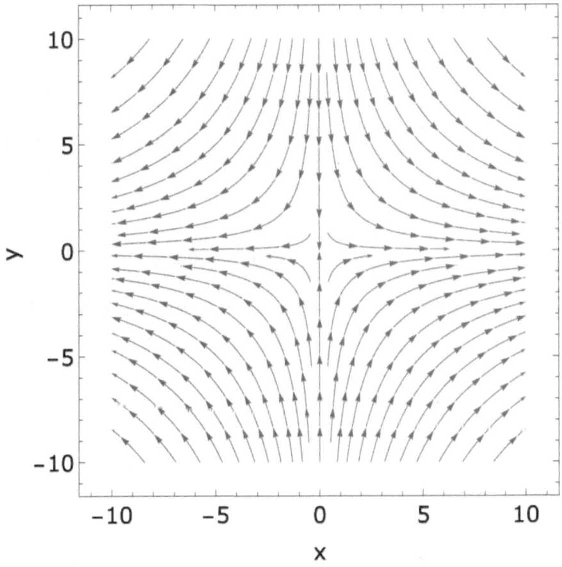

Figure 2.3 The characteristics family to the problem **2.2**(b).

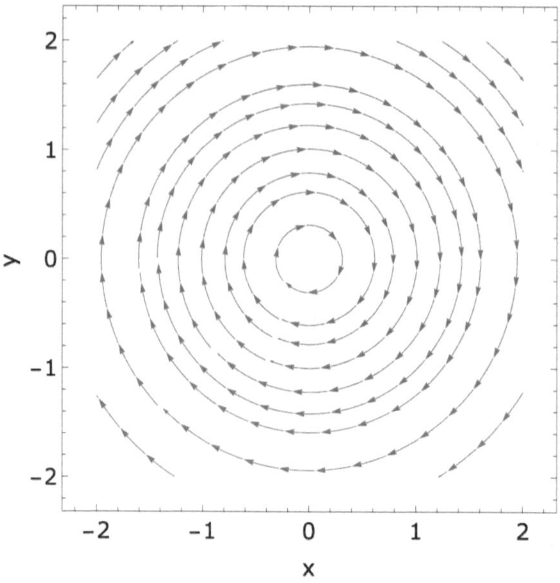

Figure 2.4 The characteristics family to the problem **2.2**(c).

Their solution

$$x(t) = A\sin(t - t_1), \quad y(t) = A\cos(t - t_1)$$

gives the parametric family of characteristics in the (x, y)-plane (circles), which are shown in Fig. 2.4. The equations for characteristics have the only first integral

$$F(x, y) = x^2 + y^2 = \text{const},$$

therefore $u(x, y) = g(x^2 + y^2)$ is the general solution, where g is an arbitrary function.

2.3. Solve the Cauchy problem for linear equation

$$\frac{\partial u}{\partial x} = y\frac{\partial u}{\partial y}, \quad u(0, y) = \cos y.$$

Solution. The equations for characteristics $\dot{x} = 1, \ \dot{y} = -y$ have the solution $(x(t) = t - t_1, \ y(t) = \exp(-t + t_2))$ and the only first integral

$$F(x, y) = ye^x = \text{const}.$$

Thus,

$$u(x, y) = g\left(y \exp(x)\right),$$

is the general solution, while g being an arbitrary function. The straight line $x = 0$, where the boundary conditions are defined, does not touch the characteristics $y = \exp(-x + x_1)$. Therefore, the function g can be found by substituting the general solution into the boundary conditions

$$u(0, y) = g(y) = \cos y.$$

Hence, finally we have the solution to the problem: $u(x, y) = \cos\left(y \exp(x)\right)$.

2.4. Solve the Cauchy problem for equation **2.2**(c) at $u(1, y) = y^2$.

Solution. The straight line $x = 1$ (where the boundary conditions are defined) touches the characteristic $x^2 + y^2 = 1$ at the point (1,0) (see the solution to the problem **2.2**(c) and Fig. 2.4). Generally speaking, the Cauchy problem solution cannot exist; nevertheless, it does exist for the given boundary conditions. One can verify this by the direct substitution of the general solution **2.2**(c) into the boundary conditions:

$$u(1, y) = g(1 + y^2) = y^2, \quad \text{hence} \quad g(z) = z - 1.$$

Thus, $u(x, y) = g(x^2 + y^2) = x^2 + y^2 - 1$. This solution is both well-defined and unique outside the unit circle $x^2 + y^2 > 1$. Inside the circle, the solution is an arbitrary function of $r^2 = x^2 + y^2$.

Note. We have found the partial solution to this problem outside the circle due to the boundary conditions' specific symmetry (to reflection $y \to -y$). There is no solution to the general Cauchy problem. For instance, the Cauchy problem has no unique solution for the initial condition $u(1, y) = y$.

2.5. (a) Show that equations for characteristics of the collisionless kinetic equation (the Liouville equation)

$$\frac{\partial f}{\partial t} + \mathbf{v}\frac{\partial f}{\partial \mathbf{r}} + e\left(\mathbf{E} + \frac{1}{c}[\mathbf{v} \times \mathbf{H}]\right)\frac{\partial f}{\partial \mathbf{p}} = 0$$

for the distribution function $f = f(\mathbf{p}, \mathbf{r}, t)$ are the equations of motion for charged particles moving in electric and magnetic fields.

(b) Show that for the Liouville equation

$$\frac{\partial f}{\partial t} + \{H, f\} = 0$$

the equations for its characteristics coincide with the Hamilton equations. Here H is the Hamilton function, and {, } means the Poisson brackets.

Solution. 2.5(a) Considering t as a parameter, one can get the equations for characteristics

$$\dot{t} = 1, \quad \dot{\mathbf{r}} = \mathbf{v}, \quad \dot{\mathbf{p}} = e\left(\mathbf{E} + [\mathbf{v} \times \mathbf{H}]/c\right),$$

which coincides with the equations of motion of charged particles in the electric $\mathbf{E}$ and magnetic $\mathbf{H}$ fields.

2.5(b). By the definition of the Poisson brackets, the Liouville equation is of the form

$$\frac{\partial f}{\partial t} + \frac{\partial H}{\partial \mathbf{p}}\frac{\partial f}{\partial \mathbf{r}} - \frac{\partial H}{\partial \mathbf{r}}\frac{\partial f}{\partial \mathbf{p}} = 0.$$

Thus, we can transform the equations for characteristics into the Hamiltonian set:

$$\dot{\mathbf{r}} = \partial H \partial \mathbf{p}, \quad \dot{\mathbf{p}} = -\frac{\partial H}{\partial \mathbf{r}}.$$

2.6. Solve the Cauchy problem for the inhomogeneous equation

$$\frac{\partial u}{\partial x} - y\frac{\partial u}{\partial y} = y, \quad u(0, y) = \sin y.$$

Solution. The general solution to the homogeneous equation has already obtained to the problem **2.3**; this is: $u(x, y) = g(ye^x)$. The particular solution of the inhomogeneous equation, written on the characteristics

$$\frac{du}{dt} = u_x\frac{dx}{dt} + u_t\frac{dy}{dt} = u_x - yu_y = y(t) = \exp(-t + t_2),$$

has the form

$$u = \int^t d\tau y(\tau) = \int^t d\tau \exp(-\tau + t_2) = -\exp(-t + t_2) + c_1 = c_1 - y.$$

Here c_1 remains constant along the characteristics. The general solution is given by a sum of the general homogeneous solution and the inhomogeneous particular one $u(x, y) = g\left(y \exp(x)\right) - y$. Its substitution into the boundary conditions leads to the equation

$$u(0, y) = g(y) - y = \sin(y), \quad \text{hence} \quad g(z) = \sin(z) + z.$$

Answer. $u(x, y) = \sin\left(y \exp(x)\right) + y\exp(x) - y.$

2.7. Find the general solution to equation

$$x\frac{\partial u}{\partial x}+y\frac{\partial u}{\partial y}+2(x^2+y^2)\frac{\partial u}{\partial z}=0$$

and solve the Cauchy problem for $u(x, y, z)|_{x^2+y^2=1}=1-z$ in the region $x^2+y^2>1$.

Solution. The equations for characteristics $\dot{x}=x,\ \dot{y}=y,\ \dot{z}=2(x^2+y^2)$ have the solution $(x(t)=\exp(t-t_1),\ y(t)=\exp(t-t_2),\ z(t)=x^2+y^2+t_3)$ with two integrals of motion:

$$F_1(x, y, z)=\frac{x}{y}=\text{const},\qquad F_2(x, y, z)=x^2+y^2-z=\text{const}.$$

The characteristics are parabolas (these are intersections of the rotation paraboloids $F_2=\text{const}$ with the planes $F_1=\text{const}$). Then

$$u(x, y, z)=g\left(\frac{x}{y}, x^2+y^2-z\right)$$

is the general solution, where g is an arbitrary function of two arguments. The cylinder $(x^2+y^2=1)$, at which the boundary conditions are given, does not touch characteristics $(y=c_1x,\ z=x^2+y^2+c_2)$. Substituting the general solution into the boundary conditions, we find the final form for the function g:

$$u(1, y, z)=g\left(\frac{1}{y}, 1+y^2-z\right)=1+y^2-z,\quad \text{hence}\quad g(\xi, \eta)=\eta.$$

Answer. $u(x, y, z)=x^2+y^2-z$ at $x^2+y^2>1$.

Note. All characteristics intersect at the z-axis, all points of which are singular. Therefore in the general case, it is impossible to find the Cauchy problem solution at the vicinity of the z-axis. Nevertheless, the obtained solution is valid in the whole space due to the axial symmetry of the boundary condition.

2.8. Find the moment of breaking the solution to the Hopf equation

$$\frac{\partial u}{\partial t}+u\frac{\partial u}{\partial x}=0,$$

describing one-dimensional flow of noninteracting particles of a dusty cloud, if $u(x, 0)=u_0(1-\tanh x)$.

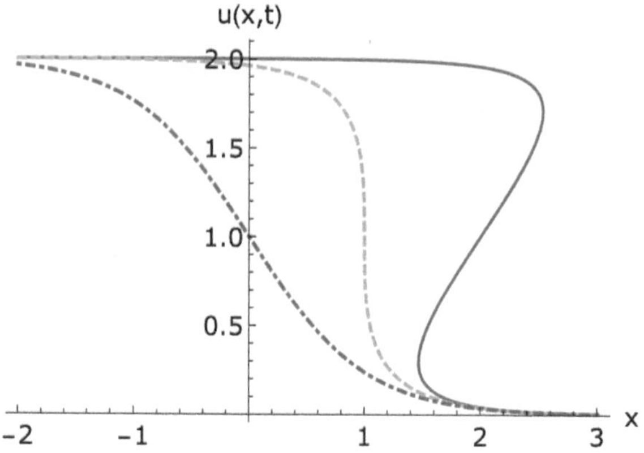

Figure 2.5 Wave front breaking of the solution to the Hopf equation for $u_0 = 1$: at $t = 0$ — before breaking (dot-dashed line), at $t = 1$ — at the moment of breaking (dashed), and at $t = 2$ — after the breaking (solid).

Solution. The equation gives $\dot{u} = 0$ at the characteristic $x = ut + C$, i.e., u is a constant at the characteristics. The general solution is given by one arbitrary function g of $\xi = x - ut$, being constant along the characteristic

$$u(x, t) = g(x - ut).. \tag{2.17}$$

For the given initial conditions at the hyperplane $t = 0$ solution is written in the implicit form $u = u_0[1 - \tanh(x - ut)]$. Differentiating u as an implicit function with respect to x one can get

$$u_x = \frac{-u_0}{\cosh^2 \xi - u_0 t}.$$

Hence it is seen that at $t < t^* = 1/u_0$ the solution is one-valued, but as $t \to t^*$, derivative u_x tends to infinity at $\xi = 0$. This is the so-called *breaking*.[2] The rapid particles of dust reach the slower ones, that correspond to the intersection of characteristics in the phase space. This is the reason why there is uniqueness for the problem solution. At first it takes place at $t = t^*$ as shown in Fig. 2.5.

2.9. For the Hopf equation

$$\frac{\partial u}{\partial t} + u\frac{\partial u}{\partial x} = 0$$

[2] In real physical systems, this process is considered the initial stage of the shock wave formation.

find a moment for the wave breaking t_0, when the solution becomes ambiguous, for

(a) $u(x, 0) = \pi/2 - \arctan(x)$;

(b)

$$u(x, 0) = \frac{1}{\sqrt{\pi}} \int_x^{\infty} d\eta \, e^{-\eta^2}.$$

Solution. 2.9(a). In accordance with (2.17) the general solution of the Hopf equation is of the form $u(x, t) = g(\xi), \xi = x - u(x, t)t$. Its substitution into the boundary conditions gives the solution in the implicit form

$$u(x, 0) = g(x) = \pi/2 - \arctan x,$$

$$u(x, t) = g(\xi) = \pi/2 - \arctan(x - tu(x, t)).$$

Its derivative against x is found with the help of the differentiation rules of composite functions:

$$u_x = -\frac{1}{1 + (x - tu)^2}(1 - tu_x) = -\frac{1}{1 - t + (x - tu)^2}.$$

As it is seen, $u_x < \infty$ for each x at $t < t^* = 1$ and the solution is one-valued. As $t \to t^*$ the derivative u_x tends to the infinity at the point $x^* = u(x^*, t^*)t^* = \pi/2$ that results in the wave *breaking* and the shock wave formation (in the real physical situation).

Solution. 2.9(b). The general solution to the Hopf equation has the form (2.17): $u(x, t) = g(\xi), \xi = x - u(x, t)t$. Satisfying the boundary conditions we get the solution in the implicit form,

$$u(x, 0) = g(x) = \frac{1}{\sqrt{\pi}} \int_x^{\infty} d\eta \, e^{-\eta^2}, \quad u(x, t) = g(\xi) = \frac{1}{\sqrt{\pi}} \int_{\xi}^{\infty} d\eta e^{-\eta^2}.$$

Its derivative with respect to x is equal to

$$u_x = -\frac{e^{-\xi^2}}{\sqrt{\pi}}(1 - tu_x) = -\frac{1}{\sqrt{\pi} e^{\xi^2} - t}. \tag{2.18}$$

Hence one can see that $u_x < \infty$ and the solution remains one-valued at $t < t^* = \sqrt{\pi}$. As $t \to t^*$ the derivative u_x tends to infinity. The breaking takes place at $\xi^* = 0$. Hence we get $u^* = g(\xi^*) = 1/2, \ x^* = u^* t^* = \sqrt{\pi}/2$.

2.10. Find general solutions to the following quasilinear equations:

(a) $u_t + uu_x = -x$;

(b) $u_t + uu_r = -1/r^3$, at $r > 0$.

Solution. 2.10(a). The equations for characteristics in the extended phase space $\dot{u} = u,\ \dot{u} = -x$ have solution $x(t) = A\sin(t-\phi),\ u(t) = A\cos(t-\phi)$. The amplitude and the phase

$$A^2 = u^2 + x^2, \quad \phi = t - \arctan\left(\frac{x}{u}\right),$$

are constant along the characteristics. Thus,

$$g\left[u^2 + x^2, t - \arctan\left(\frac{x}{u}\right)\right] = 0,$$

where $g(\xi, \eta)$ is an arbitrary function. The general solution can be represented also in a more convenient form:

$$t - \arctan\frac{x}{u} = f(u^2 + x^2), \quad \text{or} \quad u = x\cot(t - f(u^2 + x^2)).$$

Here f is an arbitrary function.

Solution. 2.10(b). The equations for characteristics $\dot{r} = u,\ \dot{u} = -1/r^3$ can be solved in the extended space in the same manner as for the problem of classical mechanics for a free particle motion in the spherical coordinates with a nonzero angular momentum:

$$u^2 - \frac{1}{r^2} = E, \quad ur = E(t - t_0),$$

where E and t_0 are constant along characteristics. The general solution is written in the implicit form and expressed by means of the derivative of the connection function E and t_0:

$$g\left(u^2 - \frac{1}{r^2},\ t - \frac{ur^3}{r^2u^2 - 1}\right) = 0.$$

2.11. Solve the Cauchy problem for equation $u_t + uu_x = 1$ at $u(x, 0) = 1 - \tanh x$.

Solution. The equations for characteristics in the extended space $\dot{x} = u$, $\dot{u} = 1$ have the solution $(u(t) = t + u_0,\ x(t) = u^2(t)/2 + x_0)$. The quantities

$$x_0 = x - \frac{1}{2}u^2 = \text{const}, \quad u_0 = u - t = \text{const}$$

are constants down a characteristic. Thus $g(u - t, u^2 - 2x) = 0$ is the general solution within $g(\xi, \eta)$ being an arbitrary function. To find the specific form of g from the boundary conditions, it is convenient to represent the general solution as

$$u^2(x,t) = 2x + f(u(x,t) - t),$$

where now $f(\xi)$ is an arbitrary function. Hence, satisfying the boundary conditions

$$u(x,0) = \sqrt{2x + f(u)} = 1 - \tanh x,$$

we get the specific form for the function f:

$$f(u) = u^2 - 2x = u^2 - 2\tanh^{-1}(1-u).$$

As a result, the Cauchy problem solution takes the form:

$$u^2(x,t) = 2x + (u(x,t) - t)^2 - 2\tanh^{-1}(1 - u(x,t) + t).$$

This solution also can be written in another simpler form.

Answer.

$$u(x,t) = 1 + t - \tanh\left(x - tu(x,t) + t^2/2\right).$$

2.12. Find characteristics, the relations on characteristics, and write down the general solution for the systems:

(a)

$$\begin{cases} u_x + v_x + u_y - 3v_y = 0, \\ u_x + v_x - 3u_y + v_y = 0. \end{cases}$$

(b)

$$\begin{cases} (x-1)u_t - (x+1)v_t + u_x = 0, \\ (x+1)u_t - (x-1)v_t - v_x = 0. \end{cases}$$

Solution. 2.12(a). Rewrite the system in matrix representation

$$A\begin{pmatrix} u_x \\ v_x \end{pmatrix} + B\begin{pmatrix} u_y \\ v_y \end{pmatrix} = 0,$$

where matrices A and B are of the form

$$A = \begin{pmatrix} 1 & 1 \\ 1 & 1 \end{pmatrix}, \quad B = \begin{pmatrix} 1 & -3 \\ -3 & 1 \end{pmatrix}.$$

Multiply the system from left by B^{-1} (here we assume B to be a non-singular matrix), then

$$E\begin{pmatrix} u_y \\ v_y \end{pmatrix} + C\begin{pmatrix} u_x \\ v_x \end{pmatrix} = 0, \quad C = B^{-1}A = -\frac{1}{2}\begin{pmatrix} 1 & 1 \\ 1 & 1 \end{pmatrix}.$$

The equations for characteristics are defined from Eq. (2.9), which is convenient to rewrite as follows

$$|Edx - Cdy| = 0.$$

Hence $dx/dy = \lambda_{\pm}$, where $\lambda_{-} = 0, \lambda_{+} = -1$ are the eigenvalues of the matrix C. Thus, the characteristics families are given by the equations $x = c_1$ and $x + y = c_2$. Because the matrix C is symmetric and its eigenvalues are different, it can be diagonalized $C \to T^{-1}CT$ utilizing substitution

$$\begin{pmatrix} u \\ v \end{pmatrix} = T\begin{pmatrix} f \\ g \end{pmatrix}, \quad T = \begin{pmatrix} \frac{1}{\sqrt{2}} & \frac{1}{\sqrt{2}} \\ -\frac{1}{\sqrt{2}} & \frac{1}{\sqrt{2}} \end{pmatrix}.$$

Here T is the matrix composing from the orthonormalized eigenvector-columns of the matrix C. As a result, the system takes the canonical form:

$$\begin{cases} f_y = 0, \\ g_x - g_y = 0. \end{cases}$$

Hence $f = f(x),\ g = g(x + y)$ is the general solution, where $f(x),\ g(z)$ are arbitrary functions. Notice that the condition $f = \text{const}$ being the relation on the characteristic $x = c_1$, for function g is equivalent to the relation $g = \text{const}$ on the characteristic $x + y = c_2$. The next step is to express u and v in terms of f and g.

Answer.

$$u(x,y) = \frac{1}{\sqrt{2}}\left(f(x) + g(x+y)\right), \quad v(x,y) = \frac{1}{\sqrt{2}}\left(-f(x) + g(x+y)\right).$$

Note. This approach based on the diagonalization of matrix C may be helpful when C depends on the variables. The simplest solvable case is when T is constant.

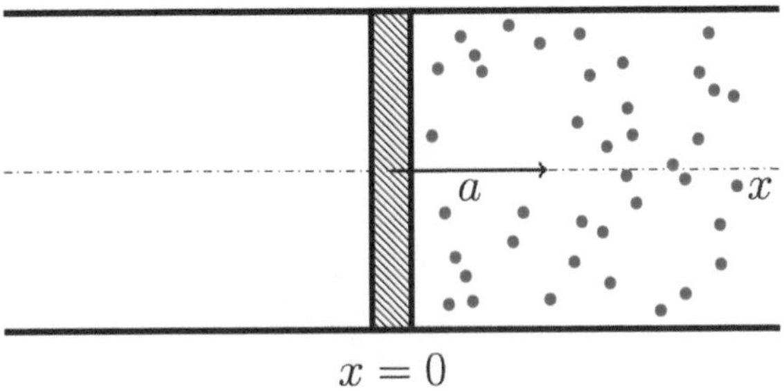

Figure 2.6 Piston in the tube with a gas.

Solution. 2.12(b). Represent the system in the matrix form

$$A\begin{pmatrix}u_t\\v_t\end{pmatrix}+B\begin{pmatrix}u_x\\v_x\end{pmatrix}=0,\quad A=\begin{pmatrix}x-1 & -x-1\\x+1 & 1-x\end{pmatrix},\quad B=\begin{pmatrix}1 & 0\\0 & -1\end{pmatrix}.$$

The equations on characteristics are defined by means of Eq. (2.9), which is convenient to rewrite as follows

$$|Adx-Bdt|=0.$$

Hence we get $dx/dt=-1/2$ for the first family of characteristics and respectively $dx/dt=1/2x$ for the second one. Integrating these equations one can obtain both the families. Two functions are constant along corresponding curves $2x+t=c_1$ and $x^2-t=c_2$. To find the relations on characteristics substitute $2dx+dt=0$ into the formula (2.10). As a result, we have $dv+du=0$ along the characteristic $2x+t=c_1$. And consequently, $v+u=f_1(2x+t)$, where f_1 is an arbitrary function. Analogously, we verify that the expression $u-v=f_2(x^2-t)$ occurs constant, along characteristics $x^2-t=c_2$ from the second family, where f_2 is another arbitrary function. The general solution is received from the conditions on characteristics:

$$u(x,t)=\frac{1}{2}\left(f_1(2x+t)+f_2(x^2-t)\right),$$

$$v(x,t)=\frac{1}{2}\left(f_1(2x+t)-f_2(x^2-t)\right).$$

2.13. A polytropic gas occupies the region $x>0$ to the right of the piston (Fig. 2.6). The piston moves with acceleration a. Find gas velocity $v(x,t)$ till the moment of the shock wave formation.

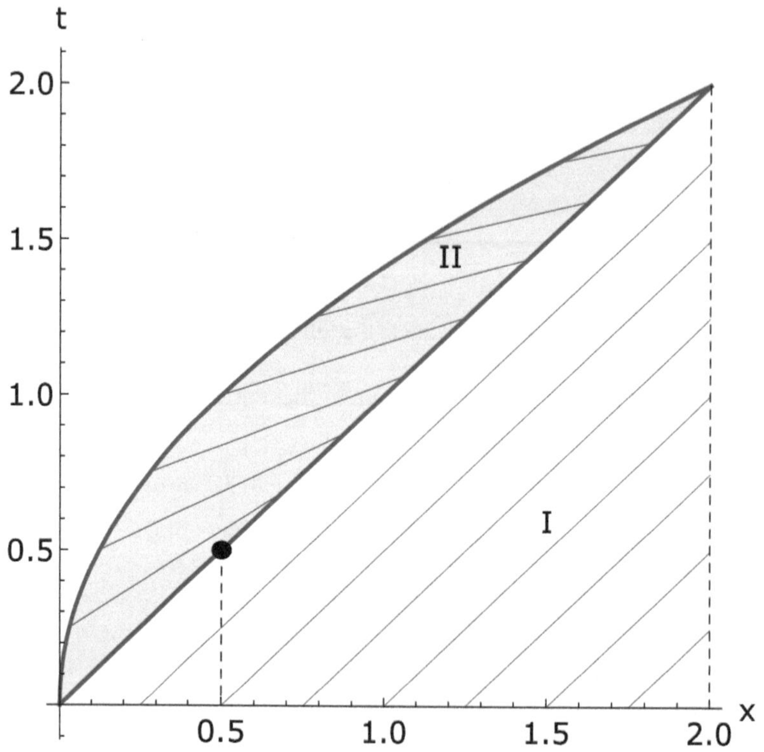

Figure 2.7 The family of characteristics $\dot{x} = c(\rho) + v$ (solid line) for the problem of polytropic gas motion due to the accelerated piston: $\gamma = 3$, $a = 1$, $c_0 = 1$: I is the rest gas; II is the moving gas. Region II from the left is bounded by the piston $x(t) = at^2/2$ (parabola), and from the right by the front of propagating perturbations $x(t) = c_0 t$ (straight line). Circle $t^* = 0.5c_0/a$, $x^* = 0.5$ marks the coordinates of breaking.

Solution. Let $a > 0$. In this case in the (x, t)-plane (Fig. 2.7) gas is occupied the region bounded from one side by semiaxis $t = 0$, $x > 0$, where the boundary conditions are given in the form

$$v(x, 0) = 0, \quad \rho(x, 0) = \rho_0.$$

In other words, along the characteristics intersecting this boundary, the Riemann invariants (2.20) are zero: $J_\pm = 0$. The equations for characteristics (2.12) become trivial and integrable in the domain, where both the invariants equal zero

$$x_+ = c_0 t + x_0, \quad x_- = -c_0 t + x_1, \quad x_0, x_1 > 0.$$

Hence one can see that this domain is bounded by straight line $x = c_0 t$, marked as I in Fig. 2.7. The solution inside this domain is

$$v(x, t) = 0, \quad \rho(x, t) = \rho_0.$$

The characteristic x_- intersects the straight line $x = c_0 t$ and advects the invariant $J_-(x, t) = 0$ from the boundary $(t = 0)$ to the region $x < c_0 t$. Thus, in the whole region II (see Fig. 2.7) the following relations take place

$$J_-(x, t) = v(x, t) - \frac{2}{\gamma - 1}\left(c(\rho(x, t)) - c_0\right) = 0,$$

with

$$c(\rho(x, t)) = c_0 + \frac{\gamma - 1}{2} v(x, t).$$

Hence

$$J_+(x, t) = v(x, t) + \frac{2}{\gamma - 1}\left(c(\rho(x, t)) - c_0\right) = 2v(x, t),$$

and $v = J_+/2$ is satisfied to the quasilinear equation

$$\frac{\partial v}{\partial t} = -\left(v + c(\rho)\right)\frac{\partial v}{\partial x} = -\left(c_0 + \frac{\gamma + 1}{2} v(x, t)\right)\frac{\partial v}{\partial x}.$$

The general solution of this equation is of the form

$$f(v(x, t)) = x - \left(c_0 + \frac{\gamma + 1}{2} v(x, t)\right) t, \tag{2.19}$$

where f is an arbitrary function.

From another side, the gas is bounded by the piston at $x = at^2/2$. The boundary conditions on the disk are

$$v(at^2/2, t) = at, \;\; x = v^2/2a.$$

Substituting the boundary conditions into the general solution (2.19)

$$f(at) = \frac{at^2}{2} - \left(c_0 + \frac{\gamma + 1}{2} at\right) t, \quad f(v) = -\left(c_0 + \frac{\gamma v}{2}\right)\frac{v}{a},$$

we get the solution for the Cauchy problem in the domain II

$$x = \left(c_0 + \frac{\gamma + 1}{2} v(x, t)\right) t - \frac{\gamma v^2(x, t)}{2a} - \frac{c_0 v(x, t)}{a},$$

$$v(x, t) = \frac{(\gamma + 1)at - 2c_0}{2\gamma} + \sqrt{\left(\frac{(\gamma - 1)at + 2c_0}{2\gamma}\right)^2 - \frac{2a}{\gamma}\left(x - \frac{at^2}{2}\right)},$$

where the sign 'plus' before the square root is chosen from the boundary condition $v = at$ at $x = at^2/2$. The density distribution $\rho(x, t)$ is found from the equation $J_- = 0$.

To find the moment of the shock wave formation one needs to differentiate the obtained solution:

$$dv/dx \to \infty \quad \text{at} \quad v^* = -\frac{c_0}{\gamma} + \frac{a(\gamma+1)t}{2\gamma}.$$

Because in the given case the gas velocity is positive and also v^* becomes positive only at $t > t^* = 2c_0/a(\gamma + 1)$, then t^* is the desired moment of time for the shock wave formation. Besides, $v^* = 0$. From this solution, one can see that the shock forms at the point $x^* = c_0 t^*$, namely, on the front of the propagating perturbations.

For $a < 0$ the characteristics on the (x, t)-plane are different. Nevertheless, all the region with gas is divided into two parts: the region of an unperturbed gas and the region with free gas. In the latter region, all the obtained formulas are applicable after changing $a \to -|a|$. Try to draw graphs for the characteristics and solution $v(x)$ at fixed t. Show that at the moment of time $t^* = 2c_0/(|a|(\gamma - 1))$ the gas density near the piston tends to zero, and at $t > t^*$ gas is absent at the region $x < x^* = 2c_0^2/|a|(\gamma - 1)^2 - 2c_0 t/(\gamma - 1)$ at the right side of the piston.

Exercises

2.14. Find the general solution to linear equation:

$$u_x \sin x + u_y \cos y = 0.$$

2.15. Find general solutions to quasilinear equations:

(a) $u_t + u_x = u$;

(b) $\frac{\partial \rho}{\partial t} + \frac{\partial}{\partial x}(\rho(1 - \rho)) = 0$.

2.16. For which initial conditions $u(x, 0) = f(x)$ does a solution of the Hopf equation remain smooth, i.e., $|u_x| < \infty$ at $t > 0$?

2.17. Find the solution for the equations of one-dimensional gas dynamics (2.11) assuming velocity v dependent on x, t only implicitly via function ρ: $v(x, t) = v(\rho(x, t))$ (the simple Riemann wave). Show that for the simple wave one of the Riemann invariants does not depend on x, t. For which initial conditions might such solutions be realized?

2.18. Find the Riemann invariants (2.13) for a polytropic gas, where power dependence $p\rho^{-\gamma} = \text{const}$ bounds the pressure and density.

2.19. Find the general solution of the system

$$\begin{cases} u_t - u_x = uv, \\ v_t + v_x = -uv. \end{cases}$$

2.20. Find characteristics, the relations on characteristics, and write down the general solution for the system:

$$\begin{cases} u_t + v_t - \frac{x}{t}(u_x + v_x) = 0, \\ u_t - v_t + t(u_x - v_x) = 0. \end{cases}$$

2.21. Define regions of hyperbolicity and find characteristics and the relations for the system:

$$\begin{cases} u_t + au_x + bv_x = 0, \\ v_t + cu_x + dv_x = 0, \end{cases}$$

where a, b, c, d are constants.

2.22. Find the characteristic normals for the system of four variables, the Dirac equation:

$$\frac{1}{c}\hat{I}\frac{\partial\psi}{\partial t} + \hat{\alpha_1}\frac{\partial\psi}{\partial x} + \hat{\alpha_2}\frac{\partial\psi}{\partial y} + \hat{\alpha_3}\frac{\partial\psi}{\partial z} + \mathrm{i}\hat{\beta}m\psi = 0,$$

where $\hat{I}$ is unit 4×4 matrix,

$$\hat{\alpha_1} = \begin{pmatrix} 0 & 0 & 0 & 1 \\ 0 & 0 & 1 & 0 \\ 0 & 1 & 0 & 0 \\ 1 & 0 & 0 & 0 \end{pmatrix}, \quad \hat{\alpha_2} = \begin{pmatrix} 0 & 0 & 0 & -\mathrm{i} \\ 0 & 0 & \mathrm{i} & 0 \\ 0 & -\mathrm{i} & 0 & 0 \\ \mathrm{i} & 0 & 0 & 0 \end{pmatrix},$$

$$\hat{\alpha_3} = \begin{pmatrix} 0 & 0 & 1 & 0 \\ 0 & 0 & 0 & -1 \\ 1 & 0 & 0 & 0 \\ 0 & -1 & 0 & 0 \end{pmatrix}, \quad \hat{\beta} = \begin{pmatrix} 1 & 0 & 0 & 0 \\ 0 & 1 & 0 & 0 \\ 0 & 0 & -1 & 0 \\ 0 & 0 & 0 & -1 \end{pmatrix}.$$

2.23. The same for Maxwell's equations

$$\frac{1}{c}\frac{\partial H_1}{\partial t} + \frac{\partial E_3}{\partial y} - \frac{\partial E_2}{\partial z} = 0, \quad \frac{1}{c}\frac{\partial H_2}{\partial t} + \frac{\partial E_1}{\partial z} - \frac{\partial E_3}{\partial x} = 0, \quad \frac{1}{c}\frac{\partial H_3}{\partial t} + \frac{\partial E_2}{\partial x} - \frac{\partial E_1}{\partial y} = 0,$$

$$\frac{1}{c}\frac{\partial E_1}{\partial t} - \frac{\partial H_3}{\partial y} + \frac{\partial H_2}{\partial z} = 0, \quad \frac{1}{c}\frac{\partial E_2}{\partial t} - \frac{\partial H_1}{\partial z} + \frac{\partial H_3}{\partial x} = 0, \quad \frac{1}{c}\frac{\partial E_3}{\partial t} - \frac{\partial H_2}{\partial x} + \frac{\partial H_1}{\partial y} = 0.$$

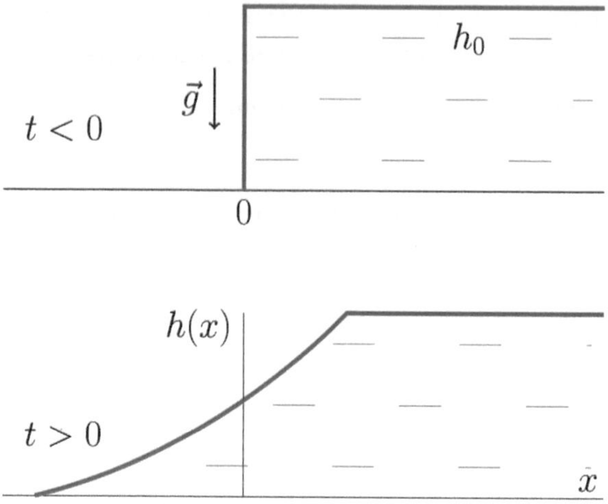

Figure 2.8 The vertical section to problem **2.24**.

2.24. Describe the spreading of a thin layer of ideal fluid after opening the damper. The process is described by the following system of equations:

$$\frac{\partial h}{\partial t} + \frac{\partial}{\partial x}(hv) = 0, \quad \frac{\partial v}{\partial t} + v\frac{\partial v}{\partial x} = -g\frac{\partial h}{\partial x}.$$

Here $h(x, t)$ is the thickness of the layer, $v(x, t)$ is the horizontal velocity of the liquid, which is assumed to be the same in the vertical section, g is the acceleration of gravity. The initial thickness of the layer is h_0, see Fig. 2.8.

Answers

2.14.

$$u(x, y) = g\left(\tan(x/2)[1 - \tan(y/2)]/[1 + \tan(y/2)]\right).$$

2.15(a). **Hint.** Make the substitution $u(x, t) = \exp(\phi(x, t))$.

Answer. $u(x, t) = g(x - t) \cdot e^x$.

2.15(b). **Answer.** $\rho = f(x - (1 - 2\rho)t)$.

2.16. **Answer.** At $df(x)/dx > 0$.

2.17. Hint. Substitute $v = f(\rho)$, get a system of two differential equations for $f(\rho)$ and write down the condition of their consistency.

2.18. Solution. In the polytropic gas ($p\rho^{-\gamma} = \text{const}$) the sound velocity is a function of density only

$$c(\rho)^2 = \frac{dp}{d\rho} = \gamma \frac{p}{\rho}.$$

Computing the integral in the expression (2.13), we get the following expression for the Riemann invariants

$$J_{\pm} = v \pm \frac{2}{\gamma - 1}(c(\rho) - c_0), \quad \gamma > 1, \tag{2.20}$$

where c_0 is an integration constant which is convenient to choose equal to the sound velocity at $v = 0$, $\rho = \rho_0$ (the rest gas).

2.19. Hint. Introduce new independent variables $\xi = (x+t)/2$, $\eta = (x-t)/2$ and potential Φ such that $u = \Phi_\xi$, $v = \Phi_\eta$. The equation for Φ will be linear.

Answer.

$$u = \frac{f'(\xi)}{f(\xi) + g(\eta)}, \quad v = \frac{g'(\eta)}{f(\xi) + g(\eta)},$$

where $f(\xi)$, $g(\eta)$ are arbitrary smooth functions.

2.20. Answer. Characteristics are $x - t^2/2 = \text{const}$, $x/t = \text{const}$, solution is

$$u = f(x - t^2/2) + g(x/t), \quad v = -f(x - t^2/2) + g(x/t).$$

2.21. Solution. Let us rewrite the system in the matrix form

$$E\begin{pmatrix} u_t \\ v_t \end{pmatrix} + A\begin{pmatrix} u_x \\ v_x \end{pmatrix} = 0, \quad A = \begin{pmatrix} a & b \\ c & d \end{pmatrix}.$$

The formula (2.9) permits to find the equations for characteristics

$$|E\,dx - A\,dt| = 0, \quad \dot{x} = \lambda_{\pm} = \frac{a+d}{2} \pm \frac{1}{2}\sqrt{(a-d)^2 + 4bc}.$$

The system occurs hyperbolic if the expression under the square root is positive:

$$D = (a-d)^2 + 4bc > 0.$$

If $D = 0$, the system is of the parabolic type, and two families degenerate into one family. In the case $D < 0$ the system belongs to the elliptic type, and the real characteristics are absent. Formula (2.10) gives the relations on a characteristic for the hyperbolic case. Their integrals are

$$vb + u(a - \lambda_{\mp}) = \text{const}, \quad \text{or} \quad uc + v(d - \lambda_{\mp}) = \text{const}.$$

2.22. Answer. The normals form a cone $\tau^2/c^2 = \xi^2 + \eta^2 + \zeta^2$.

2.23. Answer. $\tau^2/c^2 = \xi^2 + \eta^2 + \zeta^2$.

2.24. Answer. The following simple Riemann wave is the solution:

$$h(x,t) = \begin{cases} 0, & x < -2\sqrt{gh_0}\,t, \\ \dfrac{1}{9g}\left(\dfrac{x}{t} + 2\sqrt{gh_0}\right), & -2\sqrt{gh_0}\,t < x < \sqrt{gh_0}\,t, \\ h_0, & \sqrt{gh_0}\,t < x; \end{cases}$$

$$v(x,t) = \begin{cases} \dfrac{2}{3}\left(\dfrac{x}{t} - \sqrt{gh_0}\right), & -2\sqrt{gh_0}\,t < x < \sqrt{gh_0}\,t, \\ 0, & \sqrt{gh_0}\,t < x. \end{cases}$$

Chapter 3

Second-Order Linear Equations

3.1 Canonical Form

The most general linear second-order partial differential equation with two independent variables is

$$a_{11}u_{xx} + 2a_{12}u_{xy} + a_{22}u_{yy} + b_1u_x + b_2u_y + cu + f = 0. \qquad (3.1)$$

Here all coefficients a_{11}, a_{12}, a_{22}, b_1, b_2, f depend on x and y.

One can simplify the equation by a reversible transformation of variables. The sign of the discriminant

$$D = a_{12}^2 - a_{11}a_{22}$$

determines the type of equation and its canonical form. The correct change of variables leading to the *canonical form* can be found by solving equations for characteristics:

$$\frac{dy}{dx} = \frac{a_{12} + \sqrt{D}}{a_{11}}; \qquad (3.2)$$

$$\frac{dy}{dx} = \frac{a_{12} - \sqrt{D}}{a_{11}}. \qquad (3.3)$$

Equations (3.2) and (3.3) have in the general case two different integrals

$$\psi_+(x, y) = \text{const}, \quad \psi_-(x, y) = \text{const}. \qquad (3.4)$$

Mathematical Methods of Physics: Problems with Solutions
Igor V. Kolokolov, Evgeny A. Kuznetsov, Alexander I. Milstein, Evgeny V. Podivilov, Alexander I. Chernykh, David A. Shapiro, and Elena G. Shapiro

ISBN 978-981-5129-21-2 (Hardcover), 978-1-003-51360-5 (eBook)
www.jennystanford.com

For $D > 0$, the equation (3.1) is called *hyperbolic*. In the new variables

$$\xi = \psi_+(x, y), \quad \eta = \psi_-(x, y)$$

it is reduced to the first canonical form

$$u_{\xi\eta} + b_1 u_\xi + b_2 u_\eta + cu + f = 0. \tag{3.5}$$

The designations of the coefficients in the equation (3.5) are the same as in the equation (3.1), although the coefficients themselves do change after changing the variables. This remark should be taken into account further. For hyperbolic equations, the second canonical form is also adopted. By making the substitution in (3.5)

$$\alpha = \frac{\xi + \eta}{2}, \quad \beta = \frac{\xi - \eta}{2},$$

we get

$$u_{\alpha\alpha} - u_{\beta\beta} + b_1 u_\alpha + b_2 u_\beta + cu + f = 0. \tag{3.6}$$

For $D = 0$ the equation is called *parabolic*. Equations (3.2), (3.3) and their integrals ψ_+ and ψ_- in this case coincide. The change of variables

$$\xi = \psi_-(x, y), \quad \eta = \varphi(x, y),$$

where φ is any function such that φ and ψ are functionally independent, leads (3.1) to the canonical form of the parabolic type

$$u_{\eta\eta} + b_1 u_\xi + b_2 u_\eta + cu + f = 0.$$

For $D < 0$ the equation is related to the *elliptic* type. Equations (3.2) and (3.3) in this case are complex conjugated, as well as their integrals ψ_+ and ψ_- are.

By means of the change of variables

$$\xi = \operatorname{Re} \psi_+(x, y), \quad \eta = \operatorname{Im} \psi_+(x, y)$$

the equation is reduced to the canonical form of the elliptic type

$$u_{\xi\xi} + u_{\eta\eta} + b_1 u_\xi + b_2 u_\eta + cu + f = 0.$$

To further simplify equations of any type, one needs to look for a solution in the form

$$u = F(\xi, \eta)v,$$

where F is the given function, $v(\xi, \eta)$ is a new unknown function. Such a replacement does not change the canonical form, but it allows one to obtain conditions on the given function F under which the equation for v will not contain one or both derivatives of u_ξ and u_η.

If the coefficients in the equation are constant, then substitution

$$F = e^{\lambda\xi + \mu\eta},$$

with appropriate choosing λ and μ, leads to the equations of the hyperbolic, parabolic and elliptic types, respectively, of the forms

$$v_{\xi\eta} + \gamma v + f = 0,$$

$$v_{\xi\xi} + b_2 v_\eta + f = 0,$$

$$v_{\xi\xi} + v_{\eta\eta} + \gamma v + f = 0.$$

If the coefficients γ and f of the hyperbolic type equation turn out equal to zero

$$v_{\xi\eta} = 0,$$

then its general solution on the entire plane (ξ, η) has the form

$$v(\xi, \eta) = f_1(\xi) + f_2(\eta).$$

Here f_1 and f_2 are arbitrary functions. Going back to the original variables (x, y) and the function u, we get

$$u(x, y) = [f_1(\xi(x, y)) + f_2(\eta(x, y))]F(x, y).$$

A particular form of the functions f_1 and f_2 for the Cauchy problem can be defined according to the initial conditions given on the line in the plane (x, y).

3.2 Curvilinear Coordinates

In an orthogonal curvilinear coordinate system in three-dimensional space, an element of length can be written as

$$ds^2 = h_1^2 dq_1^2 + h_2^2 dq_2^2 + h_3^2 dq_3^2.$$

Quantities

$$h_i = \sqrt{\left(\frac{\partial x}{\partial q_i}\right)^2 + \left(\frac{\partial y}{\partial q_i}\right)^2 + \left(\frac{\partial z}{\partial q_i}\right)^2}$$

are called *Lame coefficients*. Gradient, divergence and rotor are expressed in the curved coordinates by formulas

$$\operatorname{grad}\varphi = \mathbf{e}_1 \frac{1}{h_1}\frac{\partial\varphi}{\partial q_1} + \mathbf{e}_2 \frac{1}{h_2}\frac{\partial\varphi}{\partial q_2} + \mathbf{e}_3 \frac{1}{h_3}\frac{\partial\varphi}{\partial q_3},$$

$$\operatorname{div}\mathbf{a} = \frac{1}{h_1 h_2 h_3}\left(\frac{\partial(h_2 h_3 a_1)}{\partial q_1} + \frac{\partial(h_1 h_3 a_2)}{\partial q_2} + \frac{\partial(h_1 h_2 a_3)}{\partial q_3}\right),$$

$$\operatorname{rot}\mathbf{a} = \begin{vmatrix} \frac{\mathbf{e}_1}{h_2 h_3} & \frac{\mathbf{e}_2}{h_1 h_3} & \frac{\mathbf{e}_3}{h_1 h_2} \\ \frac{\partial}{\partial q_1} & \frac{\partial}{\partial q_2} & \frac{\partial}{\partial q_3} \\ h_1 a_1 & h_2 a_2 & h_3 a_3 \end{vmatrix}.$$

Here $\mathbf{e}_1$, $\mathbf{e}_2$, $\mathbf{e}_3$ are local orthogonal unit vectors of the given coordinate system.

3.3 Separation of Variables

The *separation of variables* method consists in finding a solution to a boundary value problem for a partial differential equation as a product[1] of functions, each of which depends on one coordinate only. In many practically important cases, this allows ones to reduce the problem to the search for solutions to several ordinary differential equations. Variables can be separated in special some coordinate system and not separated in others. For example, in the stationary Schrödinger equation

$$-\frac{1}{2}\,\triangle\,\psi + U\,\psi = E\,\psi$$

we can separate the variables in that coordinate system where the potential U is

$$U = \sum_{m=1}^{3} \frac{U_m(q_m)}{h_m^2},$$

where h_m are the Lame coefficients.

[1]In some cases, the solution does not have to be sought in the product form. For example, in the Hamilton–Jacobi equation, variables can be separated if one looks for a solution in the form of a sum of functions of individual coordinates.

3.4 Fourier Method

The Fourier method is the expansion of variable separation to the linear equations. The main idea of the Fourier method is a decomposition of the unknown function $u(x, t)$ over the eigenfunctions of the right-side operator. For hyperbolic or parabolic equation

$$u_{tt} = \hat{L}u, \quad u_t = \hat{L}u, \tag{3.7}$$

where $\hat{L}$ is the second-order elliptic operator the method is applicable when its eigenfunctions

$$\hat{L}\psi_n = \lambda_n \psi_n \tag{3.8}$$

form the complete set according to Eq. (1.4). The sufficient conditions of the completeness are Hermiticity and negative definiteness: $\hat{L}^\dagger = \hat{L}$, $\lambda_n < 0$.

We look for solutions to Eqs. (3.7) in the form

$$u(x, t) = \sum_n C_n(t)\psi_n(x).$$

As a result, we get the diagonal system of ODE for coefficients $c_n(t)$. After the ODE solution, we obtain the integration constants from the initial conditions.

Note that the inhomogeneous equation can be solved in the same manner, if we decompose its right-hand part over the same functions. The spectral problem (3.8) has to be supplied by the same boundary conditions, as in Eq. (3.7). That means that decomposition over eigenfunctions is possible if the unknown function $u(x, t)$ belongs to the same Hilbert space.

The simplest hyperbolic type equation (the so-called one-dimensional wave or telegraph equation) has the form

$$u_{tt} - c^2 u_{xx} = 0,$$

where t has a meaning of time and x of coordinate. In particular, this equation describes plane free string vibrations. We will use this visual interpretation. Then u is the deviation of the string from its equilibrium position, $c^2 = T/\rho$ is the square of the wave velocity expressed in terms of string tension T and its linear density ρ. For this equation, we will use two types of boundary conditions. The boundary condition $u(a, t) = 0$ models the end of the string clamped at the point $x = a$. Boundary condition $u_x(a, t) = 0$ models the end of the string, fixed on a weightless ring that glides along a bar perpendicular to the equilibrium position of the string without friction. Conditions

$$u(x, 0) = \varphi(x), \quad u_t(x, 0) = \psi(x).$$

specify the initial deviation and speed of the string.

The simplest equation of the parabolic type (diffusion or heat equation)

$$u_t = a^2 u_{xx}$$

describes, for example, heat distribution in a straight rod that is thermally insulated over its entire length and exchanges heat with the environment only through its ends. In this case, u corresponds to the deviation of the core temperature from the ambient temperature. The boundary condition $u(a, t) = 0$ means that the temperature coincides at the rod point $x = a$ with ambient temperature. The boundary condition $u_x(a, t) = 0$ models the rod with a heat-insulated end, with no heat exchange with the environment. The initial temperature distribution is given by condition $u(x, 0) = \varphi(x)$.

The simplest elliptic equation (two-dimensional Laplace equation)

$$u_{xx} + u_{yy} = 0$$

describes the electrostatic potential in a domain or stationary temperature distribution. As distinct from hyperbolic or parabolic equations, we fix the boundary conditions at the closed surface. We consider the boundary problems of two types: the Dirichlet problem with the value of u set on the boundary of the domain, and the Neumann problem when we assign the value of the normal derivative of u. Physically, one poses the Dirichlet problem when sets the potential or temperature at the boundary and the Neumann when predefines the electric field strength normal component or heat flux density.

Methods for solving second-order linear partial differential equations based on the canonical form are described, for example, in (Tychonov and Samarski, 1964). The use of curvilinear coordinates is presented in (Morse and Feshbach, 1953). The Fourier method and variable separation are discussed in textbooks (Mathews and Walker, 1970; Mikhlin, 1967; Sommerfeld, 1949).

Worked Examples

3.1. Determine the type of equations and reduce them to canonical form:

(a) $u_{xx} + u_{xy} - 2u_{yy} - 3u_x - 15u_y + 27x = 0$;
(b) $u_{xx} + 2u_{xy} + 5u_{yy} - 32u = 0$;
(c) $u_{xx} - 2u_{xy} + u_{yy} + u_x + u_y - u = 0$.

Solution. 3.1(a) The equation everywhere has a hyperbolic type, since $D = 9/4 > 0$. There are two families of characteristics defined by the equations:

$$\frac{dy}{dx} = 2 \quad \text{and} \quad \frac{dy}{dx} = -1.$$

General solutions of these equations take the form:

$$y = 2x + \xi \quad \text{and} \quad y = -x + \eta$$

respectively. Here ξ, η are arbitrary constants. Expressing them in terms of x and y,

$$\xi = y - 2x, \quad \eta = y + x,$$

we see that they are integrals of the characteristic equations that can be used as characteristic variables. Expressing derivatives of u with respect to x and y through derivatives with respect to ξ and η we get the canonical form

$$u_{\xi\eta} + u_\xi + 2u_\eta + 3(\xi - \eta) = 0.$$

Solution. 3.1(b) The equation is elliptical everywhere, since $D = -4 < 0$. The complex equation for the characteristics

$$\frac{dy}{dx} = 1 + 2\mathrm{i}$$

has solution:

$$y = (1 + 2\mathrm{i})x + (\xi - \mathrm{i}\eta).$$

Choosing the real $\xi = y - x$ and imaginary $\eta = 2x$ parts of the integral as new variables, we obtain the canonical form

$$u_{\xi\xi} + u_{\eta\eta} - 8u = 0.$$

Solution. 3.1(c) The equation is of the parabolic type. There is one family of characteristics which are given by the equation

$$\frac{dy}{dx} = -1.$$

One of the new variables is its integral $\xi = x + y$. As for the second one, we can take any independent function, for example, $\eta = x$. With this choice, we get the canonical form

$$u_{\eta\eta} + 2u_\xi + u_\eta - u = 0.$$

Recall that the condition of functional independence is that the Jacobian for a transition from the variables x, y to the variables ξ, η must be nonzero. With another choice of the variable η, the canonical form will be different.

3.2. Find areas of hyperbolicity, parabolicity, and ellipticity of the equations. Reduce them in these areas to a canonical form:

(a) Tricomi equation $u_{xx} - xu_{yy} = 0$;

(b) $xu_{xx} + 2xu_{xy} + (x-1)u_{yy} = 0$.

Solution. 3.2(a) In this case, the type of equation is different in left and right half-planes. Calculating the discriminant, one can see that the equation is hyperbolic at $x > 0$ and has the canonical form

$$u_{\xi\eta} - \frac{u_\xi - u_\eta}{6(\xi - \eta)} = 0, \quad \xi = y + \frac{2}{3}x^{3/2}, \quad \eta = y - \frac{2}{3}x^{3/2}.$$

The characteristics are semicubical parabolas with cuspidal points at line $x = 0$ of type changing. At $x < 0$ the equation is elliptic:

$$u_{\xi\xi} + u_{\eta\eta} - \frac{u_\eta}{3\eta} = 0$$

in coordinates $\xi = y, \ \eta = 2(-x)^{3/2}/3$.

Solution. 3.2(b) The equation is hyperbolic at $x > 0$:

$$u_{\xi\eta} + \frac{u_\xi - u_\eta}{2(\xi - \eta)} = 0, \quad \xi = y - x + 2\sqrt{x}, \quad \eta = y - x - 2\sqrt{x};$$

the elliptic one at $x < 0$

$$u_{\xi\xi} + u_{\eta\eta} - \frac{1}{\eta}u_\eta = 0, \quad \xi = y - x, \quad \eta = 2\sqrt{-x},$$

and the parabolic one at $x = 0$ with the canonical form $u_{yy} = 0$.

3.3. Solve the Cauchy problem:

$$u_{xx} - 2u_{xy} - 3u_{yy} = 0, \quad u(x,0) = 3x^2, \quad u_y(x,0) = 0.$$

Solution. $u_{\xi\eta} = 0$. The equation is hyperbolic, in the variables $\xi = y - x$, $\eta = y + 3x$ is reduced to the canonical form $u_{\xi\eta} = 0$. Its general solution has the form $u = f(\xi) + g(\eta)$, where f, g are arbitrary functions. To determine their form in a particular case, one needs to use the boundary conditions which are conveniently rewritten to the variables ξ, η. In the plane ξ, η the boundary conditions are given on the line $\eta = -3\xi$, obtained from the condition $y = 0$; x as a function of ξ on this line is expressed by formula $x = \dfrac{\xi - \eta}{4} = \xi$. Hence we obtain the first boundary condition in the form

$$u\big|_{y=0} = f(\xi) + g(-3\xi) = 3\xi^2. \tag{3.9}$$

The second boundary condition takes the form

$$u_y\big|_{y=0} = f'(\xi) + g'(-3\xi) = 0, \tag{3.10}$$

where the prime denotes the derivative of a function with respect to its argument, and in brackets it is shown the argument value at which it is calculated.

Differentiating the condition (3.9) with respect to ξ, we have

$$f'(\xi) - 3g'(-3\xi) = 6\xi.$$

From the equations (3.9), (3.10) we obtain

$$f'(\xi) = \frac{3}{2}\xi \quad \text{and} \quad f(\xi) = \frac{3}{4}\xi^2.$$

From the equation (3.9) we find

$$g(-3\xi) = \frac{9}{4}\xi^2.$$

Returning to the variable $\eta = -3\xi$, we obtain

$$g(\eta) = \frac{\eta^2}{4}.$$

The desired partial solution has the form

$$u(x, y) = \frac{3}{4}(y - x)^2 + \frac{1}{4}(y + 3x)^2 = y^2 + 3x^2.$$

3.4. Show that in the spherical coordinates the Laplace operator can be written in the form:

$$\Delta = \frac{1}{r^2}\frac{\partial}{\partial r}\left(r^2\frac{\partial}{\partial r}\right) - \frac{1}{r^2}\hat{\mathbf{l}}^2$$

where $\hat{\mathbf{l}}$ is the angular momentum operator :

$$\hat{\mathbf{l}} = -\mathrm{i}[\mathbf{r} \times \nabla], \quad \hat{\mathbf{l}}^2 = -[\mathbf{r} \times \nabla][\mathbf{r} \times \nabla].$$

Solution. Let write the vector product $[\mathbf{r} \times \nabla]$ in the spherical coordinates using its representation in the form of determinant:

$$[\mathbf{r} \times \nabla] = \begin{vmatrix} \mathbf{e}_r & \mathbf{e}_\theta & \mathbf{e}_\varphi \\ r & 0 & 0 \\ \dfrac{\partial}{\partial r} & \dfrac{1}{r}\dfrac{\partial}{\partial \theta} & \dfrac{1}{r\sin\theta}\dfrac{\partial}{\partial \varphi} \end{vmatrix}.$$

Here

$$\begin{aligned}
\mathbf{e}_r &= (\sin\theta\cos\varphi, \sin\theta\sin\varphi, \cos\theta), \\
\mathbf{e}_\theta &= (\cos\theta\cos\varphi, \cos\theta\sin\varphi, -\sin\theta), \\
\mathbf{e}_\varphi &= (\sin\varphi, \cos\varphi, 0)
\end{aligned} \tag{3.11}$$

are unit mutually orthogonal vectors of the local basis tangent to the lines r, θ, φ. Opening the determinant, we have

$$[\mathbf{r}\times\nabla]^2 = \left(\mathbf{e}_\varphi\frac{\partial}{\partial\theta} - \mathbf{e}_\theta\frac{1}{\sin\theta}\frac{\partial}{\partial\varphi}\right)\left(\mathbf{e}_\varphi\frac{\partial}{\partial\theta} - \mathbf{e}_\theta\frac{1}{\sin\theta}\frac{\partial}{\partial\varphi}\right). \tag{3.12}$$

Here it is necessary to take into account that the differentiations of the left bracket also act to the unit vectors $\mathbf{e}_r$, $\mathbf{e}_\theta$, $\mathbf{e}_\varphi$ of the right brackets. From the definitions (3.11) we get

$$\begin{aligned}
\frac{\partial\mathbf{e}_\varphi}{\partial\theta} &= 0, \quad \frac{\partial\mathbf{e}_\varphi}{\partial\varphi} = \cos\theta\,\mathbf{e}_\theta - \sin\theta\,\mathbf{e}_r, \\
\frac{\partial\mathbf{e}_\theta}{\partial\varphi} &= \cos\theta\,\mathbf{e}_\varphi, \quad \frac{\partial\mathbf{e}_\theta}{\partial\theta} = -\mathbf{e}_r.
\end{aligned} \tag{3.13}$$

Expanding the expression (3.12) and taking into account formulas (3.13), we obtain that only three terms giving the angular part of the Laplace operator are nonzero:

$$[\mathbf{r}\times\nabla]^2 = \frac{\partial^2}{\partial\theta^2} + \frac{\cos\theta}{\sin\theta}\frac{\partial}{\partial\theta} + \frac{1}{\sin^2\theta}\frac{\partial^2}{\partial\varphi^2}.$$

3.5. Separate variables in the Schrödinger equation if potential has the form:

(a) in the cylindrical coordinates

$$U = f(\rho) + \frac{g(\varphi)}{\rho^2} + h(z);$$

(b) in the spherical coordinates

$$U = f(r) + \frac{g(\theta)}{r^2} + \frac{h(\varphi)}{r^2\sin^2\theta}.$$

Solution. 3.5(a) Finding a solution in the form

$$\psi(\rho, \varphi, z) = R(\rho)\Phi(\varphi)Z(z),$$

we divide the Schrödinger equation by $\psi(\rho, \varphi, z)$ and then transform the result to the form

$$\left(-\frac{1}{2\rho R}\frac{\partial}{\partial\rho}\rho\frac{\partial R}{\partial\rho} + f(\rho)\right) + \left(-\frac{1}{2\rho^2\Phi}\frac{\partial^2\Phi}{\partial\varphi^2} + \frac{g(\varphi)}{\rho^2}\right) =$$

$$= \left(\frac{1}{2Z} \frac{\partial^2 Z}{\partial z^2} - h(z) \right) + E = c_1.$$

It is clear that equality is possible only if c_1 is a constant. Similarly, separating the variables ρ and φ, we obtain the equations on Φ and R:

$$-\frac{1}{2} \frac{d^2 \Phi}{d\varphi^2} + g(\varphi)\Phi = c_2 \Phi,$$

$$-\frac{1}{2} \rho \frac{d}{d\rho} \rho \frac{dR}{d\rho} + f(\rho)\rho^2 R - c_1 \rho^2 R = -c_2 R,$$

where c_2 is a constant also.

Solution. 3.5(b) If one looks for a solution in the form $\psi = R(r)P(\theta)\Phi(\varphi)$, then the following system of equations for R, P and Φ arises

$$-\frac{1}{2} \frac{d^2 \Phi}{d\varphi^2} + h(\varphi)\Phi = c_1 \Phi,$$

$$-\frac{1}{2 \sin\theta} \frac{d}{d\theta} \left(\sin\theta \frac{dP}{d\theta} \right) + g(\theta)P + \frac{c_1 P}{\sin^2\theta} = c_2 P,$$

$$-\frac{1}{2r^2} \frac{d}{dr} \left(r^2 \frac{dR}{dr} \right) + \frac{c_2 R}{r^2} = ER,$$

where c_1, c_2 are constants.

3.6. Separate variables in the Schrödinger equation for oscillator with the potential

$$U = \frac{kr^2}{2}$$

(a) in Cartesian coordinates;
(b) in spherical coordinates.

Solution. 3.6(a) If one looks for a solution as product $\psi(x, y, z) = X(x)Y(y)Z(z)$ then the equations is separated into three equations for one-dimensional harmonic oscillator:

$$\frac{d^2}{dx^2} X + kx^2 X = c_1 X,$$

$$\frac{d^2}{dy^2} Y + ky^2 Y = c_2 Y,$$

$$\frac{d^2}{dz^2} Z + kz^2 Z = (2E - c_1 - c_2)Z.$$

Solution. 3.6(b) If one looks for a solution in the form

$$\psi(r,\theta,\varphi) = R(r)P(\theta)\Phi(\varphi),$$

then for $R,\ P,\ \Phi$ equations follow

$$\frac{d^2\Phi}{d\varphi^2} + c_2\Phi = 0\,,$$

$$\sin\theta\frac{d}{d\theta}\left(\sin\theta\frac{dP}{d\theta}\right) - (c_1\sin^2\theta + c_2)P = 0\,,$$

$$\frac{d}{dr}\left(r^2\frac{dR}{dr}\right) + (2Er^2 + c_1 - kr^4)R = 0\,. \qquad (3.14)$$

3.7. Separate variables in the Schrödinger equation using the parabolic coordinates, if the field is a superposition of homogeneous and Coulomb fields:

$$U = -\frac{1}{r} + az.$$

Solution. The Schrödinger equation in the parabolic coordinates has the form (see problem **3.30**(c))

$$\frac{4}{\xi+\eta}\left[\frac{\partial}{\partial\xi}\left(\xi\frac{\partial\psi}{\partial\xi}\right) + \frac{\partial}{\partial\eta}\left(\eta\frac{\partial\psi}{\partial\eta}\right)\right] + \frac{1}{\xi\eta}\frac{\partial^2}{\partial\varphi^2} + 2[E + \frac{2}{\xi+\eta} - \frac{a}{2}(\xi-\eta)] = 0.$$

Seeking for solution in the form

$$\psi = f_1(\xi)f_2(\eta)\Phi(\varphi),$$

one gets the equation

$$\frac{d^2\Phi}{d\varphi^2} + c_1\Phi = 0,$$

$$\frac{d}{d\xi}\left(\xi\frac{df_1}{d\xi}\right) + \left(\frac{E}{2}\xi - \frac{c_1}{4\xi} - a\frac{\xi^2}{4} + \frac{1}{2} + c_2\right) f_1 = 0,$$

$$\frac{d}{d\eta}\left(\eta\frac{df_2}{d\eta}\right) + \left(\frac{E}{2}\eta - \frac{c_1}{4\eta} + a\frac{\eta^2}{4} + \frac{1}{2} - c_2\right) f_2 = 0.$$

3.8. Find an expression for the Laplace operator in an arbitrary curvilinear nonorthogonal coordinate system.

Solution. Note that the eigenvalue problem for Laplace operator $\Delta\psi = \lambda\psi$ with zero boundary conditions at the infinity is equivalent to the variational principle $S[\psi] = \min$ for the functional

$$S[\psi] = \int \left[(\nabla\psi)^2 + \lambda\psi^2\right] d^n x.$$

Let us perform a non-degenerate coordinate transformation, from x variables to the new y variables: $x_i = x_i(y_1, \ldots, y_n), i = 1, 2, \ldots, n$. Here non-degeneracy means that Jacobian of the transformation takes only nonzeroth values:

$$J = \frac{\partial(x_1, x_2, \ldots, x_n)}{\partial(y_1, y_2, \ldots, y_n)} \neq 0.$$

The square of the interval length and the square of the gradient are converted as follows:

$$ds^2 = dx_i dx_i = \frac{\partial x_i}{\partial y_j}\frac{\partial x_i}{\partial y_k} dy_j dy_k, \quad (\nabla\psi)^2 = \frac{\partial y_j}{\partial x_k}\frac{\partial y_l}{\partial x_k}\frac{\partial\psi}{\partial y_j}\frac{\partial\psi}{\partial y_l}.$$

By the definition of the metric tensor g_{jk}, $ds^2 = g_{jk} dy_j dy_k$, it can be expressed as through partial derivatives of the functions x_i,

$$g_{jk} = \frac{\partial x_i}{\partial y_j}\frac{\partial x_i}{\partial y_k}.$$

Therefore, we can write the square of the function ψ gradient in terms of the inverse metric tensor:

$$(\nabla\psi)^2 = g_{jl}^{-1}\frac{\partial\psi}{\partial y_j}\frac{\partial\psi}{\partial y_l},$$

and the determinant of the metric tensor takes the form

$$g = \det\left(g_{jk}\right) = |J|^2.$$

In the new system of coordinates the functional reads as

$$S = \int \left(g_{ik}^{-1}\frac{\partial\psi}{\partial y_i}\frac{\partial\psi}{\partial y_k} + \lambda\psi^2\right)\sqrt{g}\, d^n y.$$

It remains now us to calculate variation of the functional assuming variation of the ψ-function vanishing at infinity that results in the equation:

$$\delta S = 2\int \delta\psi \left[-\frac{1}{\sqrt{g}}\frac{\partial}{\partial y_i}\left(g_{ik}^{-1}\sqrt{g}\frac{\partial\psi}{\partial y_k}\right) + \lambda\psi\right]\sqrt{g}\, d^n y = 0.$$

Derivation of the last equality assumes integration by parts with the nonintegral term vanished due to the boundary conditions. Since the variation of $\delta\psi(x)$ is arbitrary, the expression in square brackets should vanish.

This gives the desired formula for the Laplacian:

$$\Delta = \frac{1}{\sqrt{g}} \frac{\partial}{\partial y_i} g_{ik}^{-1} \sqrt{g} \frac{\partial}{\partial y_k}.$$

For the particular case of orthogonal coordinates we have

$$g_{ik} = h_i^2 \delta_{ik}, \quad g = h \equiv h_1 h_2 h_3, \quad g_{ik}^{-1} = h_i^{-2} \delta_{ik}, \quad \Delta = \frac{1}{h} \frac{\partial}{\partial y_i} \frac{h}{h_i^2} \frac{\partial}{\partial y_i}.$$

3.9. Solve the Cauchy problem

$$u_{tt} - c^2 u_{xx} = 0,$$

$$0 \leqslant x \leqslant L, \quad u(0,t) = u(L,t) = 0, \quad u(x,0) = \varphi(x), \quad u_t(x,0) = \psi(x).$$

Consider the partial case

$$u(x,0) = 0, \quad u_t(x,0) = \sin\frac{2\pi x}{L}.$$

Solution. The solution to the problem by the Fourier method splits into two steps.

The first step consists in finding of partial solutions for which only the boundary conditions along the coordinate x are satisfied. Looking for $u(x,t)$ in the form

$$u(x,t) = X(x)T(t)$$

and separating variables, we have

$$\frac{X''(x)}{X(x)} = \frac{T''(t)}{c^2 T(t)} = -\lambda, \tag{3.15}$$

where λ is an arbitrary constant. The general solution to the equation

$$\frac{X''(x)}{X(x)} + \lambda = 0 \tag{3.16}$$

for $\lambda > 0$ has the form

$$X = A\cos\sqrt{\lambda}x + B\sin\sqrt{\lambda}x. \tag{3.17}$$

Function $X(x)$ must satisfy the same boundary conditions as $u(x,t)$:

$$X(0) = X(L) = 0, \tag{3.18}$$

then we have

$$A = 0 \quad \text{and} \quad \sqrt{\lambda}L = n\pi, \qquad n = 1, 2, \ldots$$

Hence the spectrum of admissible values of λ and the spatial form of the solution with number n are obtained:

$$\lambda_n = \left(\frac{n\pi}{L}\right)^2, \qquad X_n(x) = \sin\frac{n\pi x}{L}.$$

Draw their graphs at $n = 1, 2, 3$.

Substituting λ_n in equation (3.15) we find the time dependence of the n-th solution:

$$T_n(t) = A_n \cos\frac{n\pi ct}{L} + B_n \sin\frac{n\pi ct}{L},$$

where A_n, B_n are arbitrary constants.

The solutions of $X_n(x)$ form a complete basis. The general solution to the problem can be written as linear combination of fundamental solutions:

$$u(x, t) = \sum_{n=1}^{\infty} X_n(x)T_n(t). \tag{3.19}$$

At the second step, the amplitudes A_n, B_n are expressed in terms of the initial conditions $u(x, 0) = \varphi(x)$, $u_t(x, 0) = \psi(x)$. For $t = 0$, using representation of the solution (3.19), we have

$$\varphi(x) = \sum_{n=1}^{\infty} A_n X_n(x), \tag{3.20}$$

$$\psi(x) = \sum_{n=1}^{\infty} \lambda_n B_n X_n(x). \tag{3.21}$$

The eigen solutions to the problem (3.16), (3.18) are mutually orthogonal:

$$\int_0^L X_n(x)X_m(x)dx = \frac{L}{2}\delta_{nm}, \tag{3.22}$$

Multiplying the equations (3.20), (3.21) by $X_m(x)$, integrating from zero to L and taking into account the orthogonality of the eigen solutions, finally we have

$$A_m = \frac{2}{L}\int_0^L \varphi(x)X_m(x)dx, \quad B_m = \frac{2}{\lambda_m L}\int_0^L \psi(x)X_m(x)dx. \tag{3.23}$$

The partial solution is

$$u(x, t) = \frac{L}{2\pi c}\sin\frac{2\pi ct}{L}\sin\frac{2\pi x}{L}.$$

3.10. Solve the problem

$$u_{tt} - c^2 u_{xx} = 0,$$

$$0 \leqslant x \leqslant L, \quad u(0, t) = u_x(L, t) = 0,$$

$$u(x, 0) = \sin\frac{5\pi x}{2L}, \quad u_t(x, 0) = \sin\frac{\pi x}{2L}.$$

Try to solve this problem under the condition $u_t(x, 0) = \cos\frac{\pi x}{2L}$.

Solution. The same scheme performs the solution to the problem as for problem **3.9**. The difference is only with solving the spectral problem (3.16) on function $X(x)$, which changed due to boundary conditions:

$$X(0) = X'(L) = 0.$$

From the general solution (3.17) we get the equation for λ:

$$\sqrt{\lambda}\cos\sqrt{\lambda}L = 0.$$

Its solution is

$$\lambda_n = \left(\frac{2n-1}{2L}\pi\right)^2, \quad n = 1, 2, \ldots$$

The corresponding eigenfunctions are

$$X_n(x) = \sin\frac{2n-1}{2L}\pi x.$$

Draw their graphs for $n = 1, 2, 3$.

Note that $\lambda = 0$ is not an eigenvalue, since it corresponds to the trivial solution $X = 0$. The solution is given by formulas (3.19)–(3.21). In the partial case, calculating the coefficients we obtain

$$u(x, t) = \cos\frac{5\pi ct}{2L}\sin\frac{5\pi x}{2L} + \frac{2L}{\pi}\sin\frac{\pi ct}{2L}\sin\frac{\pi x}{2L}.$$

At $u_t(x, 0) = \cos(\pi x/(2L))$ the problem has no solution. The reason is that such an initial condition is not consistent with the boundary condition $u(0, t) = 0$, which implies that there must be $u_t(0, t) = 0$ for any moment.

3.11. Solve the problem

$$u_{tt} - u_{xx} = 0,$$

$$0 \leqslant x \leqslant L, \quad u_x(0, t) = u_x(L, t) = 0,$$

$$u(x, 0) = x, \quad u_t(x, 0) = 1.$$

Describe the string motion.

Solution. Spectral problem

$$X'' + \lambda X = 0, \quad X'(0) = X'(L) = 0$$

has non-trivial solutions

$$X_n(x) = \cos\frac{n\pi}{L}x, \quad \lambda_n = \left(\frac{n\pi}{L}\right)^2, \quad n = 0, 1, \ldots$$

In this case, $\lambda = 0$ is the eigenvalue, with the eigenfunction $X_0 = 1$. The time dependence of this solution has non-oscillatory behavior since from $T'' = 0$ follows $T = A_0 + B_0 t$. We write the general solution in the form:

$$u(x, t) = A_0 + B_0 t + \sum_{n=1}^{\infty}\left(A_n \cos\frac{n\pi}{L}t + B_n \sin\frac{n\pi}{L}t\right)\cos\frac{n\pi x}{L}.$$

Using the initial conditions one can see that A_n, B_n are the Fourier coefficients of the expansion of these initial conditions. Calculating them we arrive at

$$u(x, t) = t + \frac{L}{2} - \frac{4L}{\pi^2}\sum_{n=0}^{\infty}\frac{1}{(2n+1)^2}\cos\frac{(2n+1)\pi}{L}t\cos\frac{(2n+1)\pi}{L}x.$$

The string is oscillating relative to its mean position which drifts with the constant velocity. Note that the initial condition does not satisfy the boundary condition: $u_x(0, 0) = u_x(L, 0) = 1 \neq 0$. Although the series converges to a solution, it is impossible to calculate its derivative with respect to x, since in this case the series diverges.

3.12. A string of length L with fixed ends in the initial time has a parabolic shape with a maximum deviation in its middle equal to h. The initial string speed is zero. Find the dependence of the deviation on time.

Solution. The problem has the form

$$u_{tt} - c^2 u_{xx} = 0,$$

$$u(0, t) = u(L, t) = 0,$$

$$u(x, 0) = \frac{4hx(L-x)}{L^2}, \quad u_t(x, 0) = 0.$$

The general solution has been already presented in the solution to problem **3.9**. From the zeroth initial velocity follows that all B_n are equal zero. Coefficients A_n are calculated by the Fourier expansion of $u(x, 0)$:

$$\frac{4hx(L-x)}{L^2} = \sum_{n=1}^{\infty} A_n \sin\frac{n\pi x}{L}.$$

Hence we get

$$u(x, t) = \frac{32h}{\pi^3} \sum_{n=0}^{\infty} \frac{1}{(2n+1)^3} \cos \frac{(2n+1)\pi ct}{L} \sin \frac{(2n+1)\pi x}{L}.$$

3.13. At the initial time, a string of length L with fixed ends is shifted for distance h at point x_0 and has zero velocity. Find the solution of the Cauchy problem.

Solution. This initial condition differs from the previous one:

$$u(x, 0) = \begin{cases} \dfrac{x}{x_0} h, & x < x_0, \\ \dfrac{L-x}{L-x_0} h, & x > x_0. \end{cases}$$

The solution is obtained by expanding the initial deviation in a Fourier series,

$$u(x, t) = \frac{2hL^2}{\pi^2 x_0 (L - x_0)} \sum_{n=1}^{\infty} \frac{1}{n^2} \sin \frac{n\pi x_0}{L} \cos \frac{n\pi c}{L} t \sin \frac{n\pi}{L} x. \tag{3.24}$$

Let us now analyze qualitative characteristics of the string motion. The solution is periodic in time with a period $\tau = \dfrac{2L}{c}$. At $t = \dfrac{\tau}{4}$ deviation of the string from equilibrium position is zero. To get the string shape for $0 < t < \dfrac{\tau}{4}$ let express the product of trigonometric functions through the sum

$$\cos \frac{n\pi c}{L} t \sin \frac{n\pi}{L} x = \frac{1}{2} \left(\sin \frac{n\pi}{L}(x - ct) + \sin \frac{n\pi}{L}(x + ct) \right). \tag{3.25}$$

The Fourier series implies that we continue the solution given by the formula (3.24) as odd function to negative $-L < x < 0$ and periodic repetition to the whole axis. The function obtained in this way will hereinafter be called the initial deviation.

If we substitute the expression (3.25) into a series (3.24) and collect the terms containing $x - ct$ terms, then the resulting sum is the expansion of the initial half-amplitude deviation shifted to the right by the distance ct. The remaining terms give half the initial deviation shifted to the left by the same distance. Thus,

$$u(x, t) = \frac{1}{2} \left(u(x - ct, 0) + u(x + ct, 0) \right).$$

For example, for $x_0 = L/2$, the initial deviation represents a sawtooth function. Its profile is an isosceles triangle. Graphically, adding up two

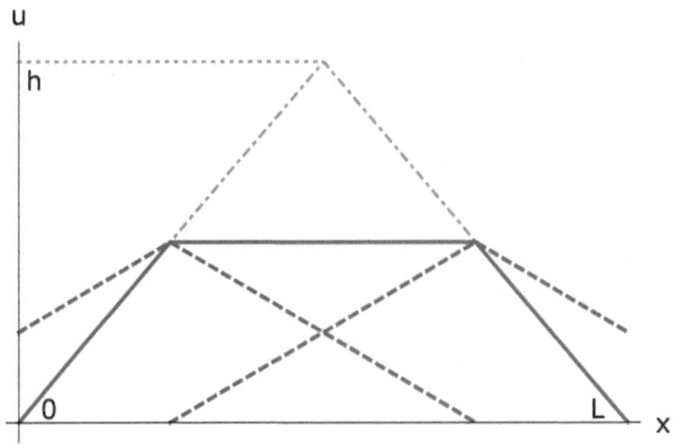

Figure 3.1 String form: dot-dashes corresponds to the initial condition $u(x, 0)$; dashed lines to $\frac{1}{2}u(x \pm L/5, 0)$; solid line is their sum $u(x, L/5c)$.

sawtooth slightly shifted curves, it is easy to obtain that for $t < \tau/4$ the string has the shape of isosceles trapezium depicted in Fig. 3.1. The horizontal part, expanding, falls with speed equal to $4h/\tau$.

3.14. Solve the boundary problem for the inhomogeneous equation

$$u_{tt} - u_{xx} = b \sinh x,$$

$$0 \leqslant x \leqslant L, \quad u(0, t) = u(L, t) = u(x, 0) = u_t(x, 0) = 0.$$

Solution. Solving an inhomogeneous equation, as well as solving a homogeneous equation, can be sought in the form of an expansion in the fundamental solution set of the (homogeneous) boundary problems that for the given boundary conditions were obtained in the problem **3.9**. The solution in the form

$$u(x, t) = \sum_{n=1}^{\infty} A_n(t) \sin \frac{n\pi x}{L} \tag{3.26}$$

satisfies the boundary conditions. To determine $A_n(t)$, we expand $b \sinh x$ in a set of fundamental solutions:

$$b \sinh x = -2b \sum_{n=1}^{\infty} \frac{(-1)^n n\pi \sinh L}{(n\pi)^2 + L^2} \sin \frac{n\pi x}{L}. \tag{3.27}$$

Substituting expressions (3.26), (3.27) to the equation with account of orthogonality of functions $\sin(n\pi x/L)$ gives the following system of differential equations,

$$\frac{d^2 A_n}{dt^2} + \left(\frac{n\pi}{L}\right)^2 A_n = -2b\frac{(-1)^n n\pi \sinh L}{(n\pi)^2 + L^2}, \quad n = 1, 2, \ldots \tag{3.28}$$

with the initial conditions which are defined from the initial conditions for $u(x, t)$:

$$A_n(0) = \frac{dA_n}{dt}|_{t=0} = 0 \quad n = 1, 2, \ldots \tag{3.29}$$

Solving the problem (3.28),(3.29) and substituting $A_n(t)$ in (3.26), yield

$$u(x, t) = 2bL^2 \sinh L \sum_{n=1}^{\infty} \frac{(-1)^n}{n\pi((n\pi)^2 + L^2)} \left(\cos\frac{n\pi t}{L} - 1\right) \sin\frac{n\pi x}{L}. \tag{3.30}$$

A different way to solve the problem is splitting into two subproblems. First, we solve the stationary boundary problem

$$-w_{xx} = b \sinh x, \quad w(0) = w(L) = 0.$$

Then, by the Fourier method, we find a solution to the boundary problem for the homogeneous equation:

$$v_{tt} - v_{xx} = 0,$$

$$v(0, t) = v(L, t) = 0, \quad v(x, 0) = -w(x), \quad v_t(x, 0) = 0.$$

By substitution, one can see that $u(x, t) = v(x, t) + w(x)$ satisfies both initial and boundary conditions for the given problem. The final answer is

$$u(x, t) = b\left(\frac{x}{L}\sinh L - \sinh x\right) +$$

$$+2bL^2 \sinh L \sum_{n=1}^{\infty} \frac{(-1)^n}{n\pi((n\pi)^2 + L^2)} \cos\frac{n\pi t}{L} \sin\frac{n\pi x}{L}. \tag{3.31}$$

From a physical point of view, this solution means that $w(x)$ defines the stationary deflection of the string under the action of the external force $b \sinh x$; $v(x, t)$ describes the oscillations around the stationary state. Note that the stationary deflection of the string in (3.31) is obtained in the explicit form, but in (3.30) as expansion in a series.

3.15. Solve the Cauchy problem at the interval $0 \leqslant x \leqslant L$:

$$u_t = a^2 u_{xx},$$

$$u(0, t) = u(L, t) = 0, \quad u(x, 0) = \begin{cases} x; & x < L/2, \\ L - x; & x > L/2 \end{cases}.$$

Compare its solution with that of **3.13**.

Solution. By separation of variables, one can see that the obtained spectral problem is the same as for problem **3.9**. The time dependence follows from equation

$$T_n' = -\lambda_n T_n$$

and now has an exponentially decaying character:

$$T_n = A_n \exp\left[-\left(\frac{n\pi}{L}\right)^2 a^2 t\right].$$

The expansion of the initial conditions in the Fourier series coincides with the obtained in the problem **3.9**. As a result, we have

$$u(x, t) = \frac{4L}{\pi^2}\sum_{n=0}^{\infty}\frac{(-1)^n}{(2n+1)^2}\exp\left[-\left(\frac{(2n+1)\pi}{L}\right)^2 a^2 t\right]\sin\frac{2n+1}{L}\pi x.$$

Note that the solution is symmetric relative to the interval middle, $u(x, t) = u(L - x, t)$ and this property is a consequence of the boundary conditions symmetry.

3.16. Solve the problem

$$u_t = a^2 u_{xx} - \beta u,$$

$$0 \leqslant x \leqslant L, \quad u(0, t) = u_x(L, t) = 0, \quad u(x, 0) = \sin\frac{\pi x}{2L}.$$

Solution. The problem can be solved, as well as the problem **3.14**, by expansion over the eigenfunctions. Another way is to make the substitution $u = e^{-\beta t} v$ and for v get a homogeneous diffusion equation with the same boundary and initial conditions, as for u. Then solving the problem for v by the Fourier method, we obtain

$$u(x, t) = \exp\left[-\beta t - \left(\frac{\pi}{2L}\right)^2 a^2 t\right]\sin\frac{\pi x}{2L}.$$

3.17. Solve the boundary problem

$$u_t = a^2 u_{xx} - \beta u + \sin\frac{\pi x}{L},$$

$$0 \leqslant x \leqslant L, \quad u(0, t) = u(L, t) = 0, \quad u(x, 0) = 0.$$

Solution. The problem solution like that of the problem **3.14** can be solved by the series expansion through the fundamental functions. A different way is to look for a solution in the form

$$u = v(x, t) + w(x),$$

where $w(x)$ is a solution for the stationary heat equation:

$$a^2 w_{xx} - \beta w + \sin \frac{\pi x}{L} = 0, \quad w(0) = w(L) = 0.$$

$w(x)$ defines the stationary temperature distribution along the rod. Partial solution to the linear inhomogeneous equation satisfying boundary conditions, has the form

$$w = \frac{1}{\left(\frac{\pi a}{L}\right)^2 + \beta} \sin \frac{\pi x}{L}.$$

Substituting $v + w$ in the initial equations and satisfying the boundary conditions, for $v(x, t)$ we arrive at the problem

$$v_t = a^2 v - \beta v,$$

$$v(0, t) = v(L, t) = 0, \quad v(x, 0) = -w(x),$$

which solution can be found by the same way as for the problem **3.16**:

$$u = \frac{1}{\left(\frac{\pi a}{L}\right)^2 + \beta} \left[1 - \exp\left(-\left(\beta + \left(\frac{\pi a}{L}\right)^2\right) t\right)\right] \sin \frac{\pi x}{L}.$$

3.18. At the initial moment, the ball has spherically symmetric temperature distribution $u(r, 0) = f(r)$. Find the dependence of the temperature distribution on time if the temperature of the ball surface is zero. Consider case $f(r) = T_0$.

Solution. Heat equation

$$u_t = a^2 \bigtriangleup u$$

in the spherical coordinates is written as

$$u_t = \frac{a^2}{r^2} \frac{\partial}{\partial r} r^2 \frac{\partial u}{\partial r}. \tag{3.32}$$

Note that in the Laplace operator there is absent dependence on angles. The boundary and initial conditions are

$$u(R, t) = 0, \quad u(r, 0) = f(r).$$

First, we look for fundamental solutions by separation of variables

$$u(r,t) = P(t)Q(r). \tag{3.33}$$

Substituting (3.33) in equation (3.32) leads to

$$\frac{P'(t)}{a^2 P(t)} = \frac{1}{Q}\frac{1}{r^2}\frac{d}{dr}r^2\frac{dQ}{dr} = -\lambda. \tag{3.34}$$

For the function Q we need to satisfy the boundary conditions. The first one follows directly from the boundary condition on the function u on the ball surface:

$$Q(R) = 0. \tag{3.35}$$

The second boundary condition is not so obvious and is generated by the fact that equation on $Q(r)$

$$\frac{d^2Q}{dr^2} + \frac{2}{r^2}\frac{dQ}{dr} + \lambda Q = 0 \tag{3.36}$$

has a singularity at the point $r = 0$, and therefore its general solution diverges at this point. Since we are only interested in bounded solutions, the second boundary condition is

$$|Q(0)| < \infty. \tag{3.37}$$

The solution to the equation (3.36) can be found in the form

$$Q(r) = p(r)q(r), \tag{3.38}$$

so that, by choosing $p(r)$, the equation for $q(r)$ does not contain the first derivative. Substituting (3.38) into (3.36), we get the condition on $p(r)$ and the equation on q:

$$p' + \frac{1}{r}p = 0, \quad pq'' + \left(p'' + \frac{2}{r}p'\right)q + \lambda pq = 0.$$

The equation on p is homogeneous and has the solution $p = 1/r$. The equation on q is simplified to an elementary one:

$$q'' + \lambda q = 0.$$

The general solution to the equation (3.36) for $\lambda > 0$ has the form

$$Q(r) = \frac{A\cos\sqrt{\lambda}r + B\sin\sqrt{\lambda}r}{r}.$$

By satisfying the boundary conditions (3.35), (3.37), the eigen solutions and corresponding eigenvalues (3.36) are

$$Q_n(r) = \frac{1}{r}\sin\frac{n\pi r}{R}, \quad \lambda_n = \frac{n\pi}{R}, \quad n = 1, 2, \ldots$$

Eigenvalues $\lambda < 0$ are impossible (why?). The time dependence for the eigen solution is obtained from equations (3.34):

$$T_n(t) = a_n \exp\left[-\left(\frac{n\pi a}{R}\right)^2 t\right],$$

and, respectively, the general solution is written as

$$u(r,t) = \sum_{n=1}^{\infty} a_n T_n(t) Q_n(r).$$

The coefficients a_n are determined from the initial conditions. To calculate them, we multiply the series by r and use the orthogonality of trigonometric functions. Then putting $t = 0$, we have

$$rf(r) = \sum_{n=1}^{\infty} a_n \sin\frac{n\pi r}{R} \quad \text{where} \quad a_n = \frac{2}{R}\int_0^R \rho f(\rho)\sin\frac{n\pi\rho}{R}d\rho.$$

In the particular case $f(r) = T_0$, the integral is calculated and the solution takes the form.

$$u(r,t) = \frac{2RT_0}{\pi r}\sum_{n=1}^{\infty}\frac{(-1)^{n+1}}{n}\exp\left[-\left(\frac{n\pi a}{R}\right)^2 t\right]\sin\frac{n\pi r}{R}.$$

3.19. Find the temperature distribution in the ball, if in the first moment in its center a point heat source of intensity Q is switched on. The initial temperature inside the ball is equal to zero, with its zero surface temperature.

Solution. The heat propagation in a space with distributed sources is described by the equation

$$u_t = \frac{\kappa}{c}\,\Delta\,u + \frac{F}{c}.$$

Here u is the temperature, κ is the thermal conductivity, c is the heat capacity of the substance per unit volume, F is the heat release intensity for unit volume. In our particular case, we obtain a boundary problem with the heat release density described by the Dirac δ-function:

$$u_t = \frac{\kappa}{c}\Delta u + \frac{Q\delta(\mathbf{r})}{c}, \quad u(r,0) = 0, \quad u(R,t) = 0.$$

We split the problem into two: stationary one

$$\Delta w = -\frac{Q\delta(\mathbf{r})}{\kappa}, \quad w(R) = 0, \tag{3.39}$$

and non-stationary

$$v_t = \frac{\kappa}{c}\Delta v, \quad v(R, t) = 0, \quad v(r, 0) = -w(r).$$

Their sum is a solution to the original problem.

First, we solve the stationary problem. For $r \neq 0$ the equation for w is homogeneous. Writing the radial part of the Laplace operator, we have

$$\frac{d}{dr} r^2 \frac{dw}{dr} = 0.$$

Twice integration gives

$$w = \frac{c_1}{r} + c_2.$$

Here constant c_1 is determined by integrating the equation (3.39) over a ball of radius $r < R$. The integral of its right-hand side is $-Q/\kappa$, and the integral from the left side transforms into an integral over the sphere from grad w with use of the Gauss theorem, since $\Delta = \text{div grad}$. Hence we have

$$c_1 = \frac{Q}{4\pi\kappa}.$$

c_2 is determined from the boundary condition $w(R) = 0$. Totally it gives

$$w(r) = \frac{Q}{4\pi\kappa}\frac{R-r}{Rr}.$$

The non-stationary part of the problem is solved by the separation of variables by the same procedure as in **3.18**. Finally, we obtain

$$u(r, t) = \frac{Q}{4\pi\kappa}\frac{R-r}{Rr} - \frac{Q}{2\pi^2\kappa r}\sum_{n=1}^{\infty}\frac{1}{n}\exp\left[-\left(\frac{n\pi a}{R}\right)^2 t\right]\sin\frac{n\pi r}{R}.$$

3.20. Solve the boundary value problem

$$u_{xx} + u_{yy} = 0, \quad 0 < x < L, \quad 0 < y < \infty,$$

$$u(0, y) = u(L, y) = 0, \quad u(x, 0) = A\frac{(L-x)x}{L^2}, \quad u(x, \infty) = 0.$$

Compare with problems **3.12** and **3.36**.

Solution. Finding a partial solution in the form $u(x, y) = X(x)Y(y)$ and separating the variables we get

$$\frac{X''}{X} = -\frac{Y''}{Y} = -\lambda.$$

The spectral problem along the coordinate x turns out to be the same as in **3.9**. The dependence on y is obtained from equation

$$Y_n'' = \lambda_n Y_n,$$

which solution is represented as the sum of the exponents:

$$Y_n = A_n \exp\left(-\left(\frac{n\pi}{L}\right)x\right) + B_n \exp\left(\left(\frac{n\pi}{L}\right)x\right).$$

Writing the solution in the form of the sum of a series, from the boundary condition for $y \to \infty$ we have $B_n = 0$ for all n. From the boundary conditions at $y = 0$ we calculate A_n. The final answer is

$$u(x, t) = \frac{8A}{\pi^3}\sum_{n=0}^{\infty}\frac{(-1)^n}{(2n+1)^3}\exp\left[-\frac{(2n+1)\pi}{L}y\right]\sin\frac{2n+1}{L}\pi x.$$

3.21. Find potential distribution between coaxial cylinders of radii $a < b$ for the Dirichlet problem:

$$u(a, \varphi) = c, \quad u(b, \varphi) = h\cos\varphi.$$

Solution. It is convenient to solve problems with cylindrical symmetry in the polar coordinates. The Laplace equation, with the absence of dependence on the coordinate z, has the form

$$\frac{1}{r}\frac{\partial}{\partial r}r\frac{\partial u}{\partial r} + \frac{1}{r^2}\frac{\partial^2 u}{\partial \varphi^2} = 0. \quad (3.40)$$

We are looking for partial solutions in the form

$$u = R(r)\Phi(\varphi). \quad (3.41)$$

Separating the variables, we obtain the equations for Φ and R:

$$\Phi'' + \lambda\Phi = 0, \quad (3.42)$$

$$\frac{d^2R}{dr^2} + \frac{1}{r}\frac{dR}{dr} - \lambda = 0. \quad (3.43)$$

The general solution to equation (3.42) reads as

$$\Phi = A\cos\sqrt{\lambda}\varphi + B\sin\sqrt{\lambda}\varphi \quad \text{at} \quad \lambda > 0,$$

$$\Phi = Ae^{\sqrt{-\lambda}\varphi} + Be^{-\sqrt{-\lambda}\varphi} \quad \text{at} \quad \lambda < 0.$$

Since the solution u must be one-valued, variation φ on 2π should not change the value of Φ. For $\lambda < 0$ the zero solution is only periodic. For $\lambda > 0$ the function Φ will be periodic with the period 2π if $\lambda = m^2$ where $m = 0, 1, \ldots$. Eigen solutions have the form

$$\Phi_m = A_m \cos m\varphi + B_m \sin m\varphi.$$

Equation (3.43) is homogeneous and therefore its solution for $\lambda = m^2$ is sought as $R = r^\alpha$. It gives

$$\alpha = \pm m.$$

For $m \neq 0$ there are two linearly independent solutions: r^m and r^{-m}. For $m = 0$ the general solution can be obtained by direct integration :

$$\Phi_0 = A_0 \ln r + B_0.$$

Thus, the general solution is written as sum

$$u(r,\varphi) = A_0 \ln r + B_0 + \sum_{m=1}^{\infty}\left[\left(A_m r^m + \frac{A_{-m}}{r^m}\right)\cos m\varphi + \left(B_m r^m + \frac{B_{-m}}{r^m}\right)\sin m\varphi\right].$$

We find coefficients A_m, B_m by expanding the boundary conditions in the Fourier series.

For the given boundary conditions the series is simplified up to the form

$$u(r,\varphi) = A_0 \ln r + B_0 + \left(A_1 r + \frac{A_{-1}}{r}\right)\cos\varphi.$$

At $r = a$ we have

$$A_0 \ln a + B_0 = c_1, \quad A_1 a + \frac{A_{-1}}{a} = 0.$$

At $r = b$ we obtain

$$A_0 \ln b + B_0 = 0, \quad A_1 b + \frac{A_{-1}}{b} = h.$$

Finding the coefficients from this system of equations we get the final answer

$$u(r,\varphi) = c\frac{\ln b - \ln r}{\ln b - \ln a} + h\frac{b(r^2 - a^2)}{(b^2 - a^2)r}\cos\varphi.$$

3.22. Find the stationary deflection of a rectangular membrane $a \times b$ with a fixed border and surface tension T, if homogeneous pressure P acts to the membrane.

Solution. The equation of the membrane equilibrium has the form

$$-\Delta u = \frac{P}{T}. \tag{3.44}$$

The bias u at the border of the rectangle is zero. Let the axis x be along the side of length a and the axis y along the second side. To solve this boundary value problem, we expand u in a series over the Laplace operator's eigenfunctions.

Eigenfunctions and their corresponding eigenvalues are found for the problem solution by separation of variables

$$-\Delta u = \lambda u\,, \quad u|_{\Gamma} = 0\,, \tag{3.45}$$

. that gives the following expressions for both eigenfunctions and eigenvalues

$$u_{n,m} = \sin\frac{n\pi x}{a}\sin\frac{m\pi y}{b}, \quad \lambda_{n,m} = \left(\frac{n\pi}{a}\right)^2 + \left(\frac{m\pi}{b}\right)^2,$$

where n, m are positive integer numbers.

Expanding the right hand side of equation (3.44) in the double Fourier series,

$$\frac{P}{T} = \sum_{n,m=1}^{\infty} c_{n,m}u_{n,m}\,, \tag{3.46}$$

we obtain the non-zeroth coefficients

$$c_{2k+1,2l+1} = \frac{4}{\pi^2}\frac{P}{T}\frac{2}{(2k+1)}\frac{2}{(2l+1)}, \qquad k, l = 0, 1, \ldots$$

Let seek for a solution to equation (3.44) in the same expansion:

$$u = \sum_{n,m=1}^{\infty} d_{n,m}u_{n,m}\,. \tag{3.47}$$

Substituting then expansions (3.46) and (3.47) in the equation and using the mutual orthogonality of functions $u_{n,m}$, we get $d_{n,m} = c_{n,m}/\lambda_{n,m}$. Finally, we have

$$u(x,y)=\frac{16}{\pi^2}\frac{P}{T}\sum_{k,l=0}^{\infty}\frac{\sin\frac{(2k+1)\pi x}{a}\sin\frac{(2l+1)\pi y}{b}}{(2k+1)(2l+1)\left[\left(\frac{(2k+1)\pi}{a}\right)^2+\left(\frac{(2l+1)\pi}{b}\right)^2\right]}.$$

Exercises

3.23. Determine the type of equation and reduce it to canonical form:

(a) $u_{xx}-(1+y^2)^2u_{yy}-2y(1+y^2)u_y=0$;
(b) $u_{xx}-2\sin x\,u_{xy}-\cos^2 x\,u_{yy}-\cos x\,u_y=0$;
(c) $(1+x^2)^2u_{xx}+u_{yy}+2x(1+x^2)u_x=0$;
(d) $\frac{1}{x}\frac{\partial}{\partial x}\left(x\frac{\partial u}{\partial x}\right)+\frac{1}{x^2}\frac{\partial^2 u}{\partial y^2}=0$;
(e) $x^2u_{xx}+2xyu_{xy}+y^2u_{yy}-2yu_x+ye^{y/x}=0$.

3.24. Find regions of hyperbolicity, parabolicity and ellipticity of equation

$$xu_{xx}+yu_{yy}+2u_x+2u_y=0$$

and in each regions reduce it to the canonical form.

3.25. Find the ellipticity condition for Chaplygin equation

$$\frac{\partial}{\partial x}(\rho\varphi_x)+\frac{\partial}{\partial y}(\rho\varphi_y)=0,$$

where $\rho=\rho(\sqrt{\varphi_x^2+\varphi_y^2})$.

3.26. Reduce the following equations to their canonical form with excluding the first derivatives:

(a) $u_{xx}-4u_{xy}+5u_{yy}-3u_x+u_y+u=0$;
(b) $u_{xx}-u_{yy}+u_x+u_y-4u=0$.

3.27. Find the general solution for equation

$$(x-y)u_{xy}-u_x+u_y=0.$$

3.28. Solve the Cauchy problem:

$$4y^2u_{xx}+2(1-y^2)u_{xy}-u_{yy}-\frac{2y}{1+y^2}(2u_x-u_y)=0,$$

$$u(x,0)=\varphi(x),\quad u_y(x,0)=\psi(x).$$

3.29. Find the expression for the Laplace operator in an arbitrary orthogonal coordinate system.

3.30. Obtain expressions for the Laplace operator:

(a) in cylindrical coordinates (z, ρ, φ)

$$x = \rho \cos\varphi, \quad y = \rho \sin\varphi, \quad z = z;$$

(b) in spherical coordinates (r, θ, φ)

$$x = r \sin\theta \cos\varphi, \quad y = r \sin\theta \cos\varphi, \quad z = r \cos\theta;$$

(c) in parabolic coordinates $(\xi, \eta, \varphi) : 0 \leqslant \xi < \infty, \ 0 \leqslant \eta < \infty, \ 0 \leqslant \varphi < 2\pi$

$$x = \sqrt{\xi\eta} \cos\varphi, \quad y = \sqrt{\xi\eta} \sin\varphi, \quad z = \frac{1}{2}(\xi - \eta);$$

(d) in prolate spheroidal coordinates $(\xi, \eta, \varphi) : 1 \leqslant \xi < \infty, \ -1 \leqslant \eta \leqslant 1, \ 0 \leqslant \varphi < 2\pi$

$$x = \frac{1}{2}\sqrt{(\xi^2 - 1)(1 - \eta^2)} \cos\varphi, \quad y = \frac{1}{2}\sqrt{(\xi^2 - 1)(1 - \eta^2)} \sin\varphi,$$

$$z = \frac{1}{2}\xi\eta.$$

3.31. Separate variables in spheroidal coordinates $u = X(\xi)Y(\eta)\mathrm{e}^{\mathrm{i}m\varphi}$ for the Helmholtz equation $(\Delta + k^2)u = 0$.

3.32. Find expressions for gradient, divergence, and curl:

(a) in cylindrical coordinates;

(b) in spherical coordinates.

3.33. Find expressions for z-component of the angular momentum operator in cylindrical and spherical coordinates:

$$l_z = -\mathrm{i}\left(x\frac{\partial}{\partial y} - y\frac{\partial}{\partial x}\right).$$

3.34. Separate variables and find eigenfunctions of the Schrödinger equation for free motion $(U = 0)$ inside a rectangular parallelepiped with sizes $a \times$

$b \times c$ assuming the wave function being equal to zero at the parallelepiped boundary.

3.35. Find the oscillations of a string with a fixed left end. Its right end, sliding freely without friction along the bar, in the initial time is situated in the distance h.

3.36. Solve boundary value problem

$$u_t = a^2 u_{xx},$$

$$0 \leqslant x \leqslant L, \quad u(0,t) = u(L,t) = 0, \quad u(x,0) = \frac{cx(L-x)}{L^2}.$$

Compare its solution with **3.12**.

3.37. Solve the Cauchy problem

$$u_t = a^2 u_{xx},$$

$$0 \leqslant x \leqslant L, \quad u(0,t) = u_x(L,t) = 0, \quad u(x,0) = Ax.$$

3.38. Find the potential distribution in the half-strip $0 < x < \infty$, $0 < y < L$.

$$u(x,0) = u_y(x,L) = 0, \quad u(0,y) = f(y), \quad u(\infty, y) = 0.$$

3.39. Find potential distribution between coaxial cylinders of radii $a < b$ for the Dirichlet problem:

$$u(a,\varphi) = 0, \quad u(b,\varphi) = Ab^2 \sin^2 \varphi.$$

3.40. Solve the Laplace equation $\Delta u = 0$ in a rectangle $a \times b$ with boundary conditions

(a)

$$u_x|_{x=0} = 0, \quad u_y\big|_{y=0} = Q_1, \quad u_x|_{x=a} = Q_2, \quad u_y\big|_{y=b} = 0,$$

see Fig. 3.2.

(b)

$$u_x|_{x=0} = 0, \quad u_y\big|_{y=0} = Q_1 \cos^2 \frac{\pi x}{a}, \quad u_x|_{x=a} = Q_2, \quad u_y\big|_{y=b} = 0,$$

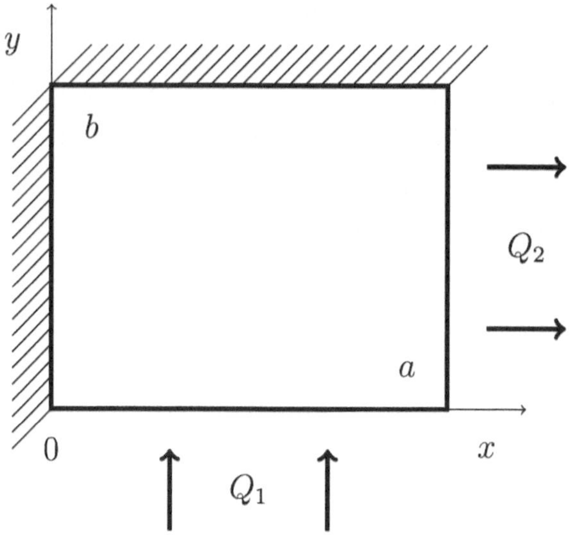

Figure 3.2 Boundary conditions for problem **3.40**(a).

Answers

3.23(a). The equation has the hyperbolic type on the entire plane and is reduced to the form $u_{\xi\eta} = 0$ in the variables $\xi = x + \operatorname{arctg} y,\ \eta = x - \operatorname{arctg} y$.

3.23(b). The equation everywhere has the hyperbolic type and is transformed to $u_{\xi\eta} = 0$ in coordinates $\xi = x + y - \cos x,\ \eta = -x + y - \cos x$.

3.23(c). The equation has the elliptic type and is reduced to $u_{\xi\xi} + u_{\eta\eta} = 0$ in variables $\xi = y,\ \eta = \operatorname{arctg} x$.

3.23(d). The equation has the elliptic type. It is transformed to the form $u_{\xi\xi} + u_{\eta\eta} = 0$ in coordinates $\xi = y,\ \eta = \ln x$.

3.23(e). The equation is parabolic everywhere and reduced to the form

$$u_{\eta\eta} + 2\frac{\xi^2}{\eta^2}u_\xi + \frac{1}{\eta}e^\xi = 0$$

in coordinates $\xi = y/x,\ \eta = y$.

3.24. The equation is hyperbolic in the second and fourth quadrants ($xy < 0$) and there takes the form

$$u_{\xi\eta} - \frac{3}{\xi^2 - \eta^2}(\eta u_\xi - \xi u_\eta) = 0$$

in coordinates

$$\xi = \sqrt{|x|} + \sqrt{|y|}, \quad \eta = \sqrt{|x|} - \sqrt{|y|}.$$

It is elliptic in the first and third quadrants ($xy > 0$) where the equation has the form

$$u_{\xi\xi} + u_{\eta\eta} + 3\left(\frac{1}{\xi}u_\xi + \frac{1}{\eta}u_\eta\right) = 0$$

in coordinates

$$\xi = \sqrt{|y|}, \quad \eta = \sqrt{|x|}.$$

The coordinate axes separate the regions of parabolic and hyperbolic behavior. The canonical form is obtained by dividing by a nonzero coordinate.

3.25. $(\rho + \rho'\sqrt{\varphi_x^2 + \varphi_y^2})\rho > 0$.

3.26(a). Reducing to the canonical form in variables $\xi = 2x + y, \quad \eta = x$ and looking for u as

$$u = w(\xi, \eta) \exp \frac{5\xi + 3\eta}{2},$$

for w we get

$$w_{\xi\xi} + w_{\eta\eta} - \frac{15}{2}w = 0.$$

3.26(b). In variables $\xi = x - y, \quad \eta = x + y$ the following change

$$u = w(\xi, \eta) \exp\left(-\frac{\xi}{2}\right),$$

leads to the equation

$$w_{\xi\eta} - w = 0.$$

3.27. Excluding the first derivatives gives $u = \dfrac{f(x) + g(y)}{x - y}$, where f and g are arbitrary functions.

3.28.

$$u(x, y) = \varphi\left(x - \frac{2y^3}{3}\right) + \frac{1}{2}\int\limits_{x-\frac{2y^3}{3}}^{x+2y} \psi(x')\,dx'$$

3.29.

$$\Delta = \frac{1}{h_1 h_2 h_3}\left[\frac{\partial}{\partial q_1}\left(\frac{h_2 h_3}{h_1}\frac{\partial}{\partial q_1}\right) + \frac{\partial}{\partial q_2}\left(\frac{h_3 h_1}{h_2}\frac{\partial}{\partial q_2}\right) + \frac{\partial}{\partial q_3}\left(\frac{h_1 h_2}{h_3}\frac{\partial}{\partial q_3}\right)\right].$$

3.30(a). $\Delta = \frac{1}{r}\frac{\partial}{\partial r}\left(r\frac{\partial}{\partial r}\right) + \frac{1}{r^2}\frac{\partial^2}{\partial \varphi^2} + \frac{\partial^2}{\partial z^2}.$

3.30(b).

$$\Delta = \frac{1}{r^2}\frac{\partial}{\partial r}\left(r^2\frac{\partial}{\partial r}\right) + \frac{1}{r^2 \sin\theta}\frac{\partial}{\partial \theta}\left(\sin\theta\frac{\partial}{\partial \theta}\right) + \frac{1}{r^2 \sin^2\theta}\frac{\partial^2}{\partial \varphi^2}.$$

3.30(c).

$$\Delta = \frac{4}{\xi + \eta}\left[\frac{\partial}{\partial \xi}\left(\xi\frac{\partial}{\partial \xi}\right) + \frac{\partial}{\partial \eta}\left(\eta\frac{\partial}{\partial \eta}\right)\right] + \frac{1}{\xi\eta}\frac{\partial^2}{\partial \varphi^2}.$$

Coordinate surfaces are confocal paraboloids.

3.30(d).

$$\Delta = \frac{4}{\xi^2 - \eta^2}\left[\frac{\partial}{\partial \xi}\left((\xi^2 - 1)\frac{\partial}{\partial \xi}\right) + \frac{\partial}{\partial \eta}\left((1 - \eta^2)\frac{\partial}{\partial \eta}\right)\right] + \frac{4}{(\xi^2 - 1)(1 - \eta^2)}\frac{\partial^2}{\partial \varphi^2}.$$

Coordinate surfaces are two-sheeted hyperboloids and prolate spheroids.

3.31.

$$(\xi^2 - 1)X'' + 2\xi X' + \left(\frac{k^2\xi^2}{4} - \frac{m^2}{\xi^2 - 1}\right) X = \lambda X,$$

$$(1 - \eta^2)Y'' - 2\eta Y' + \left(-\frac{k^2\eta^2}{4} - \frac{m^2}{1 - \eta^2}\right) Y = -\lambda Y.$$

Both the equations turn to be of the same type, but with different intervals of ξ, η.

3.32(a).

$$\operatorname{grad} U = \mathbf{e}_z\frac{\partial U}{\partial z} + \mathbf{e}_\rho\frac{\partial U}{\partial \rho} + \mathbf{e}_\varphi\frac{1}{\rho}\frac{\partial U}{\partial \varphi},$$

$$\begin{aligned}\operatorname{rot} \mathbf{A} = \mathbf{e}_z &\left(\frac{1}{\rho}A_\varphi + \frac{\partial A_\varphi}{\partial \rho} - \frac{1}{\rho}\frac{\partial A_\rho}{\partial \varphi}\right)\\ &+\mathbf{e}_\rho\left(\frac{1}{\rho}\frac{\partial A_z}{\partial \varphi} - \frac{\partial A_\varphi}{\partial z}\right)\\ &+\mathbf{e}_\varphi\left(\frac{\partial A_\rho}{\partial z} - \frac{\partial A_z}{\partial \rho}\right),\end{aligned}$$

$$\operatorname{div}\mathbf{A} = \frac{\partial A_z}{\partial z} + \frac{1}{\rho}A_\rho + \frac{\partial A_\rho}{\partial \rho} + \frac{1}{\rho}\frac{\partial A_\varphi}{\partial \varphi}.$$

3.32(b).

$$\operatorname{grad}U = \mathbf{e}_r\frac{\partial U}{\partial r} + \mathbf{e}_\theta\frac{1}{r}\frac{\partial U}{\partial \theta} + \mathbf{e}_\varphi\frac{1}{r\sin\theta}\frac{\partial U}{\partial \varphi}.$$

$$\begin{aligned}\operatorname{rot}\mathbf{A} = \mathbf{e}_r &\left(\frac{\operatorname{ctg}\theta}{r}A_\varphi + \frac{1}{r}\frac{\partial A_\varphi}{\partial \theta} - \frac{1}{r\sin\theta}\frac{\partial A_\theta}{\partial \varphi}\right)\\ +\mathbf{e}_\theta &\left(\frac{1}{r\sin\theta}\frac{\partial A_r}{\partial \varphi} - \frac{1}{r}A_\varphi - \frac{\partial A_\varphi}{\partial r}\right)\\ +\mathbf{e}_\varphi &\left(\frac{1}{r}A_\theta + \frac{\partial A_\theta}{\partial \theta} - \frac{1}{r}\frac{\partial A_r}{\partial \theta}\right),\end{aligned}$$

$$\operatorname{div}\mathbf{A} = \frac{2}{r}A_r + \frac{\partial A_r}{\partial r} + \frac{\operatorname{ctg}\theta}{r}A_\theta + \frac{1}{r}\frac{\partial A_\theta}{\partial \theta} + \frac{1}{r\sin\theta}\frac{\partial A_\varphi}{\partial \varphi}.$$

3.33. $l_z = -\mathrm{i}\dfrac{\partial}{\partial \varphi}$.

3.34. Non-normalized wave functions have the form:

$$\psi_{n,m,l} = \sin\left(\frac{n\pi x}{a}\right)\sin\left(\frac{m\pi y}{b}\right)\sin\left(\frac{l\pi z}{c}\right).$$

The corresponding energies are equal to

$$E_{n,m,l} = \frac{\pi^2}{2}\left(\frac{n^2}{a^2} + \frac{m^2}{b^2} + \frac{l^2}{c^2}\right).$$

Here n, m, l are positive integer numbers.

3.35. Standard actions may solve the problem: first, solving the spectral problem, presenting the solution as a sum of normal oscillations, and calculating the coefficients of expansion through the initial data. We get the answer from the formula (3.24) by replacing L to $2L$, and then by substituting $x_0 = L$. Why?

3.36.

$$u(x,t) = \frac{8c}{\pi^3}\sum_{n=0}^{\infty}\frac{1}{(2n+1)^3}\exp\left[-\left(\frac{(2n+1)\pi}{L}\right)^2 a^2 t\right]\sin\frac{(2n+1)\pi x}{L}.$$

3.37.

$$u(x,t)=\frac{8AL}{\pi^2}\sum_{n=0}^{\infty}\frac{(-1)^n}{(2n+1)^2}\exp\left[-\left(\frac{(2n+1)\pi}{2L}\right)^2 a^2t\right]\sin\frac{2n+1}{2L}\pi x.$$

3.38.

$$u(x,y)=\sum_{n=0}^{\infty}a_n\exp\left[-\frac{(2n+1)\pi}{2L}x\right]\sin\frac{(2n+1)\pi}{2L}y,$$

$$a_n=\frac{2}{L}\int_0^L f(y)\sin\left(\frac{(2n+1)\pi}{2L}y\right)dy.$$

3.39.

$$u(r,\varphi)=\frac{Ab^2}{2}\frac{\ln r-\ln a}{\ln b-\ln a}-\frac{Ab^4(r^4-a^4)}{2(b^4-a^4)r^2}\cos 2\varphi.$$

3.40. Solution. Sometimes the Fourier method can be applied to an elliptic operator. Consider the boundary problem to the Laplace equation $\Delta u = 0$ in a rectangular domain with dimensions $a\times b$. In particular let the boundary conditions fix the flux, the normal derivative of the unknown function (Niemann problem):

$$\left.\frac{\partial u}{\partial x}\right|_{x=0}=Q_L,\quad \left.\frac{\partial u}{\partial x}\right|_{x=a}=Q_R,\quad \left.\frac{\partial u}{\partial y}\right|_{y=0}=Q_B,\quad \left.\frac{\partial u}{\partial y}\right|_{y=b}=Q_T,$$

where Q_L, Q_R, Q_B, Q_T are the given values. The problem is split up into two auxiliary problems with zeroth at the opposite sides:

$$\left.\frac{\partial u_1}{\partial x}\right|_{x=0}=0,\quad \left.\frac{\partial u_1}{\partial x}\right|_{x=a}=0,\quad \left.\frac{\partial u_1}{\partial y}\right|_{y=0}=Q_B,\quad \left.\frac{\partial u_1}{\partial y}\right|_{y=b}=Q_T,$$

$$\left.\frac{\partial u_2}{\partial x}\right|_{x=0}=Q_L,\quad \left.\frac{\partial u_2}{\partial x}\right|_{x=a}=Q_R,\quad \left.\frac{\partial u_2}{\partial y}\right|_{y=0}=0,\quad \left.\frac{\partial u_2}{\partial y}\right|_{y=b}=0.$$

The sum $u(x,y)=u_1+u_2$ obeys the Laplace equation with given boundary conditions.

The solutions are Fourier series over eigenfunctions. For the Niemann problem the functions are cosines:

$$u_1(x,y)=a_0(y)+\sum_{n=1}^{\infty}a_n(y)\cos\frac{\pi nx}{a},\quad u_2(x,y)=b_0(x)+\sum_{n=1}^{\infty}b_n(x)\cos\frac{\pi ny}{b}.$$

Functions $a_n(y)$, $b_n(x)$, $n = 1, 2, \ldots,$ can be found from the Laplace equation:

$$a_n(y) = A_n \cosh\frac{\pi n y}{a} + B_n \sinh\frac{\pi n y}{a}, \quad b_n(x) = C_n \cosh\frac{\pi n x}{b} + D_n \sinh\frac{\pi n x}{b}.$$

For $n = 0$ they are $a_0(y) = -\alpha y^2/2$, $b_0 = \alpha x^2/2$, where α is the separation parameter.

3.40(a). **Hint.** Left and top walls are heat-insulated, whereas the heat with flux density Q_1 flows into the rectangular area through the bottom, and flux density Q_2 flows out through the right wall (Fig. 3.2). Only the zeroth harmonics enters the solution, and then the variable separation occurs in the form of a sum $u = a_0(y) + b_0(x)$. The ordinary differential equations have quadratic solutions:

$$b_0 = \alpha\frac{x^2}{2} + C_1 x, \quad a_0 = -\alpha\frac{y^2}{2} + C_2 y + C_3.$$

Substituting the solution to the boundary conditions, we find $C_1 = 0$, $C_2 = \alpha b$. The separation parameter is $\alpha = Q_2/a = Q_1/b$. The last equality gives the solvability condition $Q_1 a = Q_2 b$. Why is the problem unsolvable when this condition is not valid? One cannot find a constant C_3 from the boundary conditions.

Answer.

$$u(x, y) = \frac{Q_1}{2b}(x^2 - y^2) + Q_1 y + C_3.$$

Draw the lines of equal temperature (isotherms) and the heat streamlines.

3.40(b). **Hint.** In this case there are harmonics $n = 0$ and $n = 2$. Then the solution has a form

$$u(x, y) = \frac{\alpha}{2}\left[x^2 - (y - b)^2\right] + \left(A\cosh\frac{2\pi y}{a} + B\sinh\frac{2\pi y}{a}\right)\cos\frac{2\pi x}{a}.$$

We find coefficient A, B,, and separation parameter α from the boundary conditions.

Answer.

$$\alpha = \frac{Q_2}{a} = \frac{Q_1}{2b}, \quad A = -\frac{aQ_1}{4\pi}\coth\frac{2\pi b}{a}, \quad B = A\frac{Q_1}{4\pi}.$$

Chapter 4

Self-Similarity and Nonlinear Equations

4.1 Symmetry of Equations

Consider a second-order partial differential equation of general form with two variables

$$F(x, t, u, u_x, u_t, u_{xx}, u_{xt}, u_{tt}) = 0. \tag{4.1}$$

Suppose we find transformations depending on the real parameter a:

$$x = \psi(\bar{x}, \bar{t}, \bar{u}, a), \quad t = \phi(\bar{x}, \bar{t}, \bar{u}, a), \quad u = \xi(\bar{x}, \bar{t}, \bar{u}, a),$$

which conserve the form of Eq. (4.1).

Let us find the invariants of the transformation:

$$I(x, t, u) = \text{const},$$

which are a independent combinations of variables. Choose among them one that is independent of the solution $I_1(x, t)$. Then passing to a new independent variable $\xi = I_1$ and a new unknown function $w(\xi) = I$ we reduce the equation to an ordinary one (Polyanin and Zaitsev, 2012). If such transformations and invariants can be found for an ordinary differential equation, its order can be lowered.

Mathematical Methods of Physics: Problems with Solutions
Igor V. Kolokolov, Evgeny A. Kuznetsov, Alexander I. Milstein, Evgeny V. Podivilov, Alexander I. Chernykh, David A. Shapiro, and Elena G. Shapiro

ISBN 978-981-5129-21-2 (Hardcover), 978-1-003-51360-5 (eBook)
www.jennystanford.com

Remarks

(1) Usually, one search for a one-parameter transformations among shifts (translations), compression / tension (similarity or scaling) and rotations.
(2) Simplification of the equation depends on the number of transformations found.
(3) Very often a particular solution allows one to suggest a hypothesis about the behavior of the solution in the general case.

The derivation of an ordinary differential equation from a partial differential equation is called *self-similar* substitution in the wide sense. In a narrow sense, self-similar solutions are those that are obtained using a similarity transformation. Substitution of the form

$$u(x,t) = A(t)\, f(\xi), \quad \xi = \frac{x}{l(t)} \tag{4.2}$$

is applied in the equations of heat conduction, the amplitude $A(t)$ and the characteristic scale $l(t)$ are chosen depending on the problem. The solution to the problem is reduced to the ordinary differential equation for $f(\xi)$. In physical problems, one can usually find a self-similar substitution using the dimensional analysis.

If the equation is invariant under translations, the resulting particular solution is called a *traveling wave*. The solution in this case is chosen in the form

$$u(x,t) = f(x - Vt). \tag{4.3}$$

4.2 Nonlinear Equations

Self-similar solutions play an important role because there are no general methods for nonlinear partial differential equations. However, sometimes it is possible to find a broad class of solutions or even a general solution to a nonlinear equation by transforming it to linear by changing variables. Unfortunately, only a few of such transformations are known.

For some vital physics evolution equations, finding time dependence of integral features without explicitly constructing the solutions is possible. This fact, in turn, allows one to understand essential features of solutions, such as the appearance of singularities in a finite time.

A special place belongs to nonlinear partial differential equations of the first order. It turns out that the solution to the Cauchy problem for such the equation is reduced to finding a general solution to some system of ordinary differential equations. For simplicity, consider the case of two variables x and y. The equation of the considered type is written in the form:

$$F(u, p, q, x, y) = 0, \tag{4.4}$$

where F is some given function assumed to be sufficiently smooth, $u = u(x, y)$ is the solution, and $p = u_x$, $q = u_y$.

Differentiating Eq. (4.4) with respect to x and y and taking into account that

$$\frac{\partial}{\partial x} q = \frac{\partial^2 u}{\partial x \partial y} = \frac{\partial}{\partial y} p,$$

we obtain the relations:

$$F_p \frac{\partial p}{\partial x} + F_q \frac{\partial p}{\partial y} = -F_x - pF_u,$$
$$F_p \frac{\partial q}{\partial x} + F_q \frac{\partial q}{\partial y} = -F_y - qF_u. \tag{4.5}$$

It follows from them that along any curve $(x(\tau), y(\tau))$, where

$$\dot{x} = F_p, \quad \dot{y} = F_q, \tag{4.6}$$

and the dot denotes the derivative with respect to the parameter τ, the functions p and q satisfy the equations

$$\dot{p} = -F_x - pF_u, \quad \dot{q} = -F_y - qF_u. \tag{4.7}$$

The derivative of the function $u(x, y)$ along the curve (4.6) is also can be expressed in terms of u, p, q and x, y

$$\dot{u} = u_x \dot{x} + u_y \dot{y} = pF_p + qF_q. \tag{4.8}$$

Equations (4.6)–(4.8) form an already closed system. It defines a family of curves in five-dimensional space with coordinates (u, p, q, x, y). These curves are called characteristics of Eq. (4.4). General solution to the system (4.6)–(4.8) depends on five arbitrary constants. One of these constants reflects the arbitrariness in choosing the origin of the parameter τ, which is fixed for convenience reasons. Initial data for Eq. (4.4) are defined on some curve Γ in the plane (x, y). They are the value of the function $u(x, y)|_\Gamma$ together with the derivative of the function along this curve, which is some linear combination of p and q.

Thus, the initial conditions eliminate arbitrariness in two more constants. Finally, accounting for the initial relation (4.4) leaves only one constant free. The resulting one-parameter family of curves, when projected onto a subspace (u, x, y), forms a plot of the desired solution: the surface $u = u(x, y)$. An analytic expression can be found by exception of parameter τ and the remaining arbitrary constant. Some methods for solving equations using similarity transformations are described in (Zel'dovich and Raizer, 1966) at the consideration of problems of gas dynamics. Initial information on the analytic theory of nonlinear waves can be found in (Whitham, 1974) and the bibliography therein.

Worked Examples

4.1. Find a particular solution of the nonlinear Klein–Gordon–Fock equation:

$$u_{tt} - u_{xx} = \mathrm{e}^u. \tag{4.9}$$

Solution. We search for the translation transformation of u with the simultaneous scaling of (x, t):

$$t \to \mu t, \quad x \to \lambda x, \quad u \to u + a.$$

We find from the conservation of the equation form

$$\lambda = \mu, \quad a = -2 \ln \lambda.$$

Hence $I_1 = x/t$, $I = u + 2 \ln x$. Then we pass to the new variables,

$$u_t = -w' \frac{x}{t^2}, \quad u_{tt} = 2w' \frac{x}{t^3} + w'' \frac{x^2}{t^4}, \quad u_x = w' \frac{1}{t} - \frac{2}{x}, \quad u_{xx} = w'' \frac{1}{t^2} + \frac{2}{x^2}.$$

After that we find the ordinary differential equation from (4.9):

$$\xi^2(\xi^2 - 1)w'' + 2\xi^3 w' - 2 = \mathrm{e}^w.$$

Instead of translation and tension, one could search for a transformation of the (x, t) plane. This is the Lorentz transformation

$$t = \bar{t} \cosh \beta + \bar{x} \sinh \beta, \quad x = \bar{t} \sinh \beta + \bar{x} \cosh \beta$$

where β is the parameter. This transformation conserves the square of the interval $\xi = t^2 - x^2$. Passing to the new independent variable, we obtain another ordinary differential equation:

$$4\xi u'' + 4u' = \mathrm{e}^u.$$

The equation is invariant under simultaneous transformations $\xi \to \mu\xi$ and $u \to u + a$ if $a = -\ln\mu$. The in variants are $z = \ln\xi$ and $w(\xi) = u + \ln\xi$. In new variables we have a Newton-type equation

$$4w'' = e^{w}.$$

Now one can lower the order of equation and find a solution in quadratures.

4.2. Solve the heat conductivity equation

$$u_t = \chi u_{xx} \tag{4.10}$$

on the real axis ($-\infty < x < \infty$) for $t > 0$. We are looking for a solution satisfying the point initial condition

$$u(x, 0) = Q\delta(x) \tag{4.11}$$

and decreasing at infinity

$$u(\pm\infty, t) \to 0. \tag{4.12}$$

Solution. The solution depends on both independent variables x, t and on the parameters χ, Q. The dimensions of these quantities are following: $[t] = T$, $[\chi] = L^2T^{-1}$, $[Q] = [u]\cdot L$, $[x] = L$. Here T, L, $[u]$ are units of time, length and quantity u, respectively. Let us find the dimensionless parameter $\xi = t^n x\chi^k$ of the problem. The exponents n and k satisfies two equations: $n - k = 0$ and $1 + 2k = 0$, so that $n = k = -1/2$ and $\xi = x/\sqrt{\chi t}$. Then a solution to the heat equation can be written in the form

$$u(x, t) = \frac{Q}{\sqrt{\chi t}} f\left(\frac{x}{\sqrt{\chi t}}\right), \tag{4.13}$$

where $f(\xi)$ is a dimensionless function of dimensionless argument. After substituting (4.13) into Eq. (4.10), we find

$$f(\xi) = \frac{1}{2\sqrt{\pi}} \exp\left(-\frac{\xi^2}{4}\right).$$

The characteristic width l of the energy-containing region (where u is large) grows with time as $l(t) \sim \sqrt{\chi t}$, and the maximum value of u decreases as $A(t) \sim Q/\sqrt{\chi t}$.

Finding self-similar solutions can also be formulated in a somewhat more general form than dimensional analysis. Namely, the self-similar substitution does not contradict the equation if it is invariant to large-scale transformations of the spatial variables $\mathbf{r}$ and time t. In this case, the self-similar variable ξ is an invariant combination of $\mathbf{r}$ and t. The law of transformation of the function we are looking for is dictated both by the equation itself and by the invariance of the initial and boundary conditions.

The initial condition (4.11) is specified as δ-functions. If we set a nonzero initial condition in the region of finite width, for example, in the form of a Gaussian function

$$u(x, 0) = \frac{Q}{\sqrt{\pi}a} \exp\left(-x^2/a^2\right),$$

then the second scale a appears in the problem and the self-similar solution (4.13) will no longer be an exact solution to the The Cauchy problem but remains true asymptotically at long times, when

$$l(t) = \sqrt{\chi t} \gg a,$$

so that a can be neglected as compared to the characteristic width of the solution.

Instead of the boundary condition (4.12), one can also have the requirement that the solution is zero on the boundaries of some finite interval, $u(\pm L, t) = 0$. In this case, the problem will also have a second scale so that the self-similar solution no longer satisfies the boundary conditions. However, a self-similar solution is correct at a short time, as long as

$$l(t) = \sqrt{\chi t} \ll L.$$

Thus, in problems where the characteristic scales of the initial and boundary conditions differ significantly ($a \ll L$), the self-similar solution is *intermediate asymptotics*.

4.3. Determine a self-similar solution to the one-dimensional heat equation

$$u_t = u_{xx}$$

under the condition at $t = 0$

$$\int_{-\infty}^{\infty} u(x, 0)\, dx = 1.$$

Solution. The equation and the initial condition should be invariant to scale transformations of all variables,

$$t \longrightarrow \mu t, \quad x \longrightarrow \lambda x, \quad u \longrightarrow \nu u\,.$$

As a result,

$$\frac{\nu}{\mu} = \frac{\nu}{\lambda^2}, \tag{4.14}$$

$$\lambda\nu = 1. \tag{4.15}$$

It follows from (4.14) and (4.15) that $\lambda = \mu^{1/2}$, $\nu = \mu^{-1/2}$. Therefore, one can search for an unknown function in the form

$$u(x,t) = \frac{1}{\sqrt{t}} f\left(\frac{x}{\sqrt{t}}\right).$$

Substituting this function into the heat equation, we obtain the ordinary differential equation for the function f

$$f'' + \frac{1}{2}(\xi f)' = 0,$$

so that $f' + \xi f/2 = \text{const} = C$. The general solution has the form

$$f(\xi) = \mathrm{e}^{-\xi^2/4}\left(C_1 + C\int_0^{\xi} dy\,\mathrm{e}^{y^2/4}\right).$$

The function f is integrable only for $C = 0$; constant C_1 is found from the normalization of the initial condition. We see that in this case the self-similarity uniquely determines the solution $u(x, t)$ and, therefore, it coincides with the solution to the previous problem for $Q = 1$:

$$u(x,t) = \frac{1}{\sqrt{4\pi t}} \exp\left(-x^2/4t\right).$$

4.4. Find a self-similar substitution for a one-dimensional heat conduction equation on the semi-axis $0 \le x < \infty$ and solve the Cauchy problem for (a) $u(x, 0) = 0,\ u(0, t) = 1$; (b) $u(x, 0) = 0,\ u(0, t) = t$.

Solution. 4.4(a) Instead of (4.15), we find the second equation for λ, μ, ν from the conditions of problem. The boundary condition fixes $\nu = 1$, then the self-similar substitution

$$u(x,t) = f\left(\frac{x}{\sqrt{t}}\right)$$

leads to the equation

$$f'' + \frac{1}{2}\xi f' = 0$$

with the boundary condition $f(0) = 1$, $f(\infty) = 0$. Integration of equation gives:

$$u(x,t) = \frac{1}{\sqrt{\pi}} \int_{x/\sqrt{t}}^{\infty} \exp(-\xi^2/4)\,d\xi.$$

Solution. 4.4(b) The boundary condition gives $\nu = \mu$, so that a self-similar substitution has the form

$$u(x, t) = tf\left(\frac{x}{\sqrt{t}}\right).$$

Then we should solve the equation

$$f'' + \frac{1}{2}\xi f' - f = 0, \quad f(0) = 1, \ f(+\infty) = 0.$$

One solution, $f_1 = 1 + \xi^2/2$, is easy to guess, but it does not satisfy the condition at infinity. The second solution can be found by the substitution $f = f_1 f_2$, where f_2 is a new unknown function. In this way,

$$u(x, t) = C(x^2 + t) \int_{x/\sqrt{t}}^{\infty} \frac{\exp(-\eta^2/4)\, d\eta}{(2 + \eta^2)^2}, \quad C^{-1} = 2 \int_{0}^{\infty} \frac{\exp(-\eta^2/4)\, d\eta}{(2 + \eta^2)^2}.$$

4.5. Find a self-similar substitution and self-similar solution to the nonlinear heat equation

$$\frac{\partial u}{\partial t} = \frac{\partial}{\partial x}\left(u^n \frac{\partial u}{\partial x}\right), \quad \int_{-\infty}^{\infty} u(x, 0)\, dx = 1, \quad u(-x, t) = u(x.t).$$

Solution. Unlike (4.14), it follows from invariance of the nonlinear heat equation (or diffusion) to the scaling transformations that

$$\frac{\nu}{\mu} = \frac{\nu^{n+1}}{\lambda^2}.$$

Therefore, taking into account (4.15), we obtain

$$\lambda = \mu^{1/(n+2)}, \quad \nu = \mu^{-1/(n+2)},$$

so that the self-similar substitution has the form

$$u(x, t) = t^{-1/(n+2)} \cdot f\left(xt^{-1/(n+2)}\right).$$

The equation for the function f has the form

$$(f^n f')' + \frac{1}{n+2}(\xi f)' = 0$$

and can be integrated,

$$f^n f' + \frac{1}{n+2}\xi f = C_1.$$

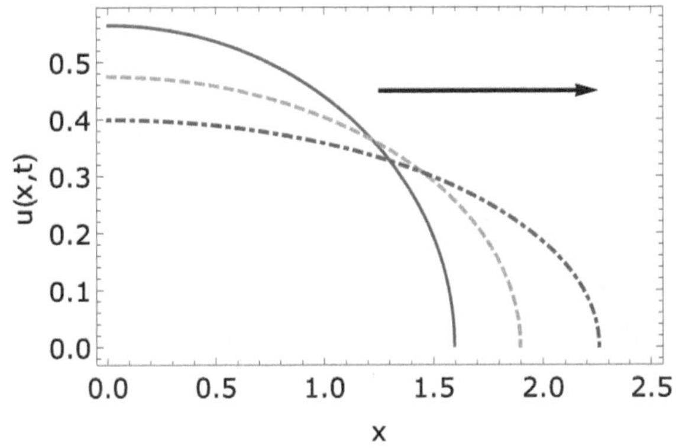

Figure 4.1 Self-similar solution to a nonlinear heat equation at $n = 2$. The arrow shows direction of wave propagation

Due to the boundary condition $f'(0) = 0$, we have $C_1 = 0$, variables are separated, and the second constant of integration is found from the normalization. For $n > 0$, we have a solution having the form of the thermal shock wave with a sharp leading edge (Fig. 4.1).

$$u(x,t) = \begin{cases} \frac{1}{t^{1/(n+2)}}\left[C - \frac{nx^2}{2(n+2)t^{\frac{2}{n+2}}}\right]^{1/n}, & x^2 < 2C\frac{n+2}{n}t^{\frac{2}{n+2}}, \\ 0, & x^2 > 2C\frac{n+2}{n}t^{\frac{2}{n+2}}; \end{cases}$$

$$C = \left[\frac{n(n+2)}{8\pi}\frac{\Gamma^2\left(\frac{1}{n}+\frac{1}{2}\right)}{\Gamma^2\left(\frac{1}{n}\right)}\right]^{\frac{n}{n+2}}.$$

In particular, for $n = 2$ we have $C = 1/\pi$. For $n = 0$ see problem **4.3**.

4.6. In the half-plane $x > 0$, there is a constant temperature $u(0, x) = u_0 > 0$. At time $t \geq 0$ the temperature on the left boundary $x = 0$ equals $u(0, x) = u_0 > 0$. The evolution equation for the temperature $u(x, t)$ has the form

$$u_t - \chi u_{xx} = Q\frac{dy(t)}{dt}\delta(x - y(t)), \tag{4.16}$$

where the time dependence of the function $y(t)$ is determined by the condition $u(y(t), t) = 0$. Let us find $u(x, t)$ and $y(t)$ for all $t > 0$. Consider, in particular, the asymptotics at $Q \to +\infty$.[1]

[1]This is a particular case of the Stefan problem on motion of crystallization front of a liquid cooled on the surface $x = 0$. The function $y(t)$ determines the position of the border between phases, and the quantity Q is proportional to the specific heat of fusion.

Solution. As a result of similarity transformation $x \to \lambda x, t \to \lambda^2 t$ and $u \to u$, which does not change the boundary and initial conditions, the common factor $1/\lambda^2$ appears in the left side of Eq. (4.16). The right side of Eq. (4.16) is transformed according to the same law if we set

$$y(t) = 2a\sqrt{\chi t},$$

where a is a dimensionless constant, and the factor $2\sqrt{\chi}$ is included for convenience. This means that the solution to the problem has a self-similar form:

$$u(x,t) = f(z), \quad z = \frac{x}{\sqrt{\chi t}}. \tag{4.17}$$

After substitution (4.17), Eq. (4.16) reduces to the ordinary differential equation

$$f''(z) + \frac{z}{2}f' + aQ\delta(z - 2a) = 0. \tag{4.18}$$

The asymptotics of the function $f(z)$ at $z \to +\infty$ follows from the initial condition $f(z \to +\infty) \to u_0$. Another boundary condition for $f(z)$ is $f(0) = -u_0$. General solution to the Eq. (4.18) in the regions $z < 2a$ and $z > 2a$ can easily be found:

$$f(z) = A_1 + B_1 \int_0^z ds\, \mathrm{e}^{-s^2/4}; \; z < 2a, \;\; f(z) = A_2 + B_2 \int_z^\infty ds\, \mathrm{e}^{-s^2/4}, \; z > 2a.$$

The constants $A_{1,2}$, $B_{1,2}$ and a follow from the requirement of continuity of $f(z)$ at $z = 2a$ and conditions $f'(2a+0) - f'(2a-0) = -aQ$ and $u(y(t), t) = f(2a) = 0$. We have

$$A_1 = -u_0, \; A_2 = u_0, \; B_1 = \frac{u_0}{\sqrt{\pi}\,\mathrm{erf}\,a}, \; B_2 = \frac{u_0}{\sqrt{\pi}(1 - \mathrm{erf}\,a)},$$

where $\mathrm{erf}\,a = \frac{2}{\sqrt{\pi}} \int_0^a ds\, \mathrm{e}^{-s^2}$ is the probability integral and the constant a is found from the transcendental equation:

$$\frac{u_0}{\sqrt{\pi}}\left(\frac{1}{\mathrm{erf}\,a} - \frac{1}{1 - \mathrm{erf}\,a}\right) = Qae^{a^2}.$$

For $Q \to 0$ the equation reduces to $\mathrm{erf}a = 1/2$, so in this limit $a \approx 0.48$ and $y = 0.96\sqrt{\chi t}$. For $Q \to +\infty$ we have $a \approx \frac{1}{2}\sqrt{2u_0/Q}$ so that

$$y(t) \approx \sqrt{\frac{2u_0\chi}{Q}t}.$$

Large heat generation decreases the speed of the crystallization front.

4.7. The Burgers equation

$$u_t + uu_x = \mu u_{xx}, \qquad -\infty < x < +\infty \tag{4.19}$$

describes weak shock waves in a medium with energy dissipation. Find a solution of the shock wave type, i.e., satisfying the conditions

$$\lim_{x\to-\infty} u(x,t) = u_1, \qquad \lim_{x\to+\infty} u(x,t) = u_2.$$

Solution. Substituting (4.3) in the Burgers equation, we obtain the ordinary differential equation

$$\mu f'' = f'(f - V) \tag{4.20}$$

with the boundary conditions $f(-\infty) = u_1$ and $f(+\infty) = u_2$. By integrating (4.20) from $-\infty$ to $+\infty$ we obtain

$$-V(u_2 - u_1) + \frac{1}{2}(u_2^2 - u_1^2) = 0\,.$$

Thus, $V = (u_1 + u_2)/2$ and the speed of the shock wave depends only on the boundary conditions and is independent of the parameter μ (viscosity of the matter).

Then, integrating (4.20) once from $-\infty$ to ξ and taking into account the formula for the front velocity V expressed via u_1 and u_2, we come to an equation with separable variables:

$$\mu f' = \frac{1}{2}(f - u_1)(f - u_2).$$

Its solution,

$$f(\xi) = \frac{u_1 + u_2 \exp\left[(u_1 - u_2)\xi/2\mu\right]}{1 + \exp\left[(u_1 - u_2)\xi/2\mu\right]} = V - \frac{u_1 - u_2}{2}\tanh\left(\frac{(u_1 \quad u_2)\xi}{4\mu}\right),$$

satisfies the boundary conditions only for $u_1 > u_2$ (otherwise, the velocity profile would begin to " straighten out " and the shock wave disappeared). This is shown in Fig. 4.2 for $u_2 = 0$. The width of the shock wave (*kink*) front increases with increasing viscosity of the matter μ.

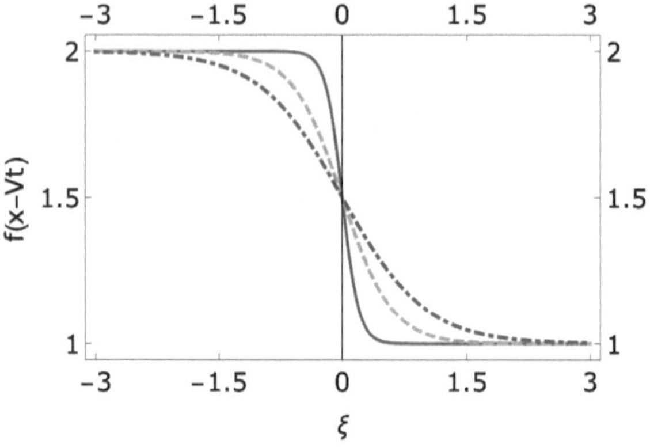

Figure 4.2 Solution to the Burgers equation of the shock wave type: $u_1 = 2$, $u_2 = 0$. Below are the values of viscosity μ. As μ decreases, the wave front becomes steeper.

4.8. Show that the solution to the Korteweg–de Vries equation,

$$u_t + 6uu_x + u_{xxx} = 0,$$

has a running wave shape. Find a particular solution vanishing at infinity together with its first and second derivatives over x.

Solution vanishing on infinity, together with its derivatives, is called *a solitary wave.*

Solution. After substitution $u(x, t) = f(x - Vt)$, we obtain the third-order equation for the function f,

$$f''' + (6f - V)f' = 0,$$

which can be integrated twice

$$f'' + 3f^2 - Vf + C_1 = 0; \quad \frac{1}{2}f'^2 + f^3 - \frac{1}{2}Vf^2 + C_1 f + C_2 = 0.$$

The first and second derivatives of the function f vanish for $x \to \infty$ if $C_1 = C_2 = 0$. Then the equation integrates in elementary functions.

$$u(x, t) = \frac{V/2}{\cosh^2\left[\frac{\sqrt{V}}{2}(x - Vt)\right]}.$$

The solution has a shape of a solitary wave (*soliton*) exponentially damping for $\xi = x - Vt \to \pm\infty$ (Fig. 4.3). Characteristic width of soliton, $\delta\xi \sim V^{-1/2}$, is uniquely related to its velocity V and amplitude $V/2$.

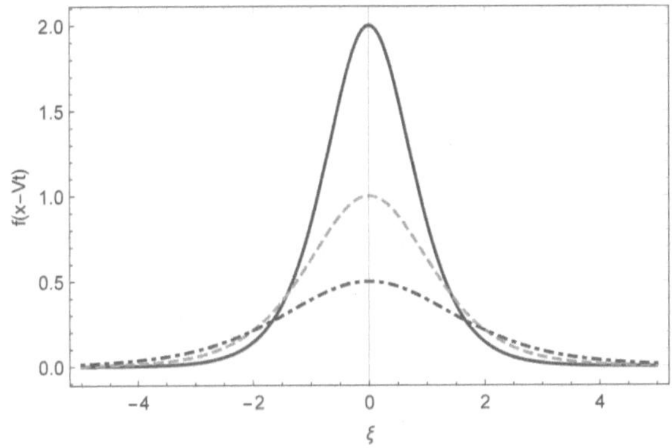

Figure 4.3 Soliton solution to the Korteweg–de Vries equation for $V = 4$ (solid line), 2 (dashed), 1 (dot-dashed).

The Korteweg–de Vries equation describes a wave in a dispersive medium, where the phase velocity depends on the wavelength. For the solution of sufficiently large amplitude, the spreading of the wave packet can be compensated by the steepening of the wave. Soliton is a solution in which the effect of dispersion compensates for nonlinearity. That is why it propagates without changing its shape.

4.9. Find a solitary-type running-wave solution to nonlinear string equation

$$u_{tt} - u_{xx} + (u^2)_{xx} + u_{xxxx} = 0.$$

Solution. Substituting the running wave $u = f(x - Vt)$ into the nonlinear string equation, we obtain an ordinary differential equation

$$f''(V^2 - 1) + (f^2)'' + f^{(IV)} = 0,$$

which we integrate twice,

$$f'' + f^2 + f(V^2 - 1) = C_1\xi + C_2.$$

The function f vanishes at infinity together with its Derivatives; therefore, $C_1 = C_2 = 0$. Then the equation can be multiplied by f'

$$\frac{1}{2}f'^2 + \frac{1}{3}f^3 + \frac{1}{2}f^2(V^2 - 1) = C_3.$$

The third constant is also zero for a solitary wave. The result is a first-order equation with separable variables. Its solution exists only for $|V| < 1$ (in dimensional units, this means that the speed of the wave should be less than the speed of sound):

$$f(\xi) = \frac{3}{2} \frac{1 - V^2}{\cosh^2 \frac{\sqrt{1-V^2}(\xi - \xi_0)}{2}}.$$

4.10. The one-dimensional motion of a polytropic gas is described system of equations

$$\begin{aligned} \frac{\partial \rho}{\partial t} + \frac{\partial}{\partial x}(\rho v) &= 0, \\ \frac{\partial v}{\partial t} + v\frac{\partial v}{\partial x} + \frac{c^2(\rho)}{\rho}\frac{\partial \rho}{\partial x} &= 0, \\ c^2(\rho) &= c_0^2 \left(\frac{\rho}{\rho_0}\right)^{\gamma - 1}. \end{aligned} \tag{4.21}$$

Find self-similar solutions for the two different initial conditions:

(a) At the initial moment, the density has a discontinuity at $x = 0$,

$$\rho(x, 0) = \begin{cases} \rho_0; \ x > 0, \\ 0; \ \ x < 0, \end{cases}$$

and the velocity is zero, $v(x, 0) = 0$.

(b) At the initial moment there is a portion of gas located at the point $x = 0$. The gas spreads without changing of the total mass.

Solution. 4.10(a) Performing a similarity transformation

$$t \longrightarrow \mu t, \quad x \longrightarrow \lambda x, \quad u \longrightarrow \alpha u, \quad \rho \longrightarrow \nu \rho,$$

we find that the equations do not change if

$$\frac{\nu}{\mu} = \frac{\nu\alpha}{\lambda}, \quad \frac{\alpha}{\mu} = \frac{\alpha^2}{\lambda} = \frac{\nu^{\gamma-1}}{\lambda}. \tag{4.22}$$

This system of equations has two free parameters. We can reduce this number to one by different ways depending on initial conditions.

Solution. 4.10(a) In this case, we choose $\nu = 1$, since $\rho = \rho_0$ on infinity. Solving (4.22) we obtain

$$\alpha = 1, \quad \lambda = \mu.$$

We are looking for a solution to the system in the form

$$\rho = f(xt^{-1}), \quad v = g(xt^{-1}).$$

The functions f and g satisfy the system of ordinary differential equations

$$\begin{aligned}(g-\xi)f' + fg' &= 0,\\ \frac{c^2(f)}{f}f' + (g-\xi)g' &= 0,\end{aligned} \tag{4.23}$$

where $\xi = xt^{-1}$ and prime denotes the derivative with respect to ξ.

We can obtain a trivial self-similar solution for a nonzero determinant of linear system of equations with respect to f' and g'. In this case

$$f' = g' = 0, \tag{4.24}$$

i.e., f and g are constant.

If determinant is zero then a relation between the functions f and g holds:

$$(g-\xi)^2 = c^2(f).$$

For nontrivial solutions

$$g = \xi \pm c(f). \tag{4.25}$$

the equations in the system (4.23) become equivalent. First, we find a solution corresponding to the sign "minus" in Eq. (4.25). Then

$$g = \xi - c_0\left(\frac{f}{\rho_0}\right)^{(\gamma-1)/2}.$$

Substituting the expression for g in Eq. (4.23) we obtain the equation

$$f - \frac{\gamma+1}{2}(f\rho_0)^{(\gamma-1)/2}\, f' = 0\,.$$

Integrating this equation, we obtain

$$\frac{\gamma+1}{\gamma-1}c_0\left(\frac{f}{\rho_0}\right)^{(\gamma-1)/2} = \frac{\gamma+1}{\gamma-1}c(f) = A + \xi, \tag{4.26}$$

where is an arbitrary constant. Substituting (4.26) in (4.25) we arrive at the solution

$$g = -\frac{\gamma-1}{\gamma+1}A + \frac{2}{\gamma+1}\xi. \tag{4.27}$$

The solution for sign "plus" in Eq. (4.25) is obtained by changing the sign in front of ξ in Eqs. (4.26) and (4.27) on the opposite.

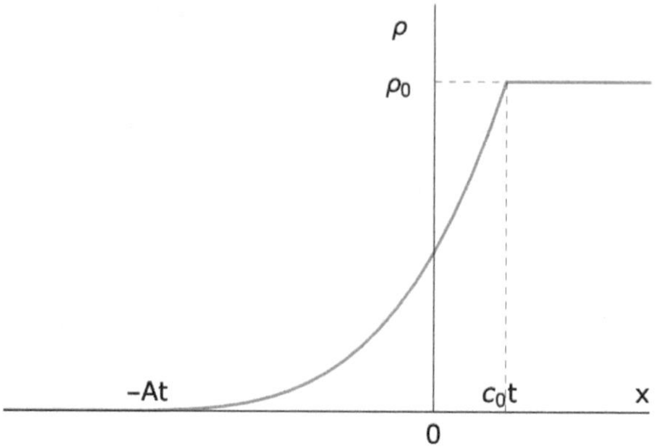

Figure 4.4 Self-similar solution to the problem **4.10** with an initial density shape in the form of step function.

Let us construct a solution to the initial problem from self-similar solutions (4.26), (4.27) and (4.24). We suppose that the solution (4.26) and (4.27) is valid up to the point ξ_0 with growing ξ and require that starting from ξ_0 it continuously transfer into the solution (4.24) with $f = \rho_0$ and $g = 0$.

The continuity condition gives a system of equations for A and ξ_0. Solving this system we get

$$\xi_0 = c_0, \quad A = \frac{2}{\gamma - 1} c_0.$$

The solution given by Eqs. (4.26 and (4.27) is valid in the interval bounded by the condition $f > 0$ from the left, which is valid for $\xi > \xi_1 = -2c_0/(\gamma - 1)$. Finally, at $t > 0$ we have:

- there is a vacuum for $x < \xi_1 t$, i.e., $v = 0$ and $\rho = 0$,
- at $\xi_1 t < x < c_0 t$ a velocity is $v = -\frac{2}{\gamma-1}\left(c_0 - \frac{x}{t}\right)$, and the density is determined from Eq. (4.26);
- for $x > c_0 t$, there is an unperturbed gas, $v = 0, \ \rho = \rho_0$ (Fig. 4.4).

At $t \to +0$ this solution tends to the initial condition.

Note that the resulting solution for the velocity has a discontinuity at the boundary with a vacuum. Formally, this feature does not contradict Eqs. (4.21) since the speed of the vacuum is undefined. However, physically, this means that near the gas-vacuum boundary, dissipative processes become significant. Therefore, they are not taken into account in the system (4.21) but could be included in the evolution equations by the terms with spatial derivatives of the second order. In this case, the gap will be smoothed out. The Burgers equation can describe such a fine structure of the shock wave for one-dimensional motion with some effective parameters. This remark is valid for all problems of this type.

Solution. 4.10(b) From the conservation of mass, we obtain

$$\nu\lambda = 1. \tag{4.28}$$

Expressing the scale factors in terms of μ we have

$$\alpha = \mu^{(1-\gamma)/(1+\gamma)}, \quad \nu = \mu^{-2/(1+\gamma)}, \quad \lambda = \mu^{2/(1+\gamma)}.$$

Looking for a solution in the form

$$\rho = t^{-2/(1+\gamma)} f(xt^{-2/(1+\gamma)}), \quad v = t^{(1-\gamma)/(1+\gamma)} g(xt^{-2/(1+\gamma)}),$$

we obtain a system of ordinary differential equations for the functions f and g,

$$-\frac{2}{\gamma+1}(f\xi)' + (fg)' = 0,$$

$$-\frac{1}{\gamma+1}[(\gamma-1)g + 2\xi g'] + gg' + f^{\gamma-2}f' = 0. \tag{4.29}$$

Integrating the first equation we find that $g(0) = 0$ for $f \neq 0$ and zero constant of integration. Then we have

$$g(\xi) = \frac{2}{\gamma+1}\xi.$$

Integrating the second equation we obtain

$$f = \left[C + \frac{(\gamma-1)^2}{(\gamma+1)^2}\xi^2\right]^{1/(\gamma-1)},$$

where C is an arbitrary constant. The second solution to the system (4.29) is trivial:

$$f = g = 0,.$$

Unfortunately, the solutions obtained under the assumption (4.28), corresponding to the gas mass conservation, describe the motion of gas with an infinite or zero mass.

4.11. Find a self-similar solution that describes gas spreading with "negative pressure,"

$$\frac{\partial \rho}{\partial t} + \frac{\partial}{\partial x}(\rho v) = 0,$$
$$\frac{\partial v}{\partial t} + v\frac{\partial v}{\partial x} - \frac{\partial \rho}{\partial x} = 0,$$

if at $t = 0$ all mass M of the gas was at the point $x = 0$.

Solution. At the beginning, the solution to our problem exactly repeats the solution to the problem (**4.10**) for $\gamma = 2$. Instead of (4.29), we obtain the system

$$-\frac{2}{3}(f\xi)' + (fg)' = 0,$$
$$-\frac{1}{3}[(g + 2\xi g'] + gg' - f' = 0, \tag{4.30}$$

It follows from this system that

$$g = \frac{2}{3}\xi, \quad f = C - \frac{1}{9}\xi^2.$$

The second solution is trivial: $f = g = 0$.

The solution that describes the spreading of a finite mass of matter has the form

$$v(x,t) = \begin{cases} \dfrac{2x}{3t}; \ |x| < 3\sqrt{C}t^{2/3}, \\ 0; \quad |x| > 3\sqrt{C}t^{2/3}; \end{cases}$$

$$\rho(x,t) = \begin{cases} t^{-2/3}\left(C - \dfrac{x^2}{9t^{4/3}}\right); \ |x| < 3\sqrt{C}t^{2/3}, \\ 0; \qquad\qquad\qquad\qquad |x| > 3\sqrt{C}t^{2/3}; \end{cases}$$

$$C = (M/4)^{2/3}.$$

Note that making the replacement $t \longrightarrow t_0 - t$ we obtain a solution in which the density goes to infinity at the time $t = t_0$. This phenomenon is called *collapse* (Fig. 4.5).

4.12. Find the asymptotics of solution $f(x,t)$ of nonuniform evolution equation

$$\frac{\partial}{\partial t}f - \frac{\partial^2}{\partial x^2}\left(1 + x^2\right) f = \frac{d^2}{dx^2}\frac{1}{1 + x^2} \tag{4.31}$$

at $t \to +\infty$ with zero initial conditions.

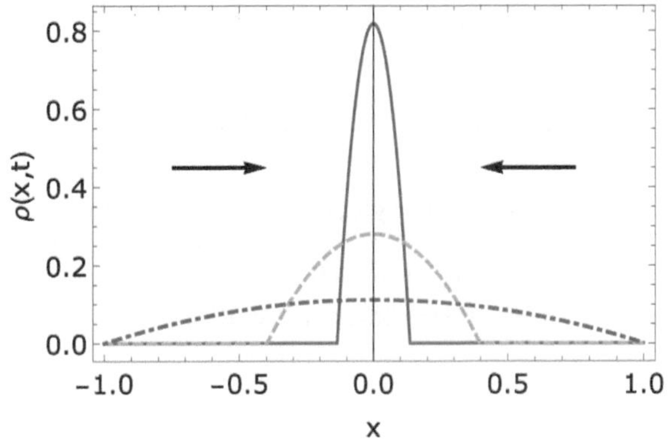

Figure 4.5 Collapse of the parabolic density profile, $t_0 = 2$

Solution. The operator $\hat{\mathcal{L}} = \frac{d^2}{dx^2}(1+x^2)$ is Hermitian in a Hilbert space with the scalar product

$$(f, g) = \int_{-\infty}^{+\infty} dx\,(1+x^2) f^* g\,, \tag{4.32}$$

and its eigenvalues are negative. If the solution to our Cauchy problem belongs to such Hilbert space, then in the limit $t \to +\infty$ it does not depend on the time t. Assuming that the function f is normalized with respect to the scalar product (4.32), we find that it satisfies the stationary equation at $t \to +\infty$,

$$-\hat{\mathcal{L}} f = \frac{d^2}{dx^2}\frac{1}{1+x^2},$$

having a general solution with two arbitrary constants,

$$f = -\frac{1}{(1 | x^2)^2} + \frac{C_1 x}{1+x^2} + \frac{C}{1+x^2}\,.$$

Using arguments of symmetry and conservation laws, the constants can be determined from the initial data without solving the real evolutionary problem. Indeed, Eq. (4.31) and the initial conditions are invariant under the reflection $x \to -x$. The solution should have the same symmetry, which immediately fixes the value $C_1 = 0$. Further, integrating both sides of (4.31) over all values of x, we find that the integral $\int_{-\infty}^{+\infty} dx\, f(x, t)$ is independent of

t. Since this integral was equal to zero at $t = 0$, then it also equals to zero at $t \to +\infty$, so that $C = \frac{1}{2}$. Finally, we arrive at the required asymptotic form:

$$f(x,t) \longrightarrow f = -\frac{1}{(1+x^2)^2} + \frac{1}{2(1+x^2)}.$$

The normalization of this function given by (4.32) is finite.

4.13. Evolution of the function $\psi(\mathbf{r}, t)$ in two-dimensional space is determined by the nonlinear Schrödinger equation

$$i\psi_t + \Delta\psi + |\psi|^2\psi = 0. \tag{4.33}$$

(a) Prove that

$$N = \int d\mathbf{r}|\psi|^2, \quad H = -\int d\mathbf{r}\left(|\nabla\psi|^2 - \frac{1}{2}|\psi|^4\right)$$

are the integrals of motion;

(b) Prove that a magnitude of localized in space initial perturbation,

$$I = \int d\mathbf{r}\,\mathbf{r}^2|\psi|^2,$$

satisfies the equation

$$\frac{d^2}{dt^2}I = -8H.$$

Using this relation, show that for positive value of the functional H a singularity in the solution $\psi(\mathbf{r}, t)$ is formed in a finite time (Talanov criterion). In other words, there is a moment t^* of time and a point in space $\mathbf{r}^*$ that $\psi(\mathbf{r}, t) \to \infty$ at $t \to t^*$ and $\mathbf{r} \to \mathbf{r}^*$.

Solution. 4.13(a) Let us write, together with Eq. (4.33), the complex conjugate equation:

$$-i\psi_t^* + \Delta\psi^* + |\psi|^2\psi^* = 0, \tag{4.34}$$

multiply (4.33) by ψ^* and (4.34) by ψ, subtract one result from the other and integrate over space. If the function $\psi(\mathbf{r}, t)$ rather quickly decreases at $\mathbf{r} \to \infty$, then we obtain $dN/dt = 0$.

Then we apply the operator ∇ to both sides of Eq. (4.33), multiply the result by $\nabla\psi^*$, subtract from the obtained equality its complex conjugation and integrate the result over space. The result of these transformations reads

$$i\frac{d}{dt}\int d\mathbf{r}|\nabla\psi|^2 = \int d\mathbf{r}\,(\psi \bigtriangleup \psi^* - \psi^* \bigtriangleup \psi)\,|\psi|^2.$$

If we multiply (4.33) by $\psi^*|\psi|^2$, subtract a complex conjugation and integrate over space, then we have

$$\frac{\mathrm{i}}{2}\frac{d}{dt}\int d\mathbf{r}|\psi|^4 = \int d\mathbf{r}\,(\psi \bigtriangleup \psi^* - \psi^* \bigtriangleup \psi)\,|\psi|^2.$$

This equation, together with the previous one, leads to the conservation law $dH/dt = 0$.

Solution. 4.13(b) Acting by operator id/dt on I and using evolution equations (4.33), we obtain the equality:

$$i\frac{d}{dt}I = \int d\mathbf{r}\,\mathbf{r}^2\,(\psi \bigtriangleup \psi^* - \psi^* \bigtriangleup \psi) = -2\int d\mathbf{r}\,\left(\psi(\mathbf{r}\nabla)\psi^* - \psi^*(\mathbf{r}\nabla)\psi\right). \tag{4.35}$$

Taking derivative over time in both sides of (4.35) and using an explicitly testable identity

$$\bigtriangleup(\mathbf{r}\nabla) = (\mathbf{r}\nabla)\bigtriangleup + 2\bigtriangleup,$$

we arrive at the relation

$$2\mathrm{i}\frac{d}{dt}\int d\mathbf{r}\,\psi(\mathbf{r}\nabla)\psi^* = 2\int d\mathbf{r}\,\left(|\psi|^2(\mathbf{r}\nabla)|\psi|^2 - 2\psi^* \bigtriangleup \psi\right).$$

Then we add to this expression its complex conjugation and obtain the result for $d^2/dt^2\, I$,

$$\frac{d^2}{dt^2}I = -8\int d\mathbf{r}\left(|\nabla\psi|^2 + \frac{1}{4}(\mathbf{r}\nabla)|\psi|^4\right) = 8H.$$

To obtain the last equality, one needs to integrate in parts over $\mathbf{r}$ and take into account that $\operatorname{div}\mathbf{r} = 2$ in two dimensions.

The quantity H has the meaning of the Hamiltonian of our continual systems, N is the number of excited degrees of freedom. Conservation of H makes it easy to find the functional I,

$$I(t) = I(0) + Ct - 4Ht^2. \tag{4.36}$$

Here C is some constant determined by the initial shape of the function ψ. We assume that the Hamiltonian is positive for a given initial condition, $H > 0$. Then for any finite value of C there is a moment when $I(t)$ vanishes. However, the functional $I(t)$ is positive by construction. The only way to resolve a contradiction is a limited time with the finite solution, i.e., the formation of a singularity, called collapse.

For the symmetric initial distribution of the function ψ to rotation, vanishing of $I(t)$ at some moment means that at this moment, the entire density $|\psi(\mathbf{r}, t)|^2$ is concentrated at the point $\mathbf{r} = 0$ where it is infinite due to conservation of the integral N.

The nonlinear Schrödinger equation describes, in particular, the propagation of light in a nonlinear medium. Talanov's criterion gives the self-focusing condition of the light beam. Nonlinearity increases the refractive index near the beam axis and acts as a converging lens. Collapse occurs when the influence of nonlinearity becomes more important than diffractive beam spread.

4.14. Equations (4.21), which describes one-dimensional barotropic gas motion, are quasilinear. Each particular solution is given by two functions ρ and v of two variables x and t,

$$\rho = \rho(x, t), \quad v = v(x, t). \tag{4.37}$$

Variables ρ and v can be used instead of x and t as new independent coordinates if the corresponding Jacobian of the transformation is not zero. This is so-called a *hodograph transformation.*

Show that the equations for the functions x and t depending on the variables ρ and v will be linear.

Solution. Differentiating Eq. (4.37) over ρ and v and assuming that x and t are functions of ρ and v, we obtain

$$\frac{\partial\rho}{\partial x}\frac{\partial x}{\partial\rho} + \frac{\partial\rho}{\partial t}\frac{\partial t}{\partial\rho} = 1, \quad \frac{\partial\rho}{\partial x}\frac{\partial x}{\partial v} + \frac{\partial\rho}{\partial t}\frac{\partial t}{\partial v} = 0,$$
$$\frac{\partial v}{\partial x}\frac{\partial x}{\partial\rho} + \frac{\partial v}{\partial t}\frac{\partial t}{\partial\rho} = 0, \quad \frac{\partial v}{\partial x}\frac{\partial x}{\partial v} + \frac{\partial v}{\partial t}\frac{\partial t}{\partial v} = 1.$$

Solving this system to the partial derivatives in Eqs. (4.21), we obtain

$$\frac{\partial\rho}{\partial x} = \frac{1}{J}\frac{\partial t}{\partial v}, \quad \frac{\partial\rho}{\partial t} = -\frac{1}{J}\frac{\partial x}{\partial v}, \quad \frac{\partial v}{\partial x} = -\frac{1}{J}\frac{\partial t}{\partial\rho}, \quad \frac{\partial v}{\partial t} = -\frac{1}{J}\frac{\partial x}{\partial\rho},$$

where $J = \partial x/\partial\rho \cdot \partial t/\partial v - \partial x/\partial v \cdot \partial t/\partial\rho$ is the Jacobian of transformation. Substituting these expressions into Eqs. (4.21), for $J \neq 0, \infty$ we obtain a system of linear equations for the functions t and x depending on the variables ρ and v,

$$\frac{\partial x}{\partial v} + \rho\frac{\partial t}{\partial\rho} - v\frac{\partial t}{\partial v} = 0,$$
$$\frac{\partial x}{\partial\rho} - v\frac{\partial t}{\partial\rho} + \frac{c^2(\rho)}{\rho}\frac{\partial t}{\partial v} = 0.$$

It is clear that in this way any quasilinear system of two equations on a function of two variables can be transformed into a linear one, if the coefficients in the original system depend only on solutions. For physically important values $\gamma = (2n+3)/(2n+1)$ the general solution to this linear system has been found analytically (Landau and Lifshitz, 1987).

4.15. (a) Show that the Burgers equation (4.19) can be transformed to a linear equation on θ by the Cole–Hopf substitution

$$u = -2\mu\frac{\partial}{\partial x}\ln\theta(x, t).$$

(b) Using this substitution, find periodic in space solution to the Burgers equation.

Solution. 4.15(a) The function θ satisfies the equation

$$\frac{\partial}{\partial x}\left(\frac{\theta_t}{\theta} - \mu\frac{\theta_{xx}}{\theta}\right) = 0.$$

Taking the integral over x, we obtain

$$\theta_t = \mu\theta_{xx} + f(t)\theta, \tag{4.38}$$

where $f(t)$ is an arbitrary function. One can set it to be equal to zero. Indeed, if one multiplies $\theta(x, t)$ by a nonzero function depending only on time t, then u does not change. On the other hand, if we look for a solution to Eq. (4.38) in the form $\theta = \varphi(t)\Phi(x, t)$ and choose φ to be the solution to the equation

$$\frac{d\varphi}{dt} = f(t)\varphi,$$

we obtain Eq. (4.38) for Φ with $f(t) = 0$.

Solution. 4.15(b) One of the simplest periodic solutions of the equation $\theta_t = \mu\theta_{xx}$ has the form $\theta(t, x) = A + \mathrm{e}^{-\mu k^2 t}\sin kx$, where A is a constant. The corresponding function,

$$u(t, x) = -2\mu\partial_x \ln\theta(t, x) = -2\mu k\frac{\mathrm{e}^{-\mu k^2 t}\cos kx}{A + \mathrm{e}^{-\mu k^2 t}\sin kx}, \tag{4.39}$$

has no singularities for $A > 1$. Therefore, at $A > 1$ the function $u(t, x)$ in (4.39) is a periodic solution to the Burgers equation for all $t \geq 1$.

4.16. Reduce the quasilinear equation

$$\Delta u - (\nabla u)^2 = 0$$

to a linear by replacement of an unknown function.

Solution. The substitution $u = \omega(v)$ leads to the equation

$$(\omega'' - \omega'^2)(\nabla v)^2 + \omega' \,\Delta\, v = 0.$$

For the function $\omega(v)$, which is the solution to the ordinary differential equation

$$\omega'' - (\omega')^2 = 0, \tag{4.40}$$

we obtain the Laplace equation for v. Equation (4.40) has an explicit solution: $\omega = c_1 - \ln|x - c_2|$.

4.17. For the system of quasilinear equations of the second-order,

$$\Delta\theta - \sin\theta\cos\theta(\nabla\varphi)^2 = 0,$$
$$\Delta\varphi + 2\cot\theta\;\nabla\theta\nabla\varphi = 0,$$

find a solution having the form $\theta = f(v)$ and $\varphi = g(v)$, where v is a new unknown function. Show that under the condition $\Delta v = 0$ the system is reduced to the ordinary differential equations for f and g. Find the corresponding solutions θ and φ.

Solution. Substitution $\theta = \theta(v)$ and $\varphi = \varphi(v)$ leads to the system

$$\theta''(\nabla v)^2 + \theta' \,\Delta\, v - \sin\theta\cos\theta\,\varphi'^2(\nabla v)^2 = 0,$$
$$\varphi''(\nabla v)^2 + \varphi' \,\Delta\, v + 2\cot\theta\,\varphi'\theta'(\nabla v)^2 = 0.$$

If v is a harmonic function, then we obtain the ordinary differential equations

$$\theta'' - \sin\theta\cos\theta\varphi'^2 = 0, \tag{4.41}$$
$$\varphi'' + 2\cot\theta\varphi'\theta' = 0. \tag{4.42}$$

After dividing by φ', Eq. (4.42) can be integrated,

$$\varphi' = \frac{c_1}{\sin^2\theta}.$$

Then we can integrate Eq. (4.41):

$$\theta'^2 + \frac{c_1^2}{\sin^2\theta} = c_2,$$

where c_1 and $c_2 > 0$ are some constants. The equation obtained can also be integrated. Finally, we have the solution

$$\theta = \arctan\sqrt{1 + \beta^2\cot^2\frac{c_1\beta v}{\beta^2 - 1}}, \quad \varphi = \arctan\frac{\tan(c_1\beta v)}{\beta},$$

where $\beta = \sqrt{c_2}/c_1 \geq 1$ and v is an arbitrary harmonic function. This system of equations is used in the theory of ferromagnetism.

4.18. Find the solution $u(x, t)$ of the equation

$$u_t u_x - u = 0, \tag{4.43}$$

satisfying the condition $u(x, 0) = x^2$.

Solution. Writing (4.43) as $F = pq - u = 0$, we obtain the equations for characteristics in the space (u, p, q, x, t):

$$\begin{aligned} \dot{x} &= F_p = q, \ \dot{t} = F_q = p, \ \dot{p} = -F_x - pF_u = p, \\ \dot{q} &= -F_t - qF_u = q, \ \dot{u} = pF_p + qF_q = 2pq. \end{aligned} \tag{4.44}$$

The general solution to this system contains five arbitrary constants C_j, where $j = 1, \ldots, 5$:

$$p = C_1 e^{\tau}, \ q = C_2 e^{\tau}, \ x = C_2 e^{\tau} + C_3, \ t = C_1 e^{\tau} + C_4, \ u = C_1 C_2 e^{2\tau} + C_5. \tag{4.45}$$

We fix the initial value of the parameter τ by the condition $t(0) = 0$. As a result, $C_4 = -C_1$. The equation $F = pq - u = 0$ gives $C_5 = 0$. Further, it follows from the initial condition that $u = x^2, \ p = u_x = 2x$ at $\tau = 0$ so that

$$C_1 C_2 = (C_2 + C_3)^2, \ C_1 = 2(C_2 + C_3).$$

Taking all these relations into account we express the constant (along characteristics!) parameters C_{2-4} through C_1:

$$\begin{aligned} p = C_1 e^{\tau}, \ q = \tfrac{1}{4} C_1 e^{\tau}, \ x = \tfrac{1}{4} C_1 e^{\tau} + \tfrac{1}{2} C_1, \\ t = C_1 (e^{\tau} - 1), \ u = \tfrac{1}{4} C_1^2 e^{2\tau}. \end{aligned} \tag{4.46}$$

Expressions (4.46) for u, t, x are nothing but parametric specifying of the surface $u = u(x, t)$ with the parameters τ and C_1. Then we exclude these parameters and obtain an explicit expression for the solution $u(x, t)$:

$$u(x, t) = \left(x + \frac{t}{4}\right)^2.$$

4.19. Find the solution $\psi(x, y, t)$ to the equation of Hamilton–Jacobi type with the condition

$$\psi_t^2 - \psi_x^2 - \psi_y^2 = 0, \quad \psi(x, y, 0) = \sqrt{x^2 + y^2}. \tag{4.47}$$

Solution. Introducing notations $p_0 = \psi_t, \ p_1 = \psi_x$, and $p_2 = \psi_y$, we rewrite the equation in the form

$$F \equiv p_0^2 - p_1^2 - p_2^2 = 0.$$

Its characteristics are curves in seven-dimensional space. They are defined by

$$\dot{p}_0 = \dot{p}_1 = \dot{p}_2 = 0,\ \dot{t} = 2p_0, \quad \dot{x} = -2p_1,\ \dot{y} = -2p_2, \quad \dot{\psi} = 0.$$

By requirement that the value $\tau = 0$ corresponds to $t = 0$, we obtain a solution to this system with six arbitrary constants:

$$p_0 = C_0,\ p_1 = C_1,\ p_2 = C_2,\ \psi = C_3$$
$$t = 2C_0\tau,\ x = -2C_1\tau + B_1,\ y = -2C_2\tau + B_2,\ .$$

At $t = 0$ we know the expressions of ψ, $p_1 = \psi_x$ and $p_2 = \psi_y$ in terms of x and y. Since $t = 0$ for $\tau = 0$, then the relations between constants C_j and B_j read

$$C_3 = \sqrt{B_1^2 + B_2^2}, \quad C_1 = \frac{B_1}{\sqrt{B_1^2 + B_2^2}}, \qquad C_2 = \frac{B_2}{\sqrt{B_1^2 + B_2^2}},$$

The condition $F = 0$ gives $C_0 = \pm\sqrt{C_1^2 + C_2^2} = \pm 1$. Thus, the solution to our Cauchy problem is not unique; there are two functions $\psi_\pm(x, y, t)$, satisfying both the equation and an initial condition (4.47). Surfaces $\psi = \psi_\pm(x, y, t)$ can be specified in the parametric form using two parameters B_1 and B_2. We exclude them and find an explicit solution:

$$\psi_\pm = \pm t + \sqrt{x^2 + y^2}.$$

Exercises

4.20. Solve the Cauchy problem for the one-dimensional heat equation in the region $0 \leq x < \infty$. The conditions for the equation are $u(x, 0) = x^3$ and $u(0, t) = 0$.

4.21. Find a self-similar substitution for KdV equation

$$u_t + 6uu_x + u_{xxx} = 0.$$

4.22. Find a self-similar substitution for the nonlinear wave equation

$$\frac{\partial^2 u}{\partial t^2} = \frac{\partial}{\partial x}\left(\varkappa(u)\frac{\partial u}{\partial x}\right).$$

4.23. Find a self-similar substitution for the equation of transonic flow

$$u_{xx} + u_y u_{yy} = 0.$$

4.24. Find a self-similar substitution to the nonlinear Schrödinger equation

$$\mathrm{i}\frac{\partial \psi}{\partial t} + \frac{\partial^2 \psi}{\partial x^2} + |\psi|^2 \psi = 0.$$

4.25. Find a substitution to the general nonlinear heat equation

$$\frac{\partial u}{\partial t} = \frac{\partial}{\partial x}\left(\varkappa(u)\frac{\partial u}{\partial x}\right)$$

in the form of running wave.

4.26. Find a solitary wave to the nonlinear Schrödinger equation

$$\mathrm{i}\left[\frac{\partial \psi}{\partial t} + V\frac{\partial \psi}{\partial x}\right] + \frac{\partial^2 \psi}{\partial x^2} + |\psi|^2 \psi = 0.$$

Hint. Search the substitution of the form $\psi(x, t) = f(x - Vt)e^{\mathrm{i}\alpha t}$.

4.27. Solve the Cauchy problem for a first-order nonlinear equation:

$$S_t + \frac{1}{2}S_x^2 = 0, \quad S(x, 0) = x^2/2.$$

4.28. Solve the Cauchy problem for a first-order nonlinear equation:

$$S_t + \frac{1}{2}S_x^2 = S_x + \frac{1}{2}, \quad S(x, 0) = x^2/2 + x.$$

4.29. Solve the Cauchy problem for a first-order nonlinear equation:

$$S_t + \frac{1}{2}S_x^2 = 1, \quad S(x, 0) = x^2/2.$$

4.30. Solve the Cauchy problem for a first-order nonlinear equation:

$$S_t + \frac{1}{2}S_x^2 = S, \quad S(x, 0) = x^2/6.$$

Answers

4.20.

$$u(x, t) = t^{3/2} f\left(\frac{x}{\sqrt{t}}\right) = x^3 + 6tx.$$

4.21.

$$u(x, t) = t^{-2/3} f(\xi), \quad \xi = xt^{-1/3}.$$

4.22.

$$u(x, t) = f(\xi), \quad \xi = x/t.$$

4.23.

$$u(x, t) = f(\xi), \quad \xi = x^{-2/3}y.$$

4.24.

$$\psi(x, t) = \frac{1}{\sqrt{t}} f\left(\frac{x}{\sqrt{t}}\right).$$

4.25.

$$u(x, t) = f(\xi), \quad \xi = x - Vt.$$

The obtained ordinary differential equation has a general solution in quadratures.

4.26.

$$\psi(x, t) = \frac{\sqrt{2\alpha}\mathrm{e}^{\mathrm{i}\alpha t}}{\cosh(\sqrt{\alpha}\xi)}, \quad \xi = x - Vt.$$

4.27

$$S(x, t) = \frac{x^2}{2(1+t)} + \text{const.}$$

4.28

$$S(x, t) = x + \frac{x^2}{2(1+t)} + \text{const.}$$

4.29

$$S(x, t) = t + \frac{x^2}{2(1+t)} + \text{const.}$$

4.30

$$S(x, t) = \frac{x^2}{2(1+2\mathrm{e}^{-t})}.$$

Chapter 5

Special Functions

5.1 Singular Points

Any linear homogeneous ordinary differential equation of the second order can be reduced to the form

$$\frac{d^2y}{dz^2} + p(z)\frac{dy}{dz} + q(z)y = 0. \tag{5.1}$$

The singular points of this equation are the points where $p(z)$ or $q(z)$ tend to infinity. If $p(z)$ and $q(z)$ are analytic in circle $|z - z_1| < R$ of the complex plane of the variable z, then there are two linearly independent solutions $y_1(z)$ and $y_2(z)$ in the vicinity of the point z_1. The functions $y_1(z)$ and $y_2(z)$ form a fundamental system of solutions, i.e., the general solution to Eq. (5.1) is expressed in terms of their linear combination

$$y(z) = Ay_1(z) + By_2(z).$$

It follows from (5.1) that the Wronskian of the two solutions,

$$W\{y_1(z), y_2(z)\} = y_1(z)\frac{dy_2(z)}{dz} - y_2(z)\frac{dy_1(z)}{dz}, \tag{5.2}$$

satisfies the equation,

$$\frac{dW(z)}{dz} = -p(z)W(z), \tag{5.3}$$

Mathematical Methods of Physics: Problems with Solutions
Igor V. Kolokolov, Evgeny A. Kuznetsov, Alexander I. Milstein, Evgeny V. Podivilov, Alexander I. Chernykh, David A. Shapiro, and Elena G. Shapiro

ISBN 978-981-5129-21-2 (Hardcover), 978-1-003-51360-5 (eBook)
www.jennystanford.com

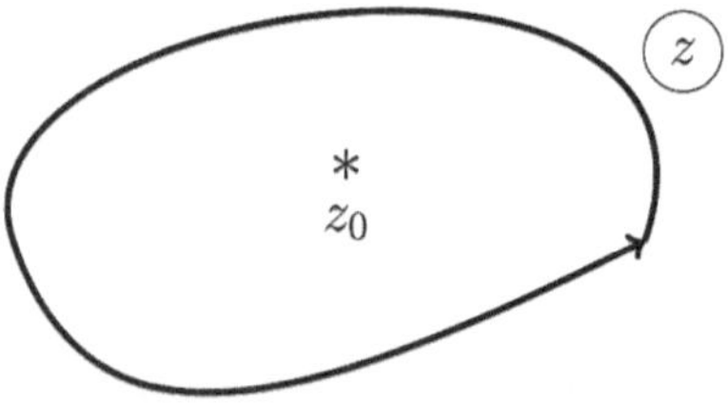

Figure 5.1 Contour in a complex plane of the variable z.

which has a solution

$$W(z') = W(z) \exp\left(- \int_{z}^{z'} dt\, p(t) \right) . \tag{5.4}$$

The function $W(z)$ vanishes only for linearly dependent $y_1(z)$ and $y_2(z)$. Both solutions can be analytically continued from z_1 along the contour C, not passing through the singular points. Moreover, it follows from (5.3) that the linear independence of solutions is conserved.

Let z_0 be an isolated singular point. Consider two linearly independent solutions $y_1(z)$ and $y_2(z)$ defined in a vicinity of a nonsingular point z located near z_0. Starting from the point z, we analytically continue these solutions along a closed contour (Fig. 5.1) going around z_0. In this case, the solutions $y_i(z)$ are transformed into new functions $y_i^+(z)$, which are linear combinations of solutions $y_i(z)$:

$$y_i(z) \to y_i^+(z) = a_{ij} y_j(z).$$

If we make the matrix a_{ij} to be diagonal by the corresponding choice of the functions $y_j(z)$,

$$y_j^+(z) = \lambda_j y_j(z),$$

then the asymptotics of the solutions at the point $z = z_0$ will have the form

$$y_i(z) = (z - z_0)^{\rho_i} \sum_{k=-\infty}^{\infty} c_k^i (z - z_0)^k . \tag{5.5}$$

Numbers $\rho_j = \dfrac{1}{2\pi \mathrm{i}} \ln \lambda_j$ are called characteristic exponents. In the nondegenerate case ($\lambda_1 \neq \lambda_2$) the difference $\rho_2 - \rho_1$ is not an integer number.

If it is impossible to diagonalize matrix a_{ij}, we reduce it to Jordan form, i.e., choose the solutions so that they are transformed, when bypassing along a closed contour (Fig. 5.1), according to the law

$$y_1^+(z) = \lambda y_1(z),$$
$$y_2^+(z) = \lambda y_2(z) + y_1(z).$$

The asymptotics of the solution y_1 at the point $z = z_0$ has the form (5.5) with $\rho_1 = \ln(\lambda)/2\pi \mathrm{i}$.

To find the asymptotics of y_2, we note that function

$$W(z) = \frac{y_2(z)}{y_1(z)} - \frac{1}{2\pi \mathrm{i}\lambda} \ln(z - z_0)$$

at analytic continuation along the contour (Fig. 5.1) transfer into itself: $W^+(z) = W(z)$, i.e., in the point $z = z_0$ the asymptotics of $W(z)$ has the form (5.5) with $\rho = 0$. The asymptotics of $y_2(z)$ contains logarithmic contributions.

If the Laurent series in the expression (5.5) terminates at the bottom (i.e., $c_k \equiv 0$ for $k \leq k_0$), then z_0 is called *regular singular point.* If the series is not terminated, then z_0 is an *irregular singular point.* An equation is called an equation of the Fuchs class if it has only regular singular points. The singular point $z_0 \neq \infty$ is regular if and only if the coefficients of Eq. (5.1) for $z \to z_0$ satisfy the conditions

$$\left|p(z)(z - z_0)\right| < \infty, \quad \left|q(z)(z - z_0)^2\right| < \infty.$$

Let $z_0 = \infty$, then conformal substitution $z = 1/t$, $d/dz = -t^2 d/dt$, transforms Eq. (5.1) to the form

$$\frac{d^2y}{dt^2} + \tilde{p}(t)\frac{dy}{dt} + \tilde{q}(t)y = 0,$$

where

$$\tilde{p}(t) = 2t - \frac{p\left(\frac{1}{t}\right)}{t^2}, \quad \tilde{q}(t) = \frac{q\left(\frac{1}{t}\right)}{t^4}.$$

Therefore, $z_0 = \infty$ is a regular singular point only for finite $|p(z)z|$ and $\left|q(z)z^2\right|$ at $z \to \infty$. The point $z_0 = \infty$ is a non-singular one only for finite $|z[2 - p(z)z]|$ and $\left|z^4 q(z)\right|$ at $z \to \infty$.

Therefore, the coefficients in Eq. (5.1) of the Fuchs class with $n+1$ regular singular points (one of which is at infinity) can be reduced to the form

$$p(z) = \sum_{k=1}^{n} \frac{A_k}{(z - z_k)}, \quad q(z) = \sum_{k=1}^{n} \frac{B_k}{(z - z_k)^2} + \frac{C_k}{(z - z_k)},$$

where A_k, B_k, C_k are fixed complex numbers with

$$\sum_{k=1}^{n} C_k = 0.$$

5.2 Hypergeometric Functions

An equation with three regular singular points is called *hypergeometric equation*. Let z_0, z_1, z_2 be singular points. The replacement

$$t = \frac{(z - z_0)(z_1 - z_2)}{(z - z_2)(z_1 - z_0)},$$

conformally maps the complex plane of the variable onto itself and transforms regular singular points to $t_0 = 0$, $t_1 = 1$, and $t_2 = \infty$. In this case the coefficients of Eq. (5.1) has the form

$$p(t) = \frac{A_0}{t} + \frac{A_1}{t-1}, \quad q(t) = \frac{C}{t(t-1)} + \frac{B_0}{t^2} + \frac{B_1}{(t-1)^2}.$$

Using the change of variable $y = t^{nu}(t-1)^{mu} F$, one can make the coefficients B_k to be equal to zero. As a result we obtain the canonical form of the hypergeometric equation

$$t(t-1)\frac{d^2F}{dt^2} + [(\alpha + \beta + 1)t - \gamma]\frac{dF}{dt} + \alpha\beta F = 0, \tag{5.6}$$

where α, β, and γ are expressed via A_k, B_k, and C in a simple form. Solutions of Eq. (5.6) are *Gaussian hypergeometric functions*. At all regular singular points, the solutions have power asymptotics. Near the singular point t', we substitute the function F in the form $F \sim (t - t')^{\nu}(1 + O(t - t'))$ into Eq. (5.6), neglect small over $t - t'$ terms, and obtain

$$\begin{aligned} F \sim t^{\nu_0}, &\quad \text{at } t \to 0: \quad \nu_0 = 0,\, 1 - \gamma, \\ F \sim (t-1)^{\nu_1}, &\quad \text{at } t \to 1: \quad \nu_1 = 0,\, \gamma - \alpha - \beta, \\ F \sim \frac{1}{t^{\nu_\infty}}, &\quad \text{at } t \to \infty: \quad \nu_\infty = \alpha,\, \beta. \end{aligned}$$

The Taylor series of the solution with $\nu_0 = 0$ has the form near $t = 0$

$$F \equiv {}_2F_1(\alpha, \beta; \gamma; t) = 1 + \frac{\alpha\beta}{\gamma}\frac{t}{1!} + \frac{\alpha(\alpha+1)\beta(\beta+1)}{\gamma(\gamma+1)}\frac{t^2}{2!} + \ldots .$$

This series shows that if α or β are non-positive integer, then the series breaks off, which means that F becomes a polynomial.

The confluent hypergeometric equation is obtained from hypergeometric one as a result of merging of two regular singular points. We make the substitution $t = x/\beta$ in the hypergeometric equation and tend β to infinity, so that $x_0 = t_0\beta = 0$, $x_1 = t_1\beta \to \infty$ and $x_2 = t_2\beta = \infty$. As a result two singular points x_1, x_2 merge into one and we arrive at the equation for a confluent hypergeometric function

$$x\frac{d^2F}{dx^2} + (\gamma - x)\frac{dF}{dx} - \alpha F = 0\,. \tag{5.7}$$

Here the point $x = \infty$ is irregular singular point. The asymptotics at $x \to 0$ has form, $F \sim x^{\nu_0}$, where $\nu_0 = 0$ or $\nu_0 = 1 - \gamma$. At $x \to \infty$ the asymptotics are exponential, $F \sim e^x$, and power, $F \sim x^{-\alpha}$. Expansion in a Taylor series of the solution with $\nu_0 = 0$ in the vicinity of $x = 0$ has the form

$${}_1F_1(\alpha;\gamma;x) = \lim_{\beta\to\infty} {}_2F_1(\alpha,\beta;\gamma;x/\beta) =$$

$$= 1 + \frac{\alpha}{\gamma}\frac{x}{1!} + \frac{\alpha(\alpha+1)}{\gamma(\gamma+1)}\frac{x^2}{2!} + \frac{\alpha(\alpha+1)(\alpha+2)}{\gamma(\gamma+1)(\gamma+2)}\frac{x^3}{3!} + \dots\,.$$

It is seen that for a non-positive integer α the series breaks off, which means that ${}_1F_1$ becomes a polynomial.

5.3 Orthogonal Polynomials

Orthogonal polynomials $P_n(x)$ are given by the generalized Rodrigues formula

$$P_n(x) = \frac{A_n}{W(x)}\frac{d^n}{dx^n}W(x)\,[R(x)]^n\,, \tag{5.8}$$

where $R(x)$ is a polynomial, n is the degree of the polynomial $P_n(x)$. The polynomials $P_n(x)$ are orthogonal with the factor $W(x)$:

$$\int_{x_1}^{x_2} dx\, W(x)P_n(x)P_m(x) = \delta_{nm}h_n, \quad W(x_1)R(x_1) = W(x_2)R(x_2) = 0\,, \tag{5.9}$$

where the interval $(x_1,\ x_2)$ is bounded by points where $W(x)R(x)$ vanishes (see problem **5.7**).

There are three general types of orthogonal polynomials (see problem **5.6**), differing by the region of definition.

(1) Polynomials defined on the whole numeric axis $(-\infty < x < \infty)$,

$$R(x) = 1, \quad W(x) = \exp[-b^2(x-a)^2]. \tag{5.10}$$

After replacement $x \to a + x/b$, these polynomials are reduced to Hermite polynomials:

$$H_n(x) = \frac{A_n}{\exp(-x^2)} \frac{d^n}{dx^n} \exp(-x^2). \tag{5.11}$$

(2) Polynomials defined on the semi axis $(a \leq x < \infty)$,

$$R(x) = x - a, \quad W^{\alpha}(x) = e^{-b(x-a)}(x-a)^{\alpha}, \quad \alpha > -1. \tag{5.12}$$

After replacement $x \to a + x/b$, these polynomials are reduced to generalized Laguerre polynomials:

$$L_n^{\alpha}(x) = \frac{A_n^{\alpha}}{x^{\alpha} e^{-x}} d^n dx^n x^{\alpha+n} e^{-x}. \tag{5.13}$$

(3) The polynomials defined on the interval $(a_1 \leq x \leq a_2)$,

$$R(x) = (x - a_1)(a_2 - x), \; W^{\alpha,\beta}(x) = (x-a_1)^{\alpha}(a_2-x)^{\beta}, \tag{5.14}$$

$$\alpha > -1, \; \beta > -1.$$

After replacement $x \to (a_1 + a_2)/2 + x\,(a_2 - a_1)/2$, these polynomials are reduced to to Jacobi polynomials:

$$P_n^{\alpha,\beta}(x) = \frac{A_n^{\alpha,\beta}}{(x+1)^{\alpha}(1-x)^{\beta}} \frac{d^n}{dx^n}(x+1)^{\alpha+n}(1-x)^{\beta+n}. \tag{5.15}$$

For $\alpha = \beta = \lambda$, the Jacobi polynomials are reduced to to Gegenbauer polynomials, and for $\lambda = 0$ they are reduced to Legendre polynomials.

Polynomials of all three types are eigenfunctions of the operator (see problem **5.8**)

$$\hat{L}P_n(x) = \frac{1}{W(x)} \frac{d}{dx}\left(W(x)R(x)\frac{d}{dx}P_n(x)\right) = -\lambda_n P_n(x), \tag{5.16}$$

$$x \in (x_1, x_2); \quad W(x_1)P_n(x_1) = W(x_2)P_n(x_2) = 0,$$

where the factor $W(x)$ and the polynomial $R(x)$ are of the form (5.10), (5.12) or (5.14). The operator $\hat{L}$ is Hermitian in a Hilbert space with scalar product, given by the formula (5.9). Equation (5.16) for Jacobi polynomials is reduced to hypergeometric, and for the Hermite and Laguerre polynomials it is reduced to a degenerate hypergeometric equation (problem **5.39**).

Books Morse and Feshbach (1953), Smirnov (1964), and Kamke (1969) include a good introduction to the analytical theory of ordinary differential equations. For readers interested in the theory of special functions, we can recommend Bateman, and Erdélyi's work (Bateman and Erdélyi, 1953-1955) with formula derivation. Finally, tables and graphs of special functions can be found in the reference books Jahnke *et al.* (1960), Abramowitz and Stegun (1965), Ovler *et al.* (2010).

Worked Examples

5.1. For the regular singular point $x_0 \neq \infty$ of Eq. (5.1), prove that the coefficients (5.1) have the form

$$p(x) = \frac{p_1(x)}{(x - x_0)}, \quad q(x) = \frac{q_1(x)}{(x - x_0)^2},$$

where $p_1(x)$ and $q_1(x)$ are functions analytic in the neighborhood of $x = x_0$.

Solution. Without loss of generality, we can put $x_0 = 0$. If $x = 0$ is a regular singular point of the equation, then the asymptotics of the solutions (5.5)and the Wronskian (5.2) for $x \to 0$ in a nondegenerate case has the form

$$y_j(x) \to x^{\rho_j}, \quad W(x) \to (\rho_2 - \rho_1)x^{\rho_1+\rho_2-1}.$$

Expressing $p(x)$ and $q(x)$ in terms of $y(x)$ and $W(x)$ using Eqs. (5.1) and (5.3), we find the asymptotics at $x \to 0$:

$$p(x) = -\frac{dW(x)/dx}{W(x)} \to \frac{A}{x}, \quad q(x) = -\frac{d^2y_j(x)/dx^2}{y_j(x)} - p(x)\frac{dy_j(x)/dx}{y_j(x)} \to \frac{B}{x^2}.$$

5.2. Find regular singular points and characteristic exponents in the Bessel equation:

$$x^2\frac{d^2y}{dx^2} + x\frac{dy}{dx} + (x^2 - \nu^2)y = 0.$$

Solution. Rewriting the equation in the form (5.1), we find $p(x) = 1/x$ and $q(x) = 1 - \nu^2/x^2$. There is one regular singular point $x = 0$ and one irregular point $x = \infty$. The asymptotics at a regular singular point must be search for in exponential form. Substituting $y(x) \sim x^{\rho}$ into the equation and neglecting small in x terms, we obtain for $x \to 0$

$$\left(\rho(\rho-1)+\rho-\nu^2\right)x^{\rho}=0\,.$$

Hence, the characteristic exponents are

$$\rho_{\pm}=\pm\nu.$$

In the degenerate case ($\rho_+ - \rho_- = 2\nu = n$, where n is an integer), the logarithmic terms can appear in the asymptotics of the second solution. Denote by $J_{n/2}(x)$ a regular at zero solution ($J_{n/2}(x) \sim x^{n/2}$ at small x, with $n \geq 0$). To find the asymptotics of the second solution, we make the following substitution

$$y_2(x)=J_{n/2}(x)\int^{x} dt f(t)\,,$$

where the function $f(x)$ satisfies the equation

$$J_{n/2}(x)\frac{df(x)}{dx}+\left(2\frac{dJ_{n/2}(x)}{dx}+\frac{J_{n/2}(x)}{x}\right)f(x)=0\,.$$

The solution to this equation has the following form:

$$f(x)=\frac{1}{xJ^2_{n/2}(x)}\,,$$

and can be represented at small x as

$$f(x)=x^{-n-1}\sum_{k=0}^{\infty}a_k x^k\,.$$

Integrating over x, we find the asymptotics of $y_2(x)$:

$$y_2(x)\to a_0\ln(x)\quad\text{for}\quad n=0,$$

$$y_2(x)\to a_0 x^{-n/2}\quad\text{for}\quad n>0.$$

If the coefficient $a_n \neq 0$, then the logarithmic terms arise in the expansion of $y_2(x)$ as a result of integration of $f(x)$.

We search for the asymptotics at an irregular singular point in the form $y(x) \to \exp(\lambda x^{\sigma})$ at $x \to \infty$. Substituting it in the equation and neglecting small in $1/x$ terms, we obtain that only three terms should be taken into account at $x \to \infty$,

$$\lambda^2\sigma^2x^{2\sigma}+\lambda\sigma^2x^{\sigma}+x^2=0.$$

If we assume that the first two terms are leading, we obtain $\sigma = 0$ which cannot be true since the last term grows faster at $x \to \infty$. We now assume that the two last terms are leading and obtain $\sigma = 2$. Since, in this case, the first term grows faster at $x \to \infty$, we conclude that our assumption is also wrong. Finally, consider the last possibility that the first and last terms are leading. Then we obtain $\sigma = 1, \quad \lambda = \pm \mathrm{i}$, and $y(x) \to \exp(\pm \mathrm{i}x)$.

5.3. Express Bessel Functions $J_\nu(x)$ in terms of the confluent hypergeometric function.

Solution. Singular points of the Bessel equation (A.14) and the confluent hypergeometric equation (5.7) coincide, but their asymptotics differ. The solutions of the Bessel equation at $x \to 0$ tend to $x^{\pm\nu}$, and at $x \to \infty$ tend to $\exp(\pm \mathrm{i}x)$ (problem **5.2**). Using this asymptotics, we write the Bessel function in the form

$$J_\nu(x) = x^\nu \exp(-\mathrm{i}x) f(x),$$

where $f(x)$ satisfies the equation

$$x\frac{d^2 f}{dx^2} + [2\nu + 1 - 2\mathrm{i}x]\frac{df}{dx} - \mathrm{i}(2\nu + 1) f = 0.$$

Comparing this equation with (5.7), we obtain

$$J_\nu(x) = \left(\frac{x}{2}\right)^\nu \frac{\exp(-\mathrm{i}x)}{\Gamma(\nu + 1)} {}_1F_1\left(\nu + \frac{1}{2}; 2\nu + 1; 2\mathrm{i}x\right).$$

We find the proportionality coefficient by a comparison of the series expansions (5.7) and (A.15) at $x \to 0$

$$\left(\frac{x}{2}\right)^{-\nu} J_\nu(x) \to \frac{1}{\Gamma(\nu + 1)}, \qquad {}_1F_1(\alpha; \gamma; 0) = 1.$$

5.4. Find characteristic exponents in regular singular points of the Legendre equation (the equation for the eigenfunctions of the angular part of the Laplace operator in spherical coordinates):

$$\frac{1}{\sin\theta}\frac{d}{d\theta}\left(\sin\theta\frac{dy}{d\theta}\right) - \frac{m^2 y}{\sin^2\theta} = -\lambda_m y. \tag{5.17}$$

Solution. Making the substitution $x = \cos\theta$ and $\lambda = l(l+1)$ we obtain the equation in the form (A.36), with three regular singular points $x = \pm 1$, $x = \infty$. We are looking for asymptotics at $x \to \pm 1$ in form:

$$y(x) \sim (1 - x)^{\rho_1}, \quad y(x) \sim (1 + x)^{\rho_2}.$$

Substituting them into Eq. (A.36) and neglecting small in $1 \pm x$ terms, we obtain

$$\rho_1 = \pm\frac{m}{2}, \quad \rho_2 = \pm\frac{m}{2}.$$

We also search for the asymptotics at $x \to \infty$ a power-law form $x^{-\rho_\infty}$ and obtain the quadratic equation for the characteristic exponent,

$$\rho_\infty(\rho_\infty + 1) = \lambda_m.$$

5.5. Express the Legendre functions $P_l^m(x)$ in terms of the Gaussian hypergeometric function.

Solution. Singular points of the Legendre equation (A.36) $x_0 = 1, x_1 = -1, x_2 = \infty$ and the hypergeometric equation (5.6) are different. Let's make the conformal change of variable $t = (1-x)/2$, which transform the regular singular points $1, -1, \infty$ into $t_0 = 0, t_1 = 1, t_2 = \infty$. In new variables the equation for the function $F(t) \equiv P_l^m(1-2t)$ has the form

$$t(t-1)\frac{d^2F}{dt^2} + [2t-1]\frac{dF}{dt} - \left(\frac{m^2}{4t(t-1)} + l(l+1)\right) F = 0\,.$$

The function

$$f(t) = \frac{F(t)}{t^{m/2}(t-1)^{m/2}}$$

satisfies the hypergeometric equation (5.6):

$$t(t-1)\frac{d^2f}{dt^2} + [2t-1](m+1)\frac{df}{dt} - [l(l+1) - m(m+1)]\, f = 0,$$

$$f(t) = A\ {}_2F_1(m-l, m+l+1; m+1; t).$$

It remains to find the proportionality coefficient A. Since hypergeometric function ${}_2F_1$ tends to unity for $t \to 0$ $(x \to 1)$ and Legendre functions (A.37) tend to

$$\left.\frac{P_l^m(x)}{(1-x^2)^{m/2}}\right|_{x\to 1} \to \frac{(l+m)!}{2^m m!(l-m)!},$$

we finally obtain

$$P_l^m(x) = \frac{(l+m)!}{2^m(l-m)!m!}(1-x^2)^{m/2}{}_2F_1(m-l, m+l+1; m+1; 1-x/2).$$

5.6. Find $W(x)$ and $R(x)$ for which the functions $P_n(x)$ given by the generalized formula Rodrigue (5.8) becomes polynomials of the n degree.

Solution. We express the first degree polynomial $P_1(x) = A_1\nu(\mu - x)$ in terms of $W(x)$ and $R(x)$ using the generalized Rodrigues formula (5.8):

$$\nu(\mu - x) = \frac{P_1(x)}{A_1} = \frac{R(x)}{W(x)}\frac{dW}{dx} + \frac{dR}{dx}. \tag{5.18}$$

Consider (5.18) as a differential equation for the function $W(x)$ at various degrees k of the polynomial $R(x)$: $k = 0,\ k = 1$ and $k \geq 2$. The solutions to this equation are:

(1) For $R = 1$, up to a constant factor the function $W(x)$ equals

$$W(x) = \exp[-\nu(x-\mu)^2/2)].$$

Substituting it into Rodrigues' formula, we find that

$$P_n(x) = A_n \exp\left(\frac{\nu(x-\mu)^2}{2}\right)\frac{d^n}{dx^n}\exp\left(-\frac{\nu(x-\mu)^2}{2}\right)$$

are polynomials of n -th degree. Making the change of variable, $\sqrt{\nu/2}\,(x-\mu) \to x$, we reduce these polynomials to Hermite polynomials.

(2) For $R = x - a$, the function

$$W^\alpha(x) = (x-a)^\alpha \exp[-\nu(x-a)]$$

depends on the parameter $\alpha = \nu(\mu - a) - 1$, which we wrote as an upper index. Substituting this function into Rodrigues' formula, we find that

$$P_n^\alpha(x) = A_n^\alpha (x-a)^{-\alpha}\exp(\nu(x-a))\frac{d^n}{dx^n}(x-a)^{\alpha+n}\exp(-\nu(x-a))$$

are polynomials of n -th degree. Making the variable change, $\nu\,(x-a) \to x$, we reduce these polynomials to generalized Laguerre polynomials.

(3) For $R(x) = \prod_{r=1}^{k}(x - a_r)$, where $k \geq 2$ and all a_r are assumed to be different, Eq. (5.18) for $W(x)$ can be written as

$$\frac{1}{W(x)}\frac{dW}{dx} = \frac{\nu(\mu - x)}{R(x)} - \frac{1}{R(x)}\frac{dR}{dx} = \sum_{r=1}^{k}\frac{\alpha_r}{x - a_r}.$$

Here α_r are the coefficients of the expansion into simple fractions. The solution to this equation for the function $W(x)$ can be written as $W(x) = \prod_{r=1}^{k}(x - a_r)^{\alpha_r}$. Substituting the functions $W(x)$ and $R(x)$ into the generalized Rodrigues formula we obtain the degree of the polynomial P_n to be $\deg P_n = n(k-1)$. It follows from the requirement $\deg P_n = n$ that k is equal to two. We denote $\alpha_1 = \alpha,\ \alpha_2 = \beta$ and indicate these parameters as superscripts, then

$$W^{\alpha,\beta}(x) = (x-a_1)^{\alpha}(a_2-x)^{\beta},$$
$$P_n^{\alpha,\beta}(x) = A_n^{\alpha,\beta}(x-a_1)^{-\alpha}(a_2-x)^{-\beta}\tfrac{d^n}{dx^n}(x-a_1)^{\alpha+n}(a_2-x)^{\beta+n}.$$

After replacement $x \to (a_1+a_2)/2 + x\,(a_2-a_1)/2$, these polynomials are reduced to Jacobi polynomials.

5.7. Prove that the polynomials given by generalized Rodrigues formula (5.8), are orthogonal with the weight function $W(x)$ on the interval $(x_1,\ x_2)$ bounded by the points in which $W(x_1)R(x_1) = W(x_2)R(x_2) = 0$.

Solution. The function $W(x)R(x)$ for all three types of polynomials found in problem **5.6** vanishes at points:

1) for (5.10) $x_1 = -\infty,\ x_2 = \infty$;
2) for (5.12) $x_1 = a,\ x_2 = \infty$;
3) for (5.14) $x_1 = a_1,\ x_2 = a_2$.

We write the weight functions of all three types as $W^{\alpha,\beta}(x)$ (the weighting function can only depend on one parameter (5.12) or independent of parameters (5.10)) and note that

$$W^{\alpha,\beta}(x)R(x) = W^{\alpha+1,\beta+1}(x). \tag{5.19}$$

Let $n > m$. Assuming for simplicity the normalization constants A_n equal to one, we integrate the scalar product by parts,

$$\begin{aligned} I_{nm}^{\alpha,\beta} &= \int_{x_1}^{x_2} dx\ W^{\alpha,\beta}(x)P_n^{\alpha,\beta}(x)P_m^{\alpha,\beta}(x) \\ &= W^{\alpha,\beta}(x)R(x)\left[(W^{\alpha+1,\beta+1}(x))^{-1}\frac{d^{n-1}W^{\alpha+1,\beta+1}R^{n-1}}{dx^{n-1}}\right]P_m^{\alpha,\beta}(x)\Bigg|_{x_1}^{x_2} \\ &\quad - \int_{x_1}^{x_2} dx\left(\frac{d^{n-1}W^{\alpha,\beta}R^n}{dx^{n-1}}\right)\frac{dP_m^{\alpha,\beta}(x)}{dx}. \end{aligned} \tag{5.20}$$

Expression in square brackets is a polynomial $P_{n-1}^{\alpha+1,\beta+1}(x)$. In particular, this means that for $\alpha > -1$ and $\beta > -1$ the integrated term vanishes at the points x_1 and x_2 where $W^{\alpha,\beta}R = 0$. We repeat the integration by parts $n-1$ times (the integrated terms vanish each time due to the relation $W^{\alpha,\beta}(x_1)R(x_1) = W^{\alpha,\beta}(x_2)R(x_2) = 0$) and obtain

$$I_{nm}^{\alpha,\beta} = \int_{x_1}^{x_2} dx(-1)^n W^{\alpha,\beta}(x)R^n(x) \left[\frac{d^n P_m^{\alpha,\beta}(x)}{dx^n}\right] = 0 \quad \text{at} \quad m < n. \qquad (5.21)$$

Here we took into account that the n -th order derivative of a polynomial of degree m is zero for $m < n$. To prove that the integral is zero for $m > n$, we should replace $n \leftrightarrow m$. As a result $I_{nm}^{\alpha,\beta} = h_n^{\alpha,\beta}\delta_{nm}$ and the orthogonality of the polynomials is proved.

5.8. Prove that the polynomials given by generalized Rodrigues formula (5.8) are eigenfunctions of the operator $\hat{L}$ (5.16) in which the functions $W(x)$ and $R(x)$ are of the form (5.10), (5.12) or (5.14). Find eigenvalues λ_n of the operator $\hat{L}$.

Solution. The multiplication of the weight function by the polynomial $R(x)$ increases its superscripts by one (5.19). The operator

$$\hat{q}^{\alpha,\beta} = \frac{R(x)}{W^{\alpha,\beta}(x)}\frac{d}{dx}W^{\alpha,\beta}(x),$$

where the derivative acts on all the functions on the right, increase the degree of the polynomial $P_n^{\alpha,\beta}(x)$ by one, transforming it into an orthogonal set of polynomials with indices α, β, reduced by one:

$$\hat{q}^{\alpha,\beta} P_n^{\alpha,\beta}(x) = A_n^{\alpha,\beta}\frac{R(x)}{W^{\alpha,\beta}(x)}\frac{d^{n+1}}{dx^{n+1}}\left(\frac{W^{\alpha,\beta}(x)}{R(x)}R^{n+1}(x)\right)$$
$$= \frac{A_n^{\alpha,\beta}}{A_{n+1}^{\alpha-1,\beta-1}}P_{n+1}^{\alpha-1,\beta-1}(x). \qquad (5.22)$$

Here $A_n^{\alpha,\beta}$ is the normalization constant, and polynomials are assumed to be normalized to 1. We call the operators that increase the degree of a polynomial by one raising operator. The operator d/dx lowers the polynomial degree by one. Let us use the completeness of the set of orthogonal polynomials $P_m^{\alpha,\beta}(x)$ and expand in this basis the polynomial $dP_{n+1}^{\alpha-1,\beta-1}/dx$ of n-th degree:

$$\int_{x_1}^{x_2} dx\, W^{\alpha,\beta}(x)\left(\frac{dP_{n+1}^{\alpha-1,\beta-1}}{dx}\right)P_m^{\alpha,\beta}(x)$$

$$= -\int_{x_1}^{x_2} dx\, P_{n+1}^{\alpha-1,\beta-1}(x) W^{\alpha-1,\beta-1}(x) \hat{q}^{\alpha,\beta} P_m^{\alpha,\beta}(x) = -\delta_{nm} \frac{A_n^{\alpha,\beta}}{A_{n+1}^{\alpha-1,\beta-1}}.$$

Here we have integrated by parts. Note that the integrated term vanishes. As a result, we have

$$\frac{dP_{n+1}^{\alpha-1,\beta-1}}{dx} = -\frac{A_n^{\alpha,\beta}}{A_{n+1}^{\alpha-1,\beta-1}} P_n^{\alpha,\beta}(x).$$

Substituting this expression in (5.22), we find that polynomials satisfy the equation

$$\frac{1}{W^{\alpha,\beta}(x)} \frac{d}{dx} \left(W^{\alpha,\beta}(x) R(x) \frac{d}{dx} P_n^{\alpha,\beta}(x) \right) = -\lambda_n P_n^{\alpha,\beta}(x),$$

$$\lambda_n^{\alpha,\beta} = \left(\frac{A_{n-1}^{\alpha+1,\beta+1}}{A_n^{\alpha,\beta}} \right)^2.$$

With the chosen normalization of the polynomials, the eigenvalues $\lambda_n^{\alpha,\beta}$ are expressed in terms of normalization constants, which, in turn, are related to the normalization integrals $I_{nn}^{\alpha,\beta}$ (5.20). Then we find eigenvalues for various weight functions $W^{\alpha,\beta}(x)$ in all three cases (5.10), (5.12) and (5.14).

(1) For Hermite polynomials (5.11)

$$\hat{q} = \exp(x^2) \frac{d}{dx} \exp(-x^2),$$

and the normalization constants are

$$(A_n)^{-2} = \int_{-\infty}^{\infty} dx (-1)^n \exp(-x^2) \frac{d^n}{dx^n} \left(\exp(x^2) \frac{d^n}{dx^n} \exp(-x^2) \right) = \frac{2^n n!}{\sqrt{\pi}}.$$

Therefore $\lambda_n = 2n$.

(2) For Laguerre polynomials (5.13)

$$\hat{q}^\alpha = x^{-\alpha+1} e^x \frac{d}{dx} x^\alpha e^{-x},$$

and the normalization constants are

$$(A_n^\alpha)^{-2} = \int_0^{\infty} dx (-1)^n x^{\alpha+n} e^{-x} \frac{d^n}{dx^n} \left(x^{-\alpha} e^x \frac{d^n}{dx^n} x^{\alpha+n} e^{-x} \right) = n!\Gamma(\alpha+n+1).$$

Hence $\lambda_n^\alpha = n$.

(3) For Jacobi polynomials (5.15)

$$\hat{q}^{\alpha,\beta} = (x-a_1)^{1-\alpha}(a_2-x)^{1-\beta}\frac{d}{dx}(x-a_1)^{\alpha}(a_2-x)^{\beta},$$

and the normalization constants are

$$(A_n^{\alpha,\beta})^{-2} = \int\limits_{-1}^{1} dx(-1)^n(x+1)^{\alpha+n}(1-x)^{\beta+n}\frac{d^n}{dx^n}\frac{\frac{d^n}{dx^n}(x+1)^{\alpha+n}(1-x)^{\beta+n}}{(x+1)^{\alpha}(1-x)^{\beta}}$$
$$= n!2^{\alpha+\beta+2n+1}B(\alpha+n+1,\beta+n+1)\times$$
$$\times\sum_{l=0}^{n}\frac{n!}{l!(n-l)!}\frac{\Gamma(n+\alpha+1)\Gamma(n+\beta+1)}{\Gamma(l+\alpha+1)\Gamma(n-l+\beta+1)}$$
$$= n!\,2^{\alpha+\beta+2n+1}\frac{\Gamma(\alpha+n+1)\Gamma(\beta+n+1)}{\Gamma(\alpha+\beta+n+1)}.$$

Therefore $\lambda_n^{\alpha,\beta} = n(\alpha+\beta+n+1)$.

5.9. By extraction of the asymptotics at the singular points of the Legendre equation

$$\frac{1}{\sin\theta}\frac{d}{d\theta}\left(\sin\theta\frac{dy}{d\theta}\right)-\frac{m^2y}{\sin^2\theta} = -\lambda_m y, \tag{5.23}$$

find the coefficients of the series expansion in $x=\cos\theta$. At what λ_m a series has a finite number of terms and a limited solution?

Solution. By extraction of power asymptotics at the points $x=\pm1$ (see problem (**5.4**))

$$f(x) = \frac{y(x)}{(1-x^2)^{m/2}},$$

we obtain an equation for the function $f(x)$

$$(1-x^2)\frac{d^2f}{dx^2}-2x(m+1)\frac{df}{dx}+(\lambda_m-m(m+1))\,f=0. \tag{5.24}$$

Then we substitute $f(x) = \sum_{n=0}^{\infty} a_nx^n$ as a power series in x into the equation and set the sum of the coefficients at the same exponents of x to be zero. As a result we obtain the recurrence relations on a_n,

$$a_{n+2} = a_n\frac{(n+m)(n+m+1)-\lambda_m}{(n+1)(n+2)}.$$

The coefficient a_{k+2} vanishes, and the series breaks off only if $\lambda_m = l(l+1),\quad l\geq m$, while $n=l-m$.

5.10. Obtain the Schlfli integral representation (A.44) from Rodrigues formula (A.37) for Legendre polynomials

Solution. It follows from the formula for the residue at the pole of order $l+1$ that the residue at point x is equal to

$$\oint^{(x+)} \frac{dz}{2\pi \mathrm{i}} \frac{f(z)}{(z-x)^{l+1}} = \frac{1}{l!}\frac{d^l f(x)}{dx^l}.$$

Substituting $f(x) = 2^{-l}(x^2-1)^l$ in this equation we obtain the integral representation of Legendre polynomials:

$$P_l(x) = \oint^{x+} \frac{dz}{2^{l+1}\pi \mathrm{i}} \frac{(z^2-1)^l}{(z-x)^{l+1}},$$

which can be written as an integral over a segment by passing from the variable z to the variable ϕ:

$$z = x + \mathrm{i}\exp(\mathrm{i}\phi)\sqrt{1-x^2}, \quad dz = (z-x)id\phi,$$

$$z^2 - 1 = x^2 + 2x(z-x) + (z-x)^2 - 1 = 2(z-x)(x + i\sqrt{1-x^2}\cos(\phi)),$$

$$P_l(\cos\theta) = \int_0^{2\pi} \frac{d\phi}{2\pi}(\cos\theta + i\sin\theta\cos\phi)^l.$$

5.11. Find a generating function (A.42) for Legendre polynomials $P_l(x)$ using the integral representation (A.44).

Solution. We substitute the integral representation for Legendre polynomials in the expression for the generating function and obtain

$$F(r, \cos\theta) = \sum_{l=0}^{\infty} r^l P_l(\cos\theta)$$

$$= \int_0^{2\pi} \frac{d\phi}{2\pi}\sum_{l=0}^{\infty} r^l(\cos\theta + \mathrm{i}\sin\theta\cos\phi)^l, \quad r < 1,$$

$$F(r, \cos\theta) = \sum_{l=0}^{\infty} r^{-l-1} P_l(\cos\theta)$$

$$= \int_0^{2\pi} \frac{d\phi}{2\pi}\sum_{l=0}^{\infty} r^{-l-1}(\cos\theta + \mathrm{i}\sin\theta\cos\phi)^l, \quad r > 1.$$

Then we sum up the geometric progression and find

$$F(r,\cos\theta)=\int_0^{2\pi}\frac{d\phi}{2\pi}\frac{1}{1-r\cos\theta-ir\sin\theta\cos\phi},\qquad r<1,$$

$$F(r,\cos\theta)=\int_0^{2\pi}\frac{d\phi}{2\pi}\frac{1}{r-\cos\theta-\mathrm{i}\sin\theta\cos\phi},\qquad r>1.$$

Passing to new variable $y=i\exp(\mathrm{i}\phi)$ we obtain

$$F(r,\cos\theta)=\oint_{|y|=1}\frac{dy}{2\pi i}\frac{-2}{r\sin\theta}\left(y^2-2\frac{1-r\cos\theta}{r\sin\theta}y-1\right)^{-1},\qquad r<1,$$

$$F(r,\cos\theta)=\oint_{|y|=1}\frac{dy}{2\pi i}\frac{-2}{\sin\theta}\left(y^2-2\frac{r-\cos\theta}{\sin\theta}y-1\right)^{-1},\qquad r>1.$$

The contribution to the integral is given by the residue of pole in the integrand that is inside the circle unit radius:

$$y^{\pm}=\frac{1-r\cos\theta\pm\sqrt{1-2r\cos\theta+r^2}}{r\sin\theta},\qquad r<1,$$

$$y^{\pm}=\frac{r-\cos\theta\pm\sqrt{1-2r\cos\theta+r^2}}{\sin\theta},\qquad r>1.$$

Using the equality $y^+y^-=1$, we find that $|y^-|<1$ and $|y^+|>1$. Therefore, the residue must be taken at the point $y=y^-$. We have

$$F(r,x)=\frac{2}{r\sin\theta}\frac{1}{|y^+-y^-|}=\frac{1}{\sqrt{1-2rx+r^2}},\qquad r<1,$$

$$F(r,x)=\frac{2}{\sin\theta}\frac{1}{|y^+-y^-|}=\frac{1}{\sqrt{1-2rx+r^2}},\qquad r>1,$$

5.12. Derive recurrence relations (A.40) and differentiation formula (A.41) using the generating function (A.42).

Solution. Taking the derivative of the generating function, we obtain

$$\frac{\partial F(r,x)}{\partial r}=\frac{x-r}{1+r^2-2rx}F(r,x),\qquad \frac{\partial F(r,x)}{\partial x}=\frac{r}{1+r^2-2rx}F(r,x).$$

Multiplying these equations by $1+r^2-2rx$, expanding the result in a series in powers of r and collecting the coefficients at the same powers of r, we find

$$(l+1)P_{l+1}(x)+(l-1)P_{l-1}(x)-2xlP_l(x)=-P_{l-1}(x)+xP_l(x),$$
$$\frac{d}{dx}P_{l+1}(x)+ddx\,P_{l-1}(x)-2xddx\,P_l(x)=P_l(x).$$

We rewrite the first equation as (A.40)

$$(l+1)P_{l+1}(x)+l\,P_{l-1}(x)=x(2l+1)P_l(x).$$

Differentiating it and subtracting the result from the second equation multiplied by $l+1/2$ or $l+1$, we obtain (A.41)

$$\frac{d}{dx}P_{l+1}(x)-ddx\,P_{l-1}(x)=(2l+1)P_l(x),$$
$$x\frac{d}{dx}P_l(x)-\frac{d}{dx}P_{l-1}(x)=l\,P_l(x).$$

5.13. Prove the orthogonality of the normalized spherical functions Y_{lm} and express them through P_l^m.

Solution. Normalized orthogonal spherical functions

$$Y_{lm}(\theta,\varphi)=A_l^m P_l^{|m|}(\cos\theta)\exp(im\varphi)$$

must satisfy the relation

$$\int d\Omega Y_{lm}^*(\theta,\varphi)Y_{l'm'}(\theta,\varphi)=2\pi\delta_{mm'}A_l^m A_{l'}^m\int_{-1}^{1}dx\,P_l^{|m|}(x)P_{l'}^{|m|}(x)=\delta_{mm'}\delta_{ll'}.$$

In other words, the associated Legendre functions must be orthogonal,

$$F=\int_{-1}^{1}dx\,P_l^{|m|}(x)P_{l'}^{|m|}(x)=\delta_{ll'}\frac{1}{2\pi(A_l^m)^2}.$$

Then we put $l'>l$, use the Rodrigues formula for the functions Legendre and integrate by parts. As a result, we have

$$F=\int_{-1}^{1}dx\frac{(1-x^2)^{|m|}}{2^{l+l'}l!l'!}\left(\frac{d^{l+|m|}}{dx^{l+|m|}}(x^2-1)^l\right)\left(\frac{d^{l'+|m|}}{dx^{l'+|m|}}(x^2-1)^{l'}\right)$$
$$=\int_{-1}^{1}dx\frac{(-1)^{l'+|m|}(x^2-1)^{l'}}{2^{l+l'}l!l'!}\left[\frac{d^{l'+|m|}}{dx^{l'+|m|}}\left((1-x^2)^{|m|}\frac{d^{l+|m|}}{dx^{l+|m|}}(x^2-1)^l\right)\right].$$

The product in square brackets is nonzero only for $l=l'$ and equals to $(-1)^m(2l)!(l+|m|)!/(l-|m|)!$. For $l'<l$ it is necessary to perform the

substitution $l \leftrightarrow l'$. Thus orthogonality is proved. Calculating the remaining integral F, we obtain an expression for the coefficients A_l^m:

$$F = \frac{2}{2l+1}\frac{(l+|m|)!}{(l-|m|)!}, \quad A_l^m = \sqrt{\frac{(2l+1)(l-|m|)!}{4\pi(l+|m|)!}}.$$

5.14. Using the semiclassical approximation, plot the dependence of spherical harmonics $|Y_{lm}(\cos\theta)|$ on the angle θ for $m = l$ and for $m = 0$ at $l \gg 1$.

Solution. It follows from the Rodrigues formula at $m = l$ that $|Y_{ll}| \sim \sin^l\theta$. For $m = 0$ we transform Eq. (A.36) to the standard form of the Sturm–Liouville operator using a substitution that excludes terms with the first derivative:

$$Y_{l0}(\theta) = \frac{f(\theta)}{\sqrt{\sin\theta}},$$

$$\frac{d^2}{d\theta^2}f(\theta) + \left[\left(l+\frac{1}{2}\right)^2 - \frac{1}{4\sin^2\theta}\right]f(\theta) = 0.$$

For $\sin\theta \gg 1/l$, we can neglect the second term in square brackets. Then, except for the regions $\theta \ll 1/l$ and $\pi - \theta \ll 1/l$, the function has the form

$$Y_{l0} \sim \frac{\sin\left[\theta\left(l+\frac{1}{2}\right)+\frac{\pi}{4}\right]}{\sqrt{\sin\theta}}.$$

The phase $\pi/4$ is determined from the requirement of symmetry

$$Y_{l0}(\theta) = (-1)^l Y_{l0}(\pi - \theta).$$

For $\theta \to 0$ and π spherical function $Y_{l0}(\theta) = \sqrt{(2l+1)/4\pi}\,P_l^0(\cos\theta)$ tends to a finite value (Fig. 5.2).

5.15. Extracting the asymptotics at the singular points, transform the radial part of the 3D Schrödinger equation for a particle in the bound state ($E < 0$) in the Coulomb field $U = -1/r$ (atomic units),

$$\Delta\Psi(\mathbf{r}) + 2\left(E + 1/r\right)\Psi(\mathbf{r}) = 0,$$

to the Laguerre equation (A.59).

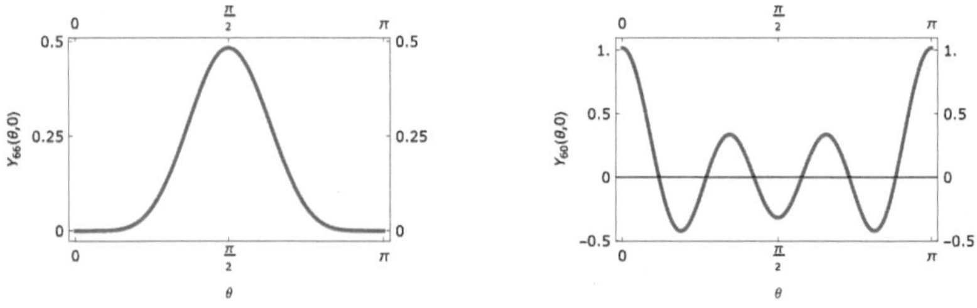

Figure 5.2 Spherical harmonics with orbital angular momentum $l = 6$: left plot $m = 6$, right plot $m = 0$.

Solution. After separating the variables, the radial part of the Schrödinger equation has the form

$$\frac{1}{r}\frac{d^2}{dr^2}r\Psi(r) + \left(\frac{2}{r} - \frac{l(l+1)}{r^2} + 2E\right)\Psi(r) = 0.$$

In this equation, the regular singular point $r = 0$ and irregular singular point $r = \infty$ coincide with the singular points of the confluent hypergeometric equation (5.7), but the corresponding asymptotics are different:

$$\text{at} \quad r \to 0: \quad \Psi(r) \sim r^l \quad \text{or} \quad r^{-l-1},$$

$$\text{at} \quad r \to \infty: \quad \Psi(r) \sim \exp(\pm r\sqrt{-2E}).$$

Extracting the asymptotics $\Psi(r) = r^l \exp(-r\sqrt{-2E})f(r)$, we obtain the equation on $f(r)$,

$$\frac{d^2 f}{dr^2} + \left(\frac{2(l+1)}{r} - 2\sqrt{-2E}\right)\frac{df}{dr} + \frac{2 - 2(l+1)\sqrt{-2E}}{r} f(r) = 0,$$

Making change of variable, $\rho = 2r\sqrt{-2E}$, we reduce this equation to the equation for a confluent hypergeometric function (5.7) with $\gamma = 2(l+1)$ and $\alpha = l + 1 - 1/\sqrt{-2E}$:

$$\rho\frac{d^2 f(\rho)}{d\rho^2} + (2(l+1) - \rho)\frac{df(\rho)}{d\rho} + \left(\frac{1}{\sqrt{-2E}} - l - 1\right) f(\rho) = 0.$$

The requirement for $\Psi(r)$ to decrease at infinity gives the condition on E,

$$E = -\frac{1}{2(-\alpha + l + 1)^2},$$

because only for integer $\alpha \leq 0$ the confluent hypergeometric equation reduces to the Laguerre equation (A.59), which satisfies the power solution (A.60), $f(\rho) = L_{-\alpha}^{2l+1}(\rho)$, growing at infinity slower than $\exp(\rho/2)$. Finally,

$$\Psi(r) = r^l \exp\left(-\frac{r}{n}\right) L_{n-l-1}^{2l+1}\left(\frac{2r}{n}\right), \quad n = \frac{1}{\sqrt{-2E}} \geq l+1.$$

5.16. Find the normalization of the radial electron wave function in a hydrogen atom:

$$R_n^l(r) = A_n^l x^l \exp(-x/2) L_{n-l-1}^{2l+1}(x), \quad x = \frac{2r}{n}.$$

Solution. By definition, the normalized wave function satisfies the equation (in atomic units),

$$1 = \int_0^\infty r^2 dr |R_n^l(r)|^2.$$

We replace the variable of integration r by $x = 2r/n$, where $n = 1/\sqrt{-2E} \geq l+1$ is integer, and use the Rodrigues formula for the Laguerre polynomials (A.60), then we have

$$(A_n^l)^{-2} = \int_0^\infty \left(\frac{n}{2}\right)^3 x^2 dx x^{-2l-2} \exp(x) \left[\frac{1}{(n-l-1)!} \frac{d^{n-l-1}}{dx^{n-l-1}} [x^{n+l} \exp(-x)]\right]^2.$$

Integrating by parts, we obtain

$$(A_n^l)^{-2} = \left(\frac{n}{2}\right)^3 \int_0^\infty dx \frac{(-1)^{n-l-1} x^{n+l} e^{-x}}{(n-l-1)!^2} \left\{\frac{d^{n-l-1}}{dx^{n-l-1}} x^{-2l} \left[e^x \frac{d^{n-l-1}}{dx^{n-l-1}} (x^{n+l} e^{-x})\right]\right\}.$$

The expression in braces is

$$(-1)^{n-l-1} \left(x(n-l)! - (n+l)(n-l-1)(n-l-1)!\right),$$

so that the integral is expressed in terms of the Γ -functions of the integer argument:

$$(A_n^l)^{-2} = \left(\frac{n}{2}\right)^3 \int_0^\infty dx \frac{\exp(-x)}{(n-l-1)!} [(n-l) x^{n+l+1} - (n+l)(n-l-1) x^{n+l}]$$

$$= \left(\frac{n}{2}\right)^3 \frac{(n-l)(n+l+1)! - (n+l)(n-l-1)(n+l)!}{(n-l-1)!}.$$

Finally, we obtain

$$R_n^l(r) = \frac{2}{n^2} \sqrt{\frac{(n-l-1)!}{(n+l)!}} \left(\frac{2r}{n}\right)^l \exp\left(-\frac{r}{n}\right) L_{n-l-1}^{2l+1}\left(\frac{2r}{n}\right).$$

5.17. For the oscillator potential, find the eigenfunctions and energies of the stationary Schrödinger equation:

$$-\frac{1}{2}(d^2dx^2 - x^2)\Psi(x) = E\Psi(x).$$

Solution. Extracting the asymptotics at $x = \infty$, one can reduce the Schrödinger equation for the oscillator to the Hermite equation. Here we consider another way of solution. By introducing the operators

$$\hat{a} = \frac{i}{\sqrt{2}}\left(\frac{d}{dx} + x\right), \quad \hat{a}^\dagger = \frac{i}{\sqrt{2}}\left(\frac{d}{dx} - x\right),$$

we write the Schrödinger equation in the form

$$\hat{H}\Psi(x) = \hat{a}^\dagger\hat{a}\Psi(x) = \left(E - \frac{1}{2}\right)\Psi(x).$$

It is easy to see that the operator $\hat{H}$ is positively defined for all functions from L^2:

$$\int dx\Psi^\dagger(x)\hat{H}\Psi(x) = \int dx\Psi^\dagger(x)\hat{a}^\dagger\hat{a}\Psi(x) = \int dx|\hat{a}\Psi(x)|^2 \geq 0.$$

The expression is zero only if

$$\hat{a}\Psi_0(x) = \frac{i}{\sqrt{2}}(\frac{d}{dx} + x)\Psi_0(x) = 0, \quad \text{i.e.,} \quad \Psi_0(x) = A\exp\left(-\frac{x^2}{2}\right).$$

In this case $E_0 = 1/2$. To find the spectrum of energies, let us consider the commutator relations

$$\begin{aligned}\hat{a}\hat{a}^\dagger\Psi(x) &= \hat{a}^\dagger\hat{a}\Psi(x) + \Psi(x),\\ \hat{a}^\dagger\hat{a}(\hat{a}^\dagger)^n\Psi(x) &= (\hat{a}^\dagger)^2\hat{a}(\hat{a}^\dagger)^{n-1}\Psi(x) + (\hat{a}^\dagger)^n\Psi(x)\\ &= n(\hat{a}^\dagger)^n\Psi(x) + (\hat{a}^\dagger)^{n+1}\hat{a}\Psi(x).\end{aligned}$$

Taking into account that $\hat{a}\Psi_0 = 0$, we find from the commutator relations that the function $\Psi_n(x) = (\hat{a}^\dagger)^n\Psi_0(x)$ satisfies the equation

$$\left(E - \frac{1}{2}\right)\Psi_n(x) = \hat{H}\Psi_n(x) = \hat{a}^\dagger\hat{a}\Psi_n(x) = n\Psi_n(x).$$

We see that the eigenfunctions $\Psi_n(x)$ with the energy

$$E_n = n + \frac{1}{2}.$$

are expressed via the Hermite polynomials $H_n(x)$ (A.51):

$$\Psi_n(x) = A_n\left(\frac{\mathrm{i}}{\sqrt{2}}\left(\frac{d}{dx} - x\right)\right)^n \mathrm{e}^{-x^2/2} = A_n\frac{i^n}{2^{n/2}}\mathrm{e}^{x^2/2}\frac{d^n}{dx^n}\mathrm{e}^{-x^2},$$

Here we used the operator identity $(\frac{d}{dx} - x)e^{x^2/2} = e^{x^2/2}\frac{d}{dx}$. From the orthogonality relation for the Hermite polynomials (A.53) we obtain the value of the normalization factor,

$$A_n = \frac{1}{\sqrt{n!\sqrt{\pi}}}.$$

5.18. Find the eigenfunctions and energies of the stationary Schrödinger equation for two-dimensional (three-dimensional) oscillator in Cartesian coordinates:

$$-\frac{1}{2}(\Delta - r^2)\Psi(\mathbf{r}) = E\Psi(\mathbf{r}).$$

Find the multiplicity of the degeneracy of the states.

Solution. In Cartesian coordinates, the variables in the equation are separated. For the two-dimensional case, we put

$$\Psi(\mathbf{r}) = \Psi^1(x_1)\Psi^2(x_2), \quad \mathbf{r} = (x_1, x_2).$$

We obtain the same equation for $\Psi^j(x_j)$

$$-\frac{1}{2}\left(\frac{d^2}{dx_j^2} - x_j^2\right)\Psi^j(x_j) = E^j\Psi^j(x_j), \quad j = 1, 2, \quad E = E^1 + E^2.$$

It is the Schrödinger equation for a one-dimensional oscillator. We found the eigenfunctions and spectrum of this equation in problem **5.17**. Thus, the energy of the state is equal to the sum of the energies $E_n^j = n_j + 1/2$ of one-dimensional oscillators,

$$E_N = n_1 + n_2 + 1 = N + 1, \quad N, n_1, n_2 = 0, 1, 2 \ldots .$$

The required eigenfunctions are the product of functions, $\Psi_{n_1}^1(x_1)\Psi_{n_2}^2(x_2)$, the explicit form of $Psi^j(x_j)$ is given in the problem **5.17**. Degeneracy multiplicity g_N of the N-th level is the number of linearly independent eigenfunctions with the same energy E_N, i.e., $n_1 + n_2 = N$. Since only one eigenfunction corresponds to each $n_{1,2}$, we find that $g_N = \sum_{n_1,n_2=0}^{N} \delta(n_1 + n_2 - N) = N + 1$.

In the three-dimensional case, we proceed in the same way. Since the energy of the state is determined by the sum of three integer numbers,

$$E_N = n_1 + n_2 + n_3 + \frac{3}{2} \equiv N + \frac{3}{2}, \quad N, n_1, n_2, n_3 = 0, 1, 2, ...,$$

then the degeneracy multiplicity is different:

$$g_N = \sum_{n_1, n_2, n_3 = 0}^{N} \delta(n_1 + n_2 + n_3 - N) = (N+1)(N+2)/2.$$

5.19. Find the eigenfunctions and the corresponding energies in spherical coordinates for the Schrödinger equation in the three-dimensional oscillator:

$$-\frac{1}{2}(\Delta - r^2)\Psi(\mathbf{r}) = E\Psi(\mathbf{r}).$$

Solution. For the Schrödinger equation in a three-dimensional oscillator, one can separate the variables in spherical coordinates,

$$\Psi(\mathbf{r}) = R(r)Y_{lm}(\theta, \varphi).$$

Here $Y_{lm}(\theta, \varphi)$ are spherical functions expressed in terms of solutions of the Legendre equation (problem **5.13**). We obtain the equation for the radial function,

$$\frac{1}{r^2}\frac{d}{dr}r^2\frac{d}{dr}R(r) + \left(2E - r^2 - \frac{l(l+1)}{r^2}\right)R(r) = 0.$$

Asymptotics of the solution of this equation for the regular singular point $r \to 0$ reads

$$R(r) \sim r^\rho, \quad \rho(\rho+1) - l(l+1) = 0, \quad \rho = l,\ \rho = -l - 1\,.$$

For irregular singular point $r \to \infty$ we have

$$R(r) \sim \exp(\pm r^2/2).$$

Extracting the asymptotics and writing $R(r) = f(r)r^l \exp(-r^2/2)$, we obtain the equation for the function $f(r)$,

$$\frac{d^2 f}{dr^2} + \left(\frac{2(l+1)}{r} - 2r\right)\frac{df}{dr} + 2\,(E - l - 3/2)\,f = 0.$$

Let us make a non-conformal change of the variable $z = r^2$:

$$z\frac{d^2 f(z)}{dz^2} + \left(l + \frac{3}{2} - z\right)\frac{df(z)}{dz} + \frac{E - l - 3/2}{2}f(z) = 0.$$

This confluent hypergeometric equation (5.7) reduces to the Laguerre equation (A.59) with indices $\nu = l + 1/2, \quad 2n = E - l - 3/2$ at a non-negative integer n. Since $R(r)$ should decrease at $r \to \infty$, we obtain that $f(z)$

is equal to the generalized Laguerre polynomial $L_n^{\nu}(z)$, and the eigenfunctions and energies are of the form:

$$\Psi_{n,l,m}(\mathbf{r}) = A_n^l r^l \exp\left(-r^2/2\right) L_n^{l+1/2}(r^2) Y_{lm}(\theta, \varphi),$$

$$E_N = \frac{3}{2} + l + 2n \equiv N + \frac{3}{2}.$$

To calculate the degeneracy multiplicity g_N of the level with energy E_N, recall that for fixed l, there are $2l + 1$ linearly independent functions Y_{lm} with different m, therefore

$$g_N = \sum_{n=0}^{[N/2]} \sum_{l=0}^{N} (2l+1)\,\delta(l + 2n - N)$$
$$= \sum_{n=0}^{[N/2]} 2(N - 2n) + 1 = \frac{(N+1)(N+2)}{2}.$$

This number coincides with the answer to a problem **5.18** for the 3D case. The wavefunctions for the N-th level are linear combinations of functions from the problem **5.18** with the same energy.

5.20. Derive the integral representation (A.67) for the Laguerre polynomials using the Laplace transform of Eq. (A.59).

Solution. Substituting the inverse Laplace transform

$$L_n^{\nu}(x) = \frac{1}{2\pi \mathrm{i}} \int_C dt\, \mathrm{e}^{tx} f(t)$$

into Eq. (A.59), we obtain

$$\int_C dt\, f(t) \mathrm{e}^{tx} \left[xt^2 + (\nu + 1 - x)t + n\right] = 0.$$

The identity $x\mathrm{e}^{tx} \equiv d\mathrm{e}^{tx}/dt$ allows one to rewrite the resulting equation in the form

$$\int_C dt\, f(t) \left[\frac{d\,\mathrm{e}^{tx}}{dt}(t^2 - t) + ((\nu+1)t + n)\mathrm{e}^{tx}\right] = 0.$$

Integrating by parts, we find that

$$\int_C dt\, \mathrm{e}^{tx} \left[-\frac{d\,(t^2 - t) f(t)}{dt} + ((\nu+1)t + n) f(t)\right] = 0,$$

under the condition that for the contour of integration C the integrated terms vanish. The expression in square brackets gives the first-order differential equation for $f(t)$:

$$\frac{d\,(t^2 - t) f(t)}{dt} = ((\nu + 1)t + n) f(t).$$

The solution of this equation reads

$$f(t) = A\frac{(1 - t)^{n+\nu}}{t^{n+1}}.$$

This function has a pole at $t = 0$. Therefore we choose the contour of integration Cas a circle of radius $t < 1$ as indicated in the explanation to the formula (A.67):

$$L_n^\nu(x) = \frac{A}{2\pi i}\int^{(0+)} \frac{(1 - t)^{n+\nu}}{t^n} e^{tx}\frac{dt}{t}.$$

A constant A cannot be determined from the equation, but is given by the additional normalization condition $L_n^\nu(0) = \Gamma(n+\nu+1)/\Gamma(\nu+1)n!$, following from the Rodrigues formula (A.60). Calculating the integral at $x = 0$ we obtain the integral representation

$$L_n^\nu(x) = \frac{(-1)^n}{2\pi i}\int^{(0+)} \frac{(1 - t)^{n+\nu}}{t^n} e^{tx}\frac{dt}{t}.$$

5.21. Prove that the Bessel function can be expressed in terms of the contour integral (Schlfli-type representation)

$$J_\nu(x) = \frac{1}{2\pi i}\int_{-\infty}^{(0+)} \frac{dz}{z^{\nu+1}} \exp\left(\frac{xz}{2} - \frac{x}{2z}\right),$$

where the integration is performed along a path starting and ending at $-\infty$, going around the point $z = 0$ in the positive direction (Fig. A.1).

Solution. Substituting the expansion of the function

$$\exp\left(-\frac{x}{2z}\right) = \sum_{n=0}^{\infty} \frac{1}{n!}\left(-\frac{x}{2z}\right)^n$$

into the contour integral, we obtain for the n th term of the series for Re $n + \nu < 0$:

$$\frac{(-1)^n (x/2)^{2n+\nu}}{n!}\frac{1}{2\pi i}\int_{-\infty}^{(0+)} dt\,\frac{\exp(t)}{t^{n+\nu+1}}$$

$$= \frac{(-1)^n (x/2)^{2n+\nu}}{\pi n!} \sin(\pi(n+\nu+1))\Gamma(-n-\nu) = \frac{(-1)^n (x/2)^{2n+\nu}}{n!\Gamma(n+\nu+1)}.$$

The latter equality follows from the formula (A.3). An analytic continuation of this expression with respect to ν gives The n -th term of the expansion of the Bessel function in a series in x (A.15).

5.22. Show that the functions $J_{-n}(x)$ and $J_n(x)$ are linearly dependent for integer n.

Solution. We use the expansion of the Bessel functions at zero (A.15) and the properties of the Γ-function, $\Gamma(m) = \infty$ for integer $m \leq 0$ and $\Gamma(m+1) = m!$ for integer $m \geq 0$. As a result we have

$$J_{-n}(x) = \left(\frac{x}{2}\right)^{-n} \sum_{k=0}^{\infty} \left(\frac{-x^2}{4}\right)^k \frac{1}{k!\Gamma(k-n+1)}$$

$$= \left(-\frac{x}{2}\right)^n \sum_{l=-n}^{\infty} \left(\frac{-x^2}{4}\right)^l \frac{1}{(l+n)!\Gamma(l+1)} = (-1)^n J_n(x).$$

Here we use the notation $k = l + n$.

5.23. Show that the Neumann function Y_ν (A.21) remains linearly independent of J_ν (A.15) at $\nu \to n$.

Solution. It is necessary to calculate the Wronskian $W\{J_\nu(x), Y_\nu(x)\}$ (5.2) and tend $\nu \to n$:

$$xW\{J_\nu(x), Y_\nu(x)\} = -\frac{xW\{J_\nu, J_{-\nu}\}}{\sin(\pi\nu)}$$

$$= \lim_{x\to 0} \frac{x}{\sin(\pi\nu)} \left(J_{-\nu}(x)\frac{dJ_\nu(x)}{dx} - J_\nu(x)\frac{dJ_{-\nu}(x)}{dx}\right)$$

$$= \frac{2\nu}{\Gamma(-\nu+1)\Gamma(\nu+1)\sin(\pi\nu)}.$$

Here we used the Wronskian of two solutions of the Bessel equation (5.4), $W(x) = \text{const}/x$, and calculated the constant at $x = 0$. Since the resulting constant (see (A.3))

$$\text{const} = \frac{2\nu}{\Gamma(-\nu+1)\Gamma(\nu+1)\sin(\pi\nu)} = \frac{2}{\Gamma(1-\nu)\Gamma(\nu)\sin(\pi\nu)} = \frac{2}{\pi}$$

is independent of ν, then the Wronskian remains finite in the limit $\nu \to n$, therefore $Y_n(x)$ and $J_n(x)$ are linearly independent.

5.24. Express Bessel functions $J_{m+1/2}$ with half-integer index in terms of elementary functions.

Solution. We have for $J_{1/2}(x)$ (A.15)

$$J_{1/2}(x) = \sqrt{x/2}\sum_{k=0}^{\infty}\frac{(-x^2/4)^k}{k!\Gamma(k+3/2)}$$
$$= \sqrt{x/2}\sum_{k=0}^{\infty}\frac{(-x^2)^k}{(2k+1)!\Gamma(3/2)} = \sqrt{2/\pi x}\sin(x),$$

where the relation (A.2)

$$\Gamma(k+3/2) = (2k+1)(2k-1)\ldots 3\cdot\Gamma(3/2)/2^k$$

and the value $\Gamma(3/2) = \sqrt{\pi}/2$ are used. In order to express $J_{m+1/2}(x)$ in terms of elementary function, we use the formulas of differentiation (A.18)

$$J_{m+1/2}(x) = -x^{m-1/2}\frac{d}{dx}\left(x^{-m+1/2}J_{m-1/2}\right)$$
$$= (-1)^m x^{m+1/2}\left(x^{-1}\frac{d}{dx}\right)^m\frac{1}{\sqrt{x}}J_{1/2}(x).$$

5.25. Using the differential equations (A.14) and the Laplace transform, obtained the integral representation of Poisson type (A.20) for the Bessel function with an integer index. What is Laplace transform of $J_0(x)$?

Solution. The function $f(x) = x^{-n}J_n(x)$ satisfies the equation

$$x\frac{d^2}{dx^2}f(x) + (2n+1)\frac{d}{dx}f(x) + xf(x) = 0,$$

where the coefficients are polynomials of the first degree. Then we substitute the inverse Laplace transform

$$f(x) = \int_C \frac{dt}{2\pi i}\exp(xt)g(t)$$

into this equation. After integration by parts, we set the integrand to be zero and obtain

$$-\frac{d}{dt}(t^2+1)g(t) + (2n+1)tg(t) = 0.$$

The integrated terms do not contribute to integration by parts if the following equation holds

$$\exp(xt_0)(t_0^2+1)g(t_0) = \exp(xt_k)(t_k^2+1)g(t_k),$$

where t_0, t_k are the initial and final points of the contour C. Finally, the differential equation for $g(t)$ has the solution

$$g(t) = A(t^2 + 1)^{n-1/2}.$$

The contour C must pass around both branch points $t^* = \pm i$ of the function $g(t)$, for example, a closed contour passing in the positive direction along the cut from $t = -i$ to $t = i$ (Fig. 8.1). As a result

$$J_n(x) = \frac{A_n x^n}{2\pi i} \int\limits_C dt \exp(xt)(t^2 + 1)^{n-1/2},$$

where A_n is a normalization constant. The expression obtained is transformed to integral representation (A.20) after replacing of variable $t \to i\tau$. Taking the integral and using the value $J_0(0) = 1$ we find $A_0 = 1$. Besides, the Laplace transform of the function $J_0(x)$ reads

$$g_0(t) = \frac{1}{\sqrt{t^2 + 1}}.$$

5.26. Using formulas of differentiation (A.18), show that the generating function

$$F(z, x) = \sum_{m=-\infty}^{\infty} z^m J_m(x)$$

is of the form (A.25)

Solution. Differentiating $F(z, x)$ over x and using formulas of differentiation (A.18), we obtain

$$\begin{aligned} \frac{\partial F(z, x)}{\partial x} &= \sum_{m=-\infty}^{\infty} z^m \frac{dJ_m(x)}{dx} \\ &= \sum_{m=-\infty}^{\infty} \frac{z^m}{2} (J_{m-1}(x) - J_{m+1}(x)) \\ &= \frac{1}{2}\left(z - \frac{1}{z}\right) F(z, x). \end{aligned}$$

Solving this differential equation, we find

$$F(z, x) = C(z) \exp\left(\frac{x}{2}\left[z - \frac{1}{z}\right]\right).$$

The quantity $C(z)$ can obtained from the condition $J_n(0) = 0$ for $n \neq 0$

$$C(z) = F(z, 0) = J_0(0) = 1.$$

5.27. Derive the Bessel integral representation

$$J_n(x) = \int_0^{\pi} \cos(n\theta - x\sin\theta)\frac{d\theta}{\pi},$$

using the generating function (A.25).

Solution. We introduce new variable $z = \exp(i\theta)$ in the contour integral, which express the Bessel function in terms of the generating function (A.25),

$$\begin{aligned} J_n(x) &= \frac{1}{2\pi \mathrm{i}} \oint \frac{dz}{z^{n+1}} \exp\left[\frac{x}{2}\left(z - \frac{1}{z}\right)\right] \\ &= \int_0^{2\pi} \frac{d\theta}{2\pi} \exp[ix\sin(\theta) - in\theta] \\ &= \int_0^{\pi} \frac{d\theta}{\pi} \cos(x\sin(\theta) - n\theta). \end{aligned}$$

In the latter equality we used the symmetry and periodicity of the functions $\sin(\theta)$ and $\exp(in\theta)$.

5.28. Derive the formula

$$J_0(|\mathbf{r_1} - \mathbf{r_2}|) = \sum_{n=-\infty}^{\infty} J_n(r_1)J_n(r_2)\exp(in\theta),$$

where θ is the angle between vectors $\mathbf{r_1}$ and $\mathbf{r_2}$.

Solution. Substituting $z = \mathrm{i}\exp(\mathrm{i}\varphi)$ in the expression for generating function (A.25), we obtain the equation

$$\sum_{m=-\infty}^{\infty} i^m \exp(\mathrm{i}m\varphi)J_m(q) = \exp(\mathrm{i}q\cos\varphi) = \exp(\mathrm{i}\mathbf{qn}),$$

where φ is the angle between vector $\mathbf{q}$ and unit vector $\mathbf{n}$, $\mathbf{n}^2 = 1$. Let $\mathbf{q} = \mathbf{r}' - \mathbf{r}$, θ is the angle between vectors $\mathbf{r}$ and $\mathbf{r}'$, α is the angle between vectors $\mathbf{r}$ and $\mathbf{n}$ (Fig. 5.3). Then we integrate the equation obtained over $\mathbf{n}$, given on a circle

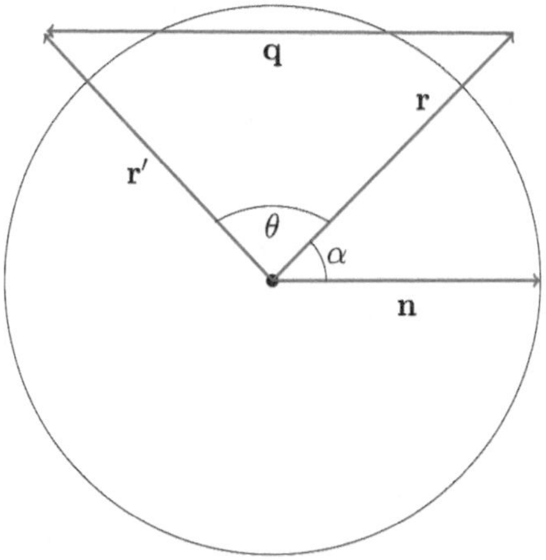

Figure 5.3 Vectors **r**, **r**′, **q** and **n**.

of unit radius:

$$J_0(|\mathbf{r}-\mathbf{r}'|) = \int_0^{2\pi} \frac{d\varphi}{2\pi} \sum_{m=-\infty}^{\infty} i^m \exp(im\varphi) J_m(|\mathbf{r}-\mathbf{r}'|)$$

$$= \int \frac{d\mathbf{n}}{2\pi} \exp[\mathrm{i}(\mathbf{r}'-\mathbf{r})\mathbf{n})]$$

$$= \int_0^{2\pi} \frac{d\alpha}{2\pi} \sum_{k=-\infty}^{\infty} \mathrm{i}^{-k} \exp(-\mathrm{i}k\alpha) J_k(r) \sum_{m=-\infty}^{\infty} \mathrm{i}^m \exp(im(\alpha+\theta) J_m(r')$$

$$= \sum_{m=-\infty}^{\infty} \exp(im\theta) J_m(r) J_m(r').$$

5.29. Find convolution $F(y) = \int_0^y dx J_0(x) J_0(y-x)$.

Solution. We perform the Laplace transform to turn the convolution into a product:

$$\int_C \frac{dt}{2\pi i} g(t) \exp(xt) = f(x)\theta(x), \quad \theta(x) = \begin{cases} 1, & x > 0, \\ 0, & x < 0. \end{cases}$$

The contour C starts and ends at $-\infty$, bypassing in a positive direction all the singularities in the complex plane of the Laplace transform $g(t)$ of the function $f(x)$,

$$F(y)=\int_0^\infty dxJ_0(x)J_0(y-x)\theta(y-x)$$

$$=\int_C\frac{dt}{2\pi i}e^{yt}g_0(t)\int_0^\infty dxJ_0(x)e^{-xt}=\int_C\frac{dt}{2\pi i}g_0^2(t)e^{yt}.$$

Laplace transform $g_0(t)$ of the Bessel function is found in the problem **5.25**. Substituting it, we find

$$F(y)=\int_C\frac{dt}{2\pi i}\frac{e^{yt}}{t^2+1}=\sin y,$$

where the contour integral is given by residues at points $t=\pm i$.

5.30. Prove the orthogonality of the Bessel functions (A.26).

Solution. Let us write the Bessel equation for the function $J_k(\lambda_m t)$ and multiply it by $J_k(\lambda_n t)$:

$$J_k(\lambda_n t)\left(\frac{1}{t}\frac{d}{dt}t\frac{d}{dt}J_k(\lambda_m t)+(\lambda_m^2-\frac{k^2}{t^2})J_k(\lambda_m t)\right)=0.$$

Then we write this equation with exchange $n\leftrightarrow m$, subtract the equations from each other and integrate the result from zero to unity over $t\,dt$:

$$(\lambda_m^2-\lambda_n^2)\int_0^1 tdtJ_k(\lambda_n t)J_k(\lambda_m t)$$

$$=-t\left(J_k(\lambda_n t)\frac{d}{dt}J_k(\lambda_m t)-J_k(\lambda_m t)\frac{d}{dt}J_k(\lambda_n t)\right)\Bigg|_0^1.$$

If $\lambda_{n,m}$ are zeros of the Bessel function, $J_k(\lambda_{n,m})=0$, then the integral vanishes for $m\neq n$, since the right-hand side of the equation vanishes. The integral also vanishes when $\lambda_{n,m}$ are zeros of the derivative $dJ_k(\lambda_m)/d\lambda_m=0$. To find the value of the integral at $n=m$ one should evaluate the limit

$$\int_0^1 tdtJ_k(\lambda_n t)J_k(\lambda_n t)=\lim_{n\to m}\frac{1}{\lambda_n^2-\lambda_m^2}\left(\lambda_m J_k(\lambda_n)\frac{dJ_k(\lambda_m)}{d\lambda_m}-\lambda_n J_k(\lambda_m)\frac{dJ_k(\lambda_n)}{d\lambda_n}\right)$$

$$=\frac{1}{2}\left(\frac{dJ_k(\lambda_m)}{d\lambda_m}\right)^2+\frac{1}{2}\left(1-\frac{k^2}{\lambda_m^2}\right)J_k^2(\lambda_m).$$

The first or second formula in (A.26) follows from this equation depending on whether λ_n is zero of the Bessel function or its derivative.

5.31. Find a solution to the heat equation in a cylinder of unit radius with heat-insulating walls, the initial condition is $U_{t=0} = (1-r)^2$:

$$\frac{\partial}{\partial t}U(t, r, \varphi, z) = \triangle U(t, r, \varphi, z).$$

Solution. In a cylindrical coordinate system, the boundary and initial conditions are independent of φ and z so that we will search for the solution as a function of r and t. Applying the variable separation method, we write

$$U(r, t) = \sum_{i=0}^{\infty} A_i R_i(r) T_i(t),$$

where the functions $R_i(r)$ and $T_i(t)$ satisfy the ordinary differential equations:

$$\frac{d}{dt}T_i(t) = -\lambda_i^2 T_i(t), \quad \frac{1}{r}\frac{d}{dr}r\frac{d}{dr}R_i(r) = -\lambda_i^2 R_i(r).$$

The solutions to these equations are as follows:

$$T_i(t) = \exp(-\lambda_i^2 t), \quad R_i(r) = J_0(\lambda_i r),$$

where $J_0(x)$ is the Bessel function. Constants λ_i are determined from the condition that the heat flux through the walls must be absent, $\partial U/\partial r|_{r=1} = 0$, so that

$$0 = \frac{d}{dr}R_i(r)_{|r=1} = \lambda_i \frac{d}{dx}J_0(x)_{|x=\lambda_i} = -\lambda_i J_1(\lambda_i),$$

In other words, λ_i are zeros of the function $dJ_0(\lambda)/d\lambda = -J_1(\lambda)$. We find the coefficients A_i from the initial condition, using the orthogonality of the Bessel functions (A.26). Taking the integral

$$\int_0^1 rdrU(r, 0)J_0(\lambda_j r) = \sum_{i=0}^{\infty} A_i \int_0^1 rdrJ_0(\lambda_i r)J_0(\lambda_j r) = A_j \frac{J_0^2(\lambda_j)}{2},$$

we find the coefficients

$$A_i = \frac{2}{J_0^2(\lambda_i)} \int_0^1 rdr(1-r)^2 J_0(\lambda_i r) = \frac{4}{\lambda_i J_0^2(\lambda_i)} \int_0^{\lambda_i} dxJ_0(x).$$

and the solution

$$U(r, t) = \sum_{i=0}^{\infty} A_i \exp(-\lambda_i^2 t) J_0(\lambda_i r).$$

Exercises

5.32. Express ${}_1F_1(a; a; x)$ in terms of elementary functions.

5.33. Show that an equation with two regular singular points can be reduced to the form

$$\frac{d^2y}{dx^2} + \frac{A}{x}\frac{dy}{dx} + \frac{B}{x^2}y = 0,$$

Solve this equation.

5.34. Prove the Kummer transfomation

$${}_1F_1(\alpha; \gamma; x) = e^x \, {}_1F_1(\gamma - \alpha; \gamma; -x).$$

5.35. Express the solution to Airy equation

$$\psi'' - x\psi = -k^2\psi$$

in terms of a confluent hypergeometric function.

5.36. Express the Laguerre polynomials $L_n^m(x)$ in terms of a confluent hypergeometric function.

5.37. Express the Hermite polynomials $H_n(x)$ in terms of a confluent hypergeometric function.

Hint. Perform a non-conformal replacement $x = \sqrt{t}$. In new variables, there is a regular singular point $x = 0$ in the equation. Consider separately the polynomials of even and odd degrees.

5.38. Reduce the one-dimension Schrödinger equation with reflectionless potential

$$\psi'' + \left(-\varkappa^2 + \frac{a}{\cosh^2 x}\right)\psi = 0, \, a = N(N+1)$$

to the hypergeometric equation. **Hint.** The nonlinear variable changing $z = \tanh x$ reduces the Schrödinger equation to the equation with polynomial coefficients

$$(1 - z^2)\frac{d}{dz}(1 - z^2)\frac{d\psi}{dz} + \left[-\varkappa^2 + a(1 - z^2)\right]\psi = 0.$$

5.39. Reduce the equation

$$\frac{d^2y}{dx^2} + \frac{r(x)}{s(x)}\frac{dy}{dx} + \frac{\sigma(x)}{s^2(x)}y = 0,$$

to hypergeometric form (5.6), (5.7). In this equation, $r(x)$ is a polynomial of a degree less or equal one, $s(x)$ and $\sigma(x)$ are polynomials of degrees less or equal two.

5.40. Show that the associated Legendre polynomials P_l^m, given by the Rodrigues formula (A.37), satisfy the Legendre equation (A.36).

5.41. For the two-dimensional oscillator, find the eigenfunctions and energies of the Schrödinger equation in polar coordinates:

$$-\frac{1}{2}(\Delta - r^2)\Psi(\mathbf{r}) = E\Psi(\mathbf{r}).$$

Hint. Extract the asymptotics of the solution at the singular points of the radial equation, make the substitution $r^2 = x$, and reduce the result to the Laguerre equation. **5.42**. Derive the recurrent relation (A.55) and the formula of differentiation (A.56) for Hermite polynomials (A.51).

5.43. Derive a generating function (A.57) for Hermite polynomials using recurrence relations (A.55).

5.44. Derive generating function (A.66) for Laguerre polynomials using the integral representation (A.67).

5.45. Find a generating function

$$F(z, x) = \sum_{n=0}^{\infty} \frac{z^n}{n!} L_n(x),$$

where$L_n(x) \equiv L_n^0(x)$ are Laguerre polynomials with $\nu = 0$.

5.46. In the vicinity of the point $x = 0$, find the expansion of the functions satisfying the equation

$$x^2\frac{d^2y}{dx^2} + x\frac{dy}{dx} + (x^2 - \nu^2)y = 0$$

with non-integer ν.

5.47. Derive formulas of differentiation for the Bessel functions (A.18):

$$\frac{d}{dx}(x^{\pm\nu}J_\nu) = \pm x^{\pm\nu}J_{\nu\mp1}.$$

Show that the Neumann function (A.21)

$$Y_\nu = (J_\nu \cos(\pi\nu) - J_{-\nu})/\sin(\pi\nu)$$

satisfies the same formulas.

Answers

5.32. ${}_1F_1(a; a; x) = \exp(x)$

5.33. The equation has two linearly independent solutions $y_{1,2} = x^{\lambda_\pm}$, where $\lambda_\pm = (1/2)(1 - A \pm \sqrt{(1-A)^2 - 4B})$.

5.34. Hint. Substitute the equality into the equation (5.7). Take into account that the right and left sides of the equation are function regular at zero.

5.35. Hint. Parameter k^2 can be removed by shifting the coordinate origin. The asymptotic behavior at infinity can be found in the form $\psi = \exp(\lambda x^\sigma)$. The result is $\sigma = 3/2$, $\lambda = \pm 2/3$. Then the following nonlinear changing the variable is available $t = x^{3/2}$.

Answer.

$$\psi = \mathrm{e}^{-2x^{3/2}/3}\, {}_1F_1(1/6, 1/3; 4x^{3/2}/3).$$

The parameters are two times different $\alpha = \gamma/2$ then the Airy equation can be reduced to the Bessel one.

5.36.

$$L_n^m = \frac{(n+m)!}{m!n!}\, {}_1F_1(-n; m+1; x).$$

5.37.

$$H_n(x) = (-1)^k \frac{n!}{k!}\, {}_1F_1\left(-k, \frac{1}{2}, x^2\right) \quad \text{at } n = 2k\,,$$

$$H_n(x) = x\,(-1)^k \frac{(n+1)!}{(k+1)!}\, {}_1F_1\left(-k, \frac{3}{2}, x^2\right) \quad \text{at } n = 2k+1\,.$$

5.38. Hint. Exclude the asymptotics at $z \to \pm 1$ by changing the unknown function $\psi = (1 - z^2)^{\varkappa/2} u(z)$, then

$$(1 - z^2)u'' - 2(1+\varkappa)zu' + [a - \varkappa(\varkappa+1)]u = 0.$$

This equation have regular points $x = \pm 1, \infty$, consequently reduces to the hypergeometric after the next changing $\xi = (z+1)/2$.

Answer.

$$u = {}_2F_1(\alpha, \beta; \gamma; z), \;\; \gamma = 1 + \varkappa, \;\; \alpha, \beta$$

$$= \varkappa + \frac{1}{2} \pm \sqrt{\left(\varkappa + \frac{1}{2}\right)^2 - \varkappa(\varkappa+1) + a}.$$

Find the series break condition. Show that at $a = N(N+1)$ there are exactly N levels in this well.

5.39. Hint. Let's consider three cases.

(1) If equation $s(x) = 0$ has two different roots (e.g., (A.36))

$$s(x) = a(x - s_1)(x - s_2),$$

then the equation has three regular singular points. It is reduced to the form (5.6) after the change of variables

$$t = (x - s_1)/(s_2 - s_1)$$

and substitution

$$y(t) = t^{\nu}(t-1)^{\mu} F(t),$$

where ν and μ are characteristic degrees in regular singular points $t = 0$ and $t = 1$.

(2) If the roots of the equation $s(x) = 0$ coincide, i.e., $s_1 = s_2 = x$, then it is necessary to transfer the irregular singular point $x = s_1$ to infinity using conformal change of variables $x - s_1 = 1/z$. In new variables, the equation retains its form; however, $s(z) = z$ is now a polynomial of first degree.

(3) If $s(x)$ is polynomial of a first degree, we make the linear change of variables $z = s(x)$. As a result the equation has one regular singular point $z = 0$ and one irregular singular point $z = \infty$ (if $r(z)$ and $\sigma(z)$ are constant, then the point $z = \infty$ is regular, but in this case the equation has power-law solutions). Making the substitution $y(z) = z^{\nu} \exp(\mu z) F(z)$ we reduce the equation to the form

$$z\frac{d^2}{dz^2}F + [\gamma - bz]\frac{d}{dz}F - \alpha F = 0.$$

For $b \neq 0$ this is the equation for confluent hypergeometric function (5.7) in variables $t = bz$. For $b = 0$ asymptotics at infinity is $F \sim \exp(\pm 2\sqrt{\alpha z})$. Comparing this asymptotics with the asymptotics (5.7), we conclude that in this case the equation can be reduced to the form (5.7) by replacing $t = 4\sqrt{\alpha z}$ and the substitution $F(t) = f(t)\exp(-t/2)$.

(4) If $s(x)$ is a constant, than we perform the substitution

$$y(x) = f(x)\exp\left(-\int^{x} d\tau \frac{r(\tau)}{2s}\right)$$

and reduce the coefficients of the equation to the form $r(x) = 0$, $s(x) = 1$. Extracting complete square in $\sigma(x) = -(a(x - x_1))^2 + a\epsilon$ and introducing a new variable $z = \sqrt{a}(x - x_1)$, we obtain the equation for the quantum oscillator

$$\frac{d^2 f}{dz^2} + (\epsilon - z^2) f = 0 .$$

This equation has one irregular singular point $z = \infty$ with asymptotics $f \sim \exp(\pm z^2/2)$. To reduce the equation to the form (5.7), it is necessary to make a non-conformal change of variable $z^2 = t$, which leads to appearance of a regular singular point $t = 0$, and to perform a substitution $f(t) = F(t) \exp(-t/2)$.

5.40. Hint. Substitute Rodrigues' formula (A.37) for the associated Legendre functions into Eq. (A.36) and obtain Eq. (5.24) for the function $f = (d/dx)^m P_l$. Show that Eq. (5.24) can be obtained from Eq. (A.35) for Legendre polynomials after applying m -fold differentiation to (A.35).

5.41.

$$\Psi(\mathbf{r}) = \sqrt{\frac{n!}{\pi(n+m)!}} r^{|m|} \exp(-r^2/2) L_n^{|m|}(r^2) \exp(im\varphi),$$

$$E_N = 2n + |m| + 1 \equiv N + 1, \quad n = 0, 1, 2, \ldots, \quad m \quad \text{is integer.}$$

5.42. Hint. Express the first derivative of a Hermite polynomial in terms of a linear combination of polynomials using (A.51). In the same way, express the second derivative of the Hermite polynomial in terms of a linear combination of polynomials. Then use Eq. (A.50).

5.43. Hint. Express the derivative over z of the generating function

$$F(z, x) = \sum_{n=0}^{\infty} (z^n/n!) H_n(x)$$

via $F(z, x)$ using (A.55). Solve the differential equation for $F(z, x)$ and use the normalization condition $F(0, x) = H_0(x) = 1$.

5.44. Hint. Substitute the integral representation (A.67) in the definition of the generating function $F(z, x) = \sum_{n=0}^{\infty} z^n L_n^{\nu}(x)$, sum the series and take the integral.

5.45.

$$F(z, x) = \exp(z) J_0(2\sqrt{zx}).$$

5.46. Hint. Extract the power asymptotics at $x = 0$ in the form $y(x) = x^{\rho} f(x)$. Substitute $y(x)$ in the equation and expand the analytic at zero function f into a Taylor series. Compare terms at the same powers of x and find the coefficients in expansion of $f(x)$. The function $y(x)$ is the sum of two linearly independent solutions $J_{\nu}(x)$ and $J_{-\nu}(x)$, having expansion (A.15) near the point $x = 0$.

5.47. Hint. Multiply the expansion of the Bessel functions at zero (A.15) by $x^{\pm\nu}$ and differentiate.

Chapter 6

Asymptotic Methods

6.1 Asymptotic Power Series

When solving various problems, the need for approximate computing integrals containing large (or small) parameters emerges. In doing so the result is represented in the form of the so-called *asymptotic series*. The formal expansion

$$f(z) \sim \sum_{n=0}^{\infty} a_n \phi_n(z) \tag{6.1}$$

is said to be asymptotic expansion of the function $f(z)$ at $z \to z_0$ if

$$f(z) - \sum_{n=0}^{N} a_n \phi_n(z) = o\left(\phi_N(z)\right)$$

for every N at $z \to z_0$. Here $\psi(z) = o\left(\phi_N(z)\right)$ means that the ratio $\psi(z)/\phi_N(z) \to 0$ at $z \to z_0$. This gives the relation:

$$a_N = \lim_{z \to z_0} \frac{f(z) - \sum_{n=0}^{N-1} a_n \phi_n(z)}{\phi_N(z)}. \tag{6.2}$$

The formal series expansion (6.1) can be a divergent one. It depends on the choice of the asymptotic sequence $\{\phi_n(z)\}$. Furthermore, two functions can

Mathematical Methods of Physics: Problems with Solutions
Igor V. Kolokolov, Evgeny A. Kuznetsov, Alexander I. Milstein, Evgeny V. Podivilov, Alexander I. Chernykh, David A. Shapiro, and Elena G. Shapiro

ISBN 978-981-5129-21-2 (Hardcover), 978-1-003-51360-5 (eBook)
www.jennystanford.com

have identical asymptotic expansion if their difference $\psi(z)$ is such that

$$\lim_{z \to z_0} \psi(z)/\phi_n(z) = 0.$$

for every n. We also can build up the expansion around the infinite point. For this purpose, let us change of variable $z \to z', z' = 1/(z - z_0)$. For the infinite point $\phi_n(z) = 1/z^n$, then the expansion (6.1) is called the asymptotic power series. One can perform operations with asymptotic power series: addition, multiplication, and term by term integration. However, it is similar to the usual power series.

A frequently used tool for the asymptotic expansion of functions represented by definite integrals is integrating parts. Sometimes it is combined with Taylor expansion.

6.2 A Laplace Integral

Suppose that we have an integral

$$g(\nu) = \int_a^b \mathrm{e}^{\nu f(x)} \phi(x) \, dx,$$

where $\phi(x)$ and $f(x)$ are continuous real functions and ν is a large positive parameter. If the function $f(x)$ reaches its maximal value at some endpoint (e.g., in the point a) and $f'(a) \neq 0, \phi(a) \neq 0$, then

$$g(\nu) \sim -\mathrm{e}^{\nu f(a)} \frac{\phi(a)}{\nu f'(a)} \left(1 + \mathcal{O}(\nu^{-1})\right). \tag{6.3}$$

Once the maximal value is reached in the point $x = c$ lying inside the interval: $a < c < b$, and $\phi(c) \neq 0$ then

$$g(\nu) \sim \sqrt{\frac{-2\pi}{\nu f''(c)}} \mathrm{e}^{\nu f(c)} \phi(c) \left(1 + \mathcal{O}(\nu^{-1})\right). \tag{6.4}$$

If the maximal value is reached at the border, e.g., at $x = a$, $\phi(a) \neq 0$, and $f'(a) = 0$ with $f''(a) < 0$ then

$$g(\nu) \sim \sqrt{-\frac{\pi}{2\nu f''(a)}} \mathrm{e}^{\nu f(a)} \phi(a). \tag{6.5}$$

The above formulas are obtained by expanding $f(x)$ near the point where $f(x)$ takes its maximum value and the subsequent integration in infinite limits or by parts. In the same manner the asymptotic formulas can be obtained when $\phi(a) = 0$ or $\phi(c) = 0$.

6.3 Method of Stationary Phase

Consider the integral of the form:

$$g(x) = \int_a^b e^{i\nu f(x)} \phi(x)\, dx,$$

where $f(x)$ is a real function and $\nu \to +\infty$. The main contribution to the integral is given either by the stationary phase points where $f'(x) = 0$ or by neighborhoods of the points a and b if $f'(x) \neq 0$ on (a, b). In the latter case

$$g(\nu) \sim \frac{\phi(b)}{i\nu f'(b)} e^{i\nu f(b)} - \frac{\phi(a)}{i\nu f'(a)} e^{i\nu f(a)} + \mathcal{O}\left(\nu^{-2}\right), \quad f'(b) \neq 0, \;\; f'(a) \neq 0.$$

If the interval (a, b) contains only one stationary phase point c at which $f''(c) \neq 0$ and $\phi(c) \neq 0$ then

$$g(\nu) \sim \sqrt{\frac{2\pi}{\nu |f''(c)|}} \phi(c) \exp\left(i\nu f(c) \pm i\pi/4\right) \left[1 + \mathcal{O}(\nu^{-1})\right]. \tag{6.6}$$

We obtain this formula by expanding $f(x)$ up to second order in $x - c$ terms in the vicinity of the point $x = c$ and after subsequent computation of the Fresnel integral. The sign of the phase $i\pi/4$ coincides with the sign of the second derivative sign $(f''(c))$. If in the interval (a, b) there are several stationary phase points where the second derivative of the function $f(x)$ and values of $\phi(x)$ are not equal to zero, then $g(\nu)$ is given by the sum of contributions of the form (6.6).

6.4 Method of Steepest Descents

Let us pose the problem of estimating an integral of the form:

$$g(\nu) = \int_\gamma e^{\nu w(z)} \phi(z)\, dz, \quad \nu \to \infty. \tag{6.7}$$

Here $w(z)$ and $\phi(z)$ are ν-independent functions of z analytic in some domain Ω which contains the integration contour γ together with the stationary

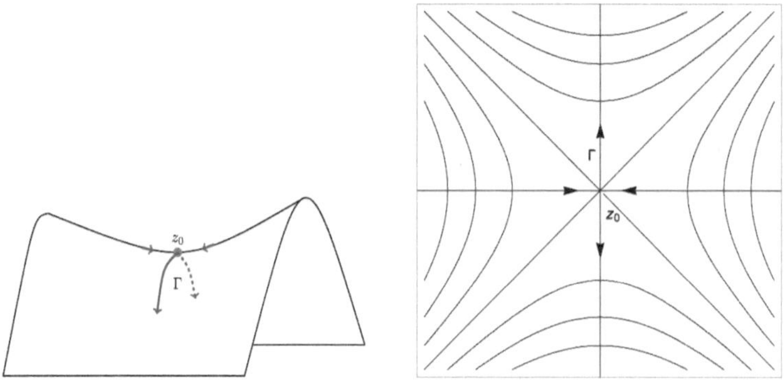

Figure 6.1 Topography of the real part of function $w(z)$ near saddle point $z = z_0$. Arrows show the directions of $u(z) = \mathrm{Re}\, w(z)$ decaying.

point z_0 where $w'(z_0) = 0$. The point z_0 may not belong to the integration contour. Let us consider the case when $w''(z_0) \neq 0$ and $\phi(z) = 1$. Then there exists a contour $\Gamma \in \Omega$, $z_0 \in \Gamma$ along which the imaginary part of the function $w(z) = u(x, y) + \mathrm{i}v(x, y)$ does not change:

$$v(x, y) = v(x_0, y_0) = \text{const},$$

where $z = x + \mathrm{i}y$. Function $w(z)$ can be represented on the contour Γ in the form $w(z^\star) = w(z_0) - \tau(z^\star)$ where $z^\star \in \Gamma$ is a point on the contour and $\tau(z^\star)$ is the real non-negative function. The gradients of the functions $u(x, y)$ and $v(x, y)$ are orthogonal by Cauchy–Riemann relations for analytic functions:

$$\frac{\partial u}{\partial x} = \frac{\partial v}{\partial y}, \ \frac{\partial u}{\partial y} = -\frac{\partial v}{\partial x} \ \Rightarrow \ (\nabla v, \nabla u) = \frac{\partial v}{\partial x}\frac{\partial u}{\partial x} + \frac{\partial v}{\partial y}\frac{\partial u}{\partial y} = 0.$$

This means that contour Γ is the *steepest-descent* path for the function $u(x, y)$. The latter decreases along Γ in both the directions from point z_0 which is called the *saddle point* of function $u(x, y)$ (Fig. 6.1).

The idea of the saddle point method is to deform the integration contour γ. The deformation aims to get the main contribution to the integral at the shortest part of the contour. The optimal contour is the line of steepest descent. We will consider only such deformations in the domain of analyticity of the function $w(x)$ for which the endpoints of the integration contour remain fixed. Contours γ' resulting from such deformations are called equivalent since the value of the integral (6.7) does not depend on γ'. The path γ' is called *minimax contour* if the extremum

$$\min_{\gamma' \in D} \max_{z \in \gamma'} \left| \mathrm{e}^{\nu w(z)} \right| = \min_{\gamma' \in D} \max_{z \in \gamma'} \mathrm{e}^{\nu u(x,y)} = \mathrm{e}^{\nu u(x_0, y_0)},$$

where D is the set of equivalent contours γ', is reached on it.

In others words, from all equivalent contours γ' we select one which passes through the saddle point z_0 of the function $u(x, y)$ and coincides in the neighborhood of z_0 with the contour Γ. Moreover, the function $u(x, y)$ must have at the point z_0 not only local, but also global maximum on the contour γ'.

If a minimax (saddle) contour exists, then z_0 called the *saddle point* for the integral (6.7). Unfortunately, the general search algorithm for such contours does not exist. Contour searching is a complex topological problem. Facilitate its solution is possible if the following three rules:

(1) Draw the level curves of the function $u(x, y)$ (as in the topographic map, Fig. 6.1).
(2) Choose such a path from start to endpoints of the integration in which the maximal value of $u(x, y)$ on the path is as small as possible (comparing with other paths). (Imagine that you are a traveler afraid of heights, and $u(x, y)$ is the height at the point x, y above sea level.)
(3) Near the point with the maximum value along the selected path $u(x, y)$ draw a contour along the line of steepest descent. (To minimize the walk at high altitude.)

As a result, we reduce the integral to a standard one.

Let's consider the two most straightforward cases:

I. There exists a saddle contour, and at the stationary point z_0 the condition for the highest derivatives $|\nu^{(n-2)/2} w''(z_0)| \gg |w^{(n)}(z_0)|$ is satisfied. Then it is enough to expand $w(z)$ in a neighborhood of the point z_0 up to quadratic terms only:

$$w(z) = w(z_0) + \frac{w''(z_0)}{2}(z - z_0)^2 + \ldots$$

The main (exponentially large) contribution to the integral is gained along contour Γ. To estimate this contribution, we replace variable z by the real along the contour Γ variable $s = \mathrm{e}^{-\mathrm{i}\psi}(z - z_0)$ and rewrite the second derivative in the form: $w''(z_0) = \rho e^{i\theta}$. Then the leading term of the asymptotic expansion of the integral in $1/\nu$ is

$$g(\nu) \sim \phi(z_0) e^{\nu w(z_0)} \int_{\gamma'} ds\, e^{i\psi} \exp\left(\frac{\nu}{2}\rho s^2 e^{2i\psi + i\theta}\right)$$

$$\sim \phi(z_0) e^{\nu w(z_0)} e^{i\psi^\star} \int_{\Gamma} \exp\left(-\frac{\nu}{2}\rho s^2\right) ds. \tag{6.8}$$

The condition determines the direction of the steepest descent are $\exp(2i\psi + i\theta) = -1$, whence $\psi_{1,2} = (\pm\pi - \theta)/2$. The angles between the tangents to Γ at the point z_0 and the positive direction of the real axis. The phase $\psi^\star$ takes one of the values $\psi_{1,2}$ which corresponds to the direction of integration along the contour γ' at point z_0. The last integral in (6.8) is the Laplace integral (6.4) thus $g(\nu)$ has the following asymptotics:

$$g(\nu) \sim \phi(z_0) e^{\nu w(z_0)} \exp(i\psi^\star) \sqrt{\frac{2\pi}{\nu |w''(z_0)|}}. \tag{6.9}$$

Note that the saddle contour can pass through several saddle points z_i at which the values of Re $w(z_i)$ are the same. In this case, it is necessary to sum contributions from each z_i.

II. A saddle contour does not exist, but the integration path γ can be deformed so that the maximal value of $u(x, y)$ on this path is in the initial or final point of γ. If the condition for the derivatives $|\nu^{n-1} w'(z_1)| \gg |w^{(n)}(z_1)|$ is fulfilled then we can terminate the expansion of $w(z)$ in a neighborhood of the point z_1 at the linear term:

$$w(z) = w(z_1) + w'(z_1)(z - z_1) + \dots .$$

One line of steepest descent $L \in \Omega$ starts from the point z_1 along which Im $w(z)$ = Im $w(z_1)$. Replace the variable z with variable $s = e^{-i\psi}(z - z_1)$, real along L. Let us write the derivative as $w'(z_1) = -\rho \exp(-i\theta)$ and deform the integration contour γ so that it coincided with L near the point z_1. Then the leading term of the asymptotic expansion of the function $g(\nu)$ in $1/\nu$ has the form

$$g(\nu) \sim \phi(z_1) e^{\nu w(z_1)} \int_{L} ds\, e^{i\psi} \exp\left(-\nu\rho s e^{i(\psi - \theta)}\right)$$

$$\sim -\frac{\phi(z_1) \exp(\nu w(z_1))}{\nu w'(z_1)}, \tag{6.10}$$

where the steepest descent direction is determined by the equality $\psi = \theta$. If the maximum value of $u(x, y)$ is in the final point z_2 of the contour γ, then by analogous reasoning, we obtain that the leading term of the $1/\nu$ asymptotic expansion of the function $g(\nu)$ has the form (6.10), in which z_1 should be replaced with z_2 and the minus sign before the fraction with a plus sign.

6.5 The Averaging Method

If *unperturbed system* with *n* degrees of freedom performs motion in a limited area of phase space, sometimes you can go to action-angle variables in which the differential equations are of the form:

$$\dot{I} = 0, \ \dot{\phi} = \omega(I); \quad I = (I_1, I_2, \ldots, I_n), \ \phi = (\phi_1, \phi_2, \ldots, \phi_n). \qquad (6.11)$$

In the unperturbed system I is a set of integrals of motion, and the ϕ variables change in the range $0 \leq \phi_i \leq 2\pi$. The perturbed equations contain in the right-hand sides additional terms proportional to the small parameter ϵ:

$$\dot{I} = \epsilon g(I, \phi, \epsilon), \quad \dot{\phi} = \omega(I) + \epsilon f(I, \phi, \epsilon), \qquad (6.12)$$

where f and g are 2π-periodic functions of the variables ϕ_i. Insofar as the variables I change slowly (due to the smallness of the parameter ϵ), then *the perturbed system* can be replaced by a much simpler *averaged system* for *slow variables* $J(t) = I(t) + \mathcal{O}(\epsilon)$:

$$\dot{J} = \epsilon G(J), \quad G(J) = \frac{\oint g(J, \phi, 0)\, d\phi}{\oint d\phi}. \qquad (6.13)$$

When passing from Eq. (6.12) to slow variables, evolution equations (6.13) the averaging over oscillation periods of the functions g is performed. This procedure is correct for $n = 1$. In the case of several degrees of freedom, the averaging over ϕ can become inapplicable if there are resonances in the system, i.e., frequencies ω_i which are components of the vector ω in the formula (6.11) satisfy the equation

$$\sum_{i=1}^{n} N_i \omega_i = 0,$$

where N_i are integer numbers.

Example. A perturbed one-dimensional system with a constant frequency ω is

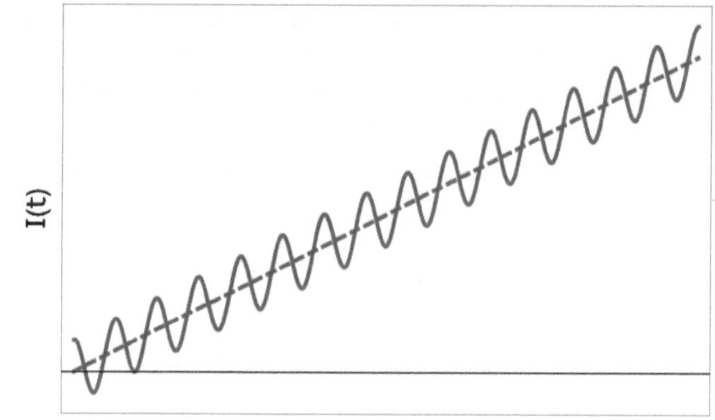

Figure 6.2 Plot of time dependence of $I(t)$ and the slow variable $J(t)$

$$\dot{I} = \epsilon(a + b\cos\phi), \quad \dot{\phi} = \omega.$$

It leads to the averaged equation

$$\dot{J} = \epsilon a,$$

which has the solution $J(t) = J_0 + \epsilon at$ (Fig. 6.2). In this case, we know the exact solution $I(t) = \epsilon at + \epsilon b \sin \omega t/\omega$ and we can make sure that the solution the averaged equation does not deviate from the exact solution if $\omega \neq 0$: $|J(t) - I(t)| \leq \text{const} \cdot \epsilon$. In the Hamiltonian system, the variables ϕ play the role of generalized coordinates, and I are generalized momenta so that the equations always look like

$$\dot{I} = -\frac{\partial H}{\partial \phi}, \quad \dot{\phi} = \frac{\partial H}{\partial I}.$$

Averaging over ϕ the derivative $\partial H/\partial\phi$, we get $\dot{J} = 0$. This equation for slow variables means that J-evolution is absent (J are *adiabatic invariants*). The main difficulty in applying the method averaging in the general case is choosing suitable variables I, which are integrals of motion of the unperturbed systems. Theory of asymptotic series and asymptotic methods of calculations of integrals are discussed in detail in books (Bender *et al.*, 1999; Erdélyi, 1952; Fedoriuk, 1977; Heading, 1962; Mathews and Walker, 1970; Olver, 1974). Books (Arnold, 1992; Bogoliubov and Mitropolsky, 1961; Cole, 1968; Nayfeh, 1985; Sanders *et al.*, 2007) describe the method of averaging over high-frequency oscillations.

Worked Examples

6.1. Find the asymptotic expansion at $x \to \infty$ of the function:

$$\Gamma(a, x) = \int_x^\infty e^{-t} t^{a-1}\, dt.$$

Solution. Integrating by parts, we get

$$\Gamma(a, x) \sim e^{-x} x^{a-1} \left(1 + \frac{a-1}{x} + \frac{(a-1)(a-2)}{x^2} + \ldots\right)$$

$$\sim \sum_{n=1}^{\infty} \frac{\Gamma(a)}{\Gamma(a+1-n)} e^{-x} x^{a-n}.$$

This result can be obtained by shifting the integration variable $t = \tau + x$ and expanding the function $(x+\tau)^{a-1}$ in a Taylor series in τ/x:

$$\Gamma(a, x) \sim e^{-x} x^{a-1} \int_0^\infty d\tau e^{-\tau} \left(1 + \frac{(a-1)\tau}{x} + \frac{(a-1)(a-2)\tau^2}{2x^2} + \ldots\right)$$

$$\sim \sum_{n=1}^{\infty} \frac{\Gamma(a)}{\Gamma(a+1-n)} e^{-x} x^{a-n}.$$

6.2 Find the asymptotics of the Euler Γ-function for $x \to +\infty$

$$\Gamma(x) = \int_0^\infty t^{x-1} \exp(-t)\, dt.$$

Derive an improved Stirling formula containing two terms of the asymptotic expansion.

Solution. Let rewrite the integral as

$$\Gamma(x) = \int_0^\infty \exp(-t + (x-1)\ln t)\, dt.$$

A multiplier for a large parameter x function $\ln t$ grows unboundedly as $t \to \infty$, therefore none of the asymptotic formulas (6.3–6.5) is not applicable. The point is that the maximum point t_0 of the integrand depends on the large parameter $t_0 = x - 1$. In this case, it is necessary make such a change to the variable (*stop* the maximum point), so that in the new variables the maximum point does not depend on x: $t = x\tau$, then

$$\Gamma(x+1) = x^{x+1} \int_0^\infty \exp(x(\ln\tau - \tau))\, d\tau\,.$$

Maximum of the function $f(\tau) = \ln\tau - \tau$ is reached at $\tau_0 = 1$ inside the integration interval; thus the leading asymptotics can be found using (6.4). To get the next term of the asymptotic expansion, it is necessary to expand the function $f(\tau)$ in a series in $\xi = \tau - 1$ up to the fourth-order terms:

$$f(\xi) = -1 - \frac{\xi^2}{2} + \frac{\xi^3}{3} - \frac{\xi^4}{4} + \mathcal{O}(\xi^5).$$

Substitute this expansion in the integral and change the variable $\xi = z\sqrt{2/x}$:

$$\Gamma(x+1) = \sqrt{2x}x^x e^{-x} \int_{-\sqrt{x/2}}^{\infty} dz \exp\left(-z^2 + \frac{2\sqrt{2}z^3}{3\sqrt{x}} - \frac{z^4}{x} + z^2\mathcal{O}(z^3/x^{3/2})\right).$$

At $x \to \infty$ the exponential function under integral sign can be expanded in a series in x^{-n}, $n > 0$ and the integration limits can be extended up to infinities:

$$\Gamma(x+1) = \sqrt{2x}x^x e^{-x} \int_{-\infty}^{\infty} dz e^{-z^2} \left(1 + \frac{2\sqrt{2}z^3}{3\sqrt{x}} + \frac{4z^6}{9x} - \frac{z^4}{x} + \mathcal{O}(x^{-3/2}).\right)$$

As a result, we get the improved Stirling formula:

$$\Gamma(x) = \frac{\Gamma(x+1)}{x} \sim \sqrt{\frac{2}{x}}\left(\frac{x}{e}\right)^x \left(\Gamma(1/2) + \frac{4\Gamma(7/2)}{9x} - \frac{\Gamma(5/2)}{x} + \mathcal{O}(x^{-2})\right)$$

$$\sim \sqrt{\frac{2\pi}{x}}\left(\frac{x}{e}\right)^x \left(1 + \frac{1}{12x} + \mathcal{O}(x^{-2})\right).$$

Here we took into account that the terms proportional to $x^{-3/2}$ contain odd degrees of the variable z. Therefore, they give zero contribution.

6.3. Find the asymptotics of the integral

$$g(\nu) = \int_{-\infty}^{\infty} e^{i\nu f(x)}\phi(x)\, dx,$$

at $\nu \to +\infty$ if $f'(c) = f''(c) = 0$, $\phi(c) \neq 0$, $f'''(c) \neq 0$ and there are no others stationary points.

Solution. Since the second derivative at the stationary point $x = c$ vanishes, the series expansion of the exponent in $\xi = x - c$ begins with the cubic term:

$$\mathrm{i}\nu f(x) = \mathrm{i}\nu \left[f(c) + \frac{1}{6} f'''(c)\xi^3 + \mathcal{O}(\xi^4) \right].$$

If $\nu \to \infty$ the function $\phi(x)$ can be substituted by its value at $x = c$. The resulting standard integral

$$g(\nu) \sim \phi(c)\mathrm{e}^{\mathrm{i}\nu f(c)} \int_{-\infty}^{\infty} \exp\left(\frac{\mathrm{i}\nu}{6} f'''(c)\xi^3\right) d\xi =$$

$$= 2\phi(c)\mathrm{e}^{\mathrm{i}\nu f(c)} \mathrm{Re} \int_{0}^{\infty} \exp\left(\frac{\mathrm{i}\nu}{6} |f'''(c)|\xi^3\right) d\xi,$$

with help of the variable changing $\nu|f'''(c)|\xi^3/6 = it$ is reduced to the Γ-function. Next we deform the contour of integration over ξ near the stationary point $\xi = 0$ in the complex plane so that the new integration variable is real:

$$d\xi = \mathrm{e}^{\mathrm{i}\pi/6}/3 \left[\frac{6}{\nu|f'''(c)|}\right]^{1/3} \frac{dt}{t^{2/3}}.$$

As a result, we obtain the superior term of the asymptotic expansion:

$$g(\nu) \sim \frac{2}{3}\phi(c)\mathrm{e}^{\mathrm{i}\nu f(c)} \cos\pi/6 \left(\frac{6}{\nu|f'''(c)|}\right)^{1/3} \Gamma(1/3).$$

6.4. Find the asymptotics of the Airy function $\mathrm{Ai}\,(x)$ at $x \to +\infty$:

$$\mathrm{Ai}\,(x) = \frac{1}{2\pi} \int_{-\infty}^{\infty} \exp\left[\mathrm{i}(t^3/3 + xt)\right] dt.$$

Solution. By changing the variable $t = \sqrt{x}z$, the integral is reduced to the form

$$\mathrm{Ai}\,(x) = \frac{\sqrt{x}}{2\pi} \int_{\gamma} \exp\left\{x^{3/2} w(z)\right\} dz, \quad w(z) = \mathrm{i}\left(\frac{z^3}{3} + z\right),$$

where the integration path γ goes along the real axis. Noting that the integral along the ray $z = R\mathrm{e}^{\mathrm{i}\theta}$, $\theta = \text{const}$, converges at $R \to \infty$ if $\mathrm{Re}\, w(z) \to R^3 \cdot \mathrm{Re}\, \mathrm{i}\mathrm{e}^{3i\theta} < 0$. From here we find the convergence sectors $\theta \in [0, \pi/3] \cup [2\pi/3, \pi] \cup [4\pi/3, 5\pi/3]$, which are shaded in Fig. 6.3.

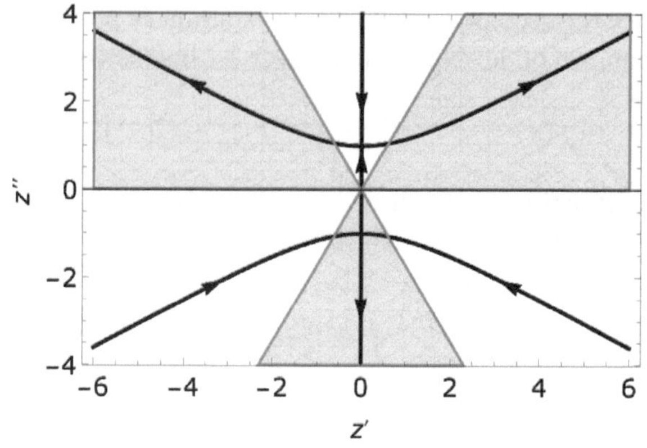

Figure 6.3 The convergence at $|z| \to \infty$ sectors of the Airy function integral representation. In z-plane there are three curves $\operatorname{Im} w = 0$: the line $\operatorname{Re} z = 0$ and two branches of hyperbola. The direction of decreasing of $\operatorname{Re} w$ is shown by the arrows

The saddle points $z_{1,2} = \pm \mathrm{i}$ are found from the condition $dw/dz = i(z^2 + 1) = 0$. At both points $\operatorname{Im} w(z_{1,2}) = 0$, therefore to find the minimax contour, it is enough to plot the zero level lines of the imaginary part of the phase. Equation $\operatorname{Im} w(z' + \mathrm{i}z'') = 0$ has three solutions $z' = 0$ and $z' = \pm\sqrt{3(z''^{\,2} - 1)}$, which are shown in Fig. 6.3. The crossing points of the zero level lines coincide with stationary ones $z_{1,2} = \pm \mathrm{i}$. Between these points the real part of the phase changes monotonically.

The integration contour can be deformed into the upper branch of the hyperbola. The maximum value of $\operatorname{Re} w(z)$ along such a contour is reached at the point $z_1 = \mathrm{i}$. The ends of our chosen path remain in the convergence sector, and the contour itself passes through the saddle point $z_1 = \mathrm{i}$ along the line of steepest descent: $\operatorname{Re} w$ near $z = \mathrm{i}$ decreases along the upper branch hyperbola and increases along the straight line $\operatorname{Re} z = 0$. Near the saddle point the series expansion of $w(z)$ in $\zeta = z - \mathrm{i}$ has the form:

$$w(\zeta) = -\frac{2}{3} - \zeta^2 + \mathcal{O}(\zeta^3).$$

Furthermore, the Laplace method can be used to calculate the integral. Therefore, we obtain the superior term of the asymptotic series expansion of the integral in $x^{-3/2}$ by formula (6.4):

$$\operatorname{Ai}(x) \sim \frac{1}{2x^{1/4}\sqrt{\pi}} \exp\left(-\frac{2}{3}x^{3/2}\right).$$

6.5. Find the asymptotics of the Γ-function

$$\Gamma(z) = \int_0^\infty t^{z-1} \exp(-t)\, dt$$

for complex value of the variable: $|z| \to +\infty, \quad \theta = |\arg z| \le \pi/2 - \epsilon < \pi/2$.

Solution. Let us change the integration variable $t = z\tau$ to stop stationary point:

$$\Gamma(z+1) = z^{z+1} \int_0^{e^{-i\theta}\infty} \exp\left(z(\ln\tau - \tau)\right) d\tau$$

Since $\operatorname{Re} z > 0$, the endpoint of integration is in the right half-plane of the complex variable τ on infinity. The integrand vanishes at both ends of the interval. Integration contour can be deformed so that it passes through the stationary point $\tau_0 = 1$. Therefore the steepest descend contour exists. It is defined by the equation

$$\operatorname{Im} z(\ln\tau - \tau) = \operatorname{Im} z(\ln\tau_0 - \tau_0) = -\operatorname{Im} z$$

and has a complicated shape. However, near the saddle point $\tau_0 = 1$ the direction of steepest descent in the complex plane τ can be easily found using the second derivative of the function $zd^2(\ln\tau - \tau)/d\tau^2 = -z$ at $\tau = \tau_0 = 1$ and it has the angle $-\theta/2$ with the positive direction of the real axis:

$$z(\ln\tau - \tau) = -z - \frac{|z|\xi^2}{2} + |z|\mathcal{O}(\xi^3), \quad \xi = (\tau - 1)e^{i\theta/2}.$$

Whence, replacing the limits of integration by infinite, for $|z| \to \infty$ we find the asymptotics of the Γ-function:

$$\Gamma(z) = \frac{\Gamma(z+1)}{z} \sim \left(\frac{z}{e}\right)^z \int_{-\infty}^{\infty} \exp\left(-\frac{|z|\xi^2}{2}\right) e^{-i\theta/2}\, d\xi \sim \sqrt{\frac{2\pi}{z}} \left(\frac{z}{e}\right)^z.$$

6.6. Derive equations for slow variables of a weakly nonlinear oscillator

$$\ddot{x} + \omega_0^2 x = -\epsilon f(x, \dot{x}), \quad \epsilon \to 0. \tag{6.14}$$

averaging over oscillations period.

Solution. Let us write the solution of the unperturbed system in the form

$$x = I\cos\phi, \quad \dot{x} = -I\omega_0 \sin\phi, \quad \phi \equiv \omega_0 t + \vartheta$$

and consider these relations as a transition to new variable action - angle $(I,\ \vartheta)$ for perturbed system (Bogolyubov–Krylov transformation). Treating these variables $(I,\ \vartheta)$ as slowly depending on time, we differentiate the function $x(t) = I\cos\phi$:

$$\dot{x} = \dot{I}\cos\phi - I(\omega_0 + \dot{\vartheta})\sin\phi.$$

Since $\dot{x} = -I\omega_0\sin\phi$ we get the first equation

$$\dot{I}\cos\phi = I\dot{\vartheta}\sin\phi\,. \tag{6.15}$$

Differentiating $\dot{x}(t) = -I\omega_0\sin\phi$ and substituting the result into (6.14), we obtain the second equation for slow variables:

$$\ddot{x} + \omega_0^2 x = -\omega_0\left(\dot{I}\sin\phi + I\dot{\vartheta}\cos\phi\right) = -\epsilon f\left(I\cos\phi, -I\omega_0\sin\phi\right). \tag{6.16}$$

The system (6.15), (6.16) can be resolved with respect to new unknown functions

$$\dot{I} = \frac{\epsilon\sin\phi}{\omega_0} f\left(I\cos\phi, -I\omega_0\sin\phi\right), \quad \dot{\vartheta} = \frac{\epsilon\cos\phi}{\omega_0 I} f\left(I\cos\phi, -I\omega_0\sin\phi\right).$$

Averaging over the fast variable ϕ and denoting the slow variables in the averaged equations as $J(t)$, $\theta(t)$, we obtain the Bogolyubov–Krylov equations:

$$\dot{J} = \frac{\epsilon}{\omega_0} G(J), \ G(J) = \frac{1}{2\pi}\int_0^{2\pi} f\left(J\cos\phi, -J\omega_0\sin\phi\right)\sin\phi\, d\phi, \tag{6.17}$$

$$\dot{\theta} = \frac{\epsilon}{\omega_0 J} F(J), \ F(J) = \frac{1}{2\pi}\int_0^{2\pi} f\left(J\cos\phi, -J\omega_0\sin\phi\right)\cos\phi\, d\phi. \tag{6.18}$$

The right-hand sides of the averaged equations contain Fourier components of the driving force.

6.7. Find and investigate for stability limiting cycle of the van der Pol equation

$$\ddot{x} + \omega_0^2 x = \epsilon(1 - x^2)\dot{x},$$

where $\epsilon \to 0$.

Solution. Averaging the driving force $f(x, \dot{x}) = -(1 - x^2)\dot{x}$ over period, we get according to the formula (6.17)

$$\dot{J} = \frac{\epsilon}{\omega_0} G(J) = \frac{\epsilon}{2} J \left(1 - \frac{J^2}{4}\right).$$

This first-order equation has three stationary points, where $\dot{J} = 0$, namely $J = 0$ and $J = \pm 2$. The first is unstable for $\epsilon > 0$, and the other two are stable. A stable stationary point corresponds to a limiting cycle in the original second-order equation. The van der Pol equation describes the establishment of a self-oscillating mode in the generator.

6.8. Find time dependence of the averaged amplitude of the linear oscillator at resonance with an external force $f(t) = f_0 \cos \omega_0 t$

$$\ddot{x} + \omega_0^2 x = f(t).$$

Solution. Let us pass to the slow variables $I,\ \theta$

$$x = I \cos \phi, \quad \dot{x} = -I\omega_0 \sin \phi, \quad \phi \equiv \omega_0 t + \theta.$$

At exact resonance, the explicit dependence of the external force on time can be expressed in terms of the fast phase of the oscillator ϕ as $f(t) = f_0 \cos(\phi - \vartheta)$. Averaging the driving force over the period, we get according to the formulas (6.17), (6.18) equations for averaged variables $(J,\ \theta)$:

$$\dot{J} = -\frac{f_0}{2\omega_0} \sin \theta, \quad \dot{\theta} = -\frac{f_0}{2\omega_0} \cos \theta\,.$$

The equation for θ has one stable stationary point $\theta = -\pi/2$. The averaged amplitude grows linearly with time: $J(t) = f_0 t/2\omega_0$. This result coincides with the asymptotics of the exact solution on large times: $x = f_0 t \sin(\omega_0 t)/2\omega_0$.

6.9. Find time dependence of the averaged amplitude of the linear oscillator for parametric resonance when its frequency changes in time according to the law $\omega^2(t) = \omega_0^2(1 + \epsilon \sin \Omega t)$ with $\Omega = 2\omega_0$:

$$\ddot{x} + \omega_0^2(1 + \epsilon \sin \Omega t)x = 0$$

Solution. Let us pass to the slow variables $I,\ \vartheta$:

$$x = I \cos \phi, \quad \dot{x} = -I\omega_0 \sin \phi, \quad \phi \equiv \omega_0 t + \vartheta.$$

At parametric resonance, the term describing perturbation can be written as $-2\epsilon\omega_0^2 x\cos(\phi-\vartheta)\sin(\phi-\vartheta)$. Then according to the formulas (6.17), (6.18) we get equations for averaged variables $(J,\ \theta)$: $(J,\ \theta)$:

$$\dot{J} = \frac{\epsilon\omega_0}{4} J\cos 2\theta, \quad \dot{\theta} = -\frac{\epsilon\omega_0}{4}\sin 2\theta\,.$$

The equation for θ has two stable stationary points $\theta = 0$ and $\theta = \pi$ (*bistability*). The averaged amplitude grows in both cases exponentially $J(t) = J_0\exp(\epsilon\omega_0 t/4)$.

6.10. Find dependence on time of the amplitude of oscillator with cubic damping:

$$\ddot{x} + \epsilon\dot{x}^3 + \omega_0^2 x = 0.$$

Solution. The averaged equation (6.17)

$$\dot{J} = \frac{\epsilon}{\omega_0}G(J), \quad G(J) = -J^3\omega_0^3\frac{1}{2\pi}\int_0^{2\pi} d\phi\,\sin^4\phi = -\frac{3}{8}J^3\omega_0^3$$

can be easily integrated. The averaged oscillation amplitude decreases in time according to the law

$$J(t) = \frac{J_0}{\sqrt{1+\frac{3}{4}J_0^2\omega_0^2 t}}.$$

6.11. Find nonlinear frequency shift of the inharmonic oscillator (Duffing equation):

$$\ddot{x} + \omega_0^2 x = -\epsilon x^3$$

Solution. Averaging according to Eq. (6.17) gives $\dot{J} = 0$, which is a consequence of the Hamiltonian nature of the system. Using (6.18), we obtain the equation for the phase:

$$\dot{\theta} = \frac{3\epsilon J^2}{8\omega_0}.$$

The quantity $\dot{\theta}$ is called as nonlinear frequency shift $\omega_0 \to \omega_{NL} = \omega_0 + 3\epsilon J^2/8\omega_0$ and is proportional to the square of the oscillation amplitude:

$$x = J\cos(\omega_{NL}t).$$

Exercises

6.12. Find the asymptotic expansion of the Fresnel integrals for $x \to +\infty$:

$$F_1(x) = \int_x^\infty \cos(\theta^2)\, d\theta, \quad F_2(x) = \int_x^\infty \sin(\theta^2)\, d\theta.$$

6.13. Find the asymptotic expansion of the integrals at $a \to \infty$:

$$F(a) = \int_0^\infty \frac{dx\, e^{-x}}{x^2 + a^2}.$$

6.14. Find the asymptotic expansion of integrals for $\nu \to +\infty$, $a = \text{const}$:

$$F(\nu, a) = \int_0^\infty \frac{dx\, \exp(-\nu x)}{x + a}.$$

6.15. Find the asymptotic expansion of integrals for $n \to +\infty$

$$S_n = \int_0^{\pi/2} \sin^n t\, dt.$$

6.16. Find the asymptotic expansion of the error function at $x \to +\infty$:

$$\text{Erfc}\, x = \int_x^\infty \exp(-t^2)\, dt.$$

6.17. Find the asymptotic expansion of the integral exponential function at $x \to +\infty$:

$$E_n(x) = \int_x^\infty \frac{\exp(-t)}{t^n}\, dt.$$

6.18. Find leading behavior of the integral at $\alpha \to +\infty$:

$$F(\alpha) = \int_0^\infty \exp\left(-t^2/2 - \alpha/t\right)\, dt.$$

Hint. Make the change of variable $t = \tau\alpha^{1/3}$.

6.19. Find leading behavior of the Legendre polynomials for $x > 1$ when $l \to +\infty$

$$P_l(x) = \frac{1}{2\pi}\int_0^{2\pi} \left(x + \sqrt{x^2 - 1}\cos t\right)^l dt.$$

6.20. Find leading behavior of the modified Bessel function at $x \to +\infty$:

$$I_n(x) = \frac{1}{\pi}\int_0^{\pi} \cos n\theta \exp(x\cos\theta)\, d\theta\,.$$

6.21. Find leading behavior of the MacDonald function for $x > 0$ and $\nu \to +\infty$:

$$K_\nu(x) = \frac{1}{2}\int_{-\infty}^{\infty} \exp(-\nu t - x\cosh t)\, dt\,.$$

6.22. Find leading behavior of the function $K_\nu(\nu/\sinh\alpha)$ at $\nu \to \infty$ and fixed $\alpha > 0$.

6.23. Find leading behavior of the Fresnel integral at $x \to +\infty$:

$$\Phi(x) = \int_x^{\infty} \exp(\mathrm{i}t^2)\, dt.$$

6.24. Find leading behavior of the Airy function at $x \to -\infty$

$$\mathrm{Ai}\,(x) = \frac{1}{2\pi}\int_{-\infty}^{\infty} \exp\left[\mathrm{i}(t^3/3 + xt)\right]\, dt.$$

Hint. Change the variable $t = \tau\sqrt{-x}$

6.25. Find leading behavior of the integer-order Bessel function at $n = \mathrm{const}$ and $x \to +\infty$

$$J_n(x) = \frac{1}{2\pi}\int_{-\pi}^{\pi} \exp(\mathrm{i}x\sin\phi - \mathrm{i}n\phi)\, d\phi\,.$$

6.26. Find leading behavior of the Bessel function $J_n(n)$ at $n \to +\infty$.

Hint. Use result of problem **6.3**.

6.27. Find asymptotic behavior of the Bessel function $J_n(n/\cos\beta)$ at $n \to +\infty$ and fixed $0 < \beta < \pi/2$.

6.28. Find the convergency sectors at infinity in a complex plane for the following integrals:

$$F_1 = \int_{-\infty}^{\infty} \exp(-x^2)\,dx; \quad F_2 = \int_{-\infty}^{\infty} \exp(\mathrm{i}x^3)\,dx;$$

$$F_3 = \int_{\gamma} \exp(x^n)\,dx, \quad n > 0 \text{ integer.}$$

6.29. Design qualitatively on complex plane the level lines of real and imaginary parts of the functions:

(a) $w(z) = z$;
(b) $w(z) = z^2 - 1$;
(c) $w(z) = z^3$;
(d) $w(z) = \ln z$;
(e) $w(z) = \ln(z^2 - 1)$;
(f) $w(z) = \mathrm{e}^z$.

6.30. Find stationary points z_0 of the functions:

$$w_1(z) = z^2 - 1; \quad w_2(z) = z^3; \quad w_3(z) = \ln(z^2 + 1).$$

In what directions away from the stationary points the real parts $\operatorname{Re} w(z)$ decrease, and in which ones increase?

6.31. Prove identity

$$\mathrm{Ai}\,(x) + \omega\,\mathrm{Ai}\,(\omega x) + \omega^2\,\mathrm{Ai}\,(\omega^2 x) = 0\,,$$

where $\omega = \exp(2\pi\mathrm{i}/3)$ for Airy functions

$$\mathrm{Ai}\,(x) = \frac{1}{2\pi}\int_{-\infty}^{\infty} \exp\left[\mathrm{i}(t^3/3 + xt)\right]\,dt\,.$$

6.32. Find leading behavior of the integral

$$F(x) = \frac{1}{2\pi}\int_{-\infty}^{\infty} \exp\left[\mathrm{i}(t^3/3 + xt)\right]\frac{t^2\,dt}{1 + t^2},$$

when $x \to +\infty$ and compare with the result of problem **6.4**.

Hint. Take into account the contribution of the residue at the pole $t = \mathrm{i}$ which arises upon deformation of the original contour of integration into the saddle contour (see Fig. 6.3).

6.33. Find leading behavior of the Legendre polynomials at $l \to \infty$ using the Schlaefli representation:

$$P_l(\cos\theta) = \frac{1}{2^l 2\pi \mathrm{i}} \oint_C \frac{(z^2-1)^l}{(z-\cos\theta)^{l+1}}\,dz,$$

where the contour C bypass the pole $z = \cos\theta$ in a positive direction.

6.34. Find the behavior in time of the amplitude of an oscillator with viscous friction:

$$\ddot{x} + 2\gamma\dot{x} + \omega_0^2 x = 0.$$

6.35. Find the behavior in time of the amplitude of an oscillator with dry friction:

$$\ddot{x} + 2\gamma\,\mathrm{sign}\,(\dot{x}) + \omega_0^2 x = 0.$$

6.36. Find the nonlinear frequency shift of the pendulum

$$l\ddot{\alpha} + g\sin\alpha = 0$$

for a small oscillations amplitude a.

6.37. Find and check for stability a limiting cycle of an equation

$$\ddot{x} - \epsilon(1-|x|)\dot{x} + \omega_0^2 x = 0.$$

Answers

6.12. At $x \to +\infty$

$$F_1 \sim -\sum_{n=0}^{\infty} \frac{\Gamma(n+1/2)}{2\Gamma(1/2)} x^{-2n-1} \sin(x^2 - n\pi/2),$$

$$F_2 \sim \sum_{n=0}^{\infty} \frac{\Gamma(n+1/2)}{2\Gamma(1/2)} x^{-2n-1} \cos(x^2 - n\pi/2).$$

6.13. At $a \to \infty$

$$F(a) \sim \sum_{n=0}^{\infty} (-1)^n \frac{(2n)!}{a^{2n+2}}.$$

6.14. At $\nu \to +\infty$ and fixed a

$$F(\nu, a) \sim \sum_{n=0}^{\infty} (-1)^n \frac{n!}{(\nu a)^{n+1}}.$$

6.15. At $n \to +\infty$

$$S_n \sim \sqrt{\frac{\pi}{2n}} \left(1 + \mathcal{O}(n^{-1})\right).$$

6.16. At $x \to +\infty$

$$\operatorname{Erfc} x \sim \frac{\exp(-x^2)}{2x} \sum_{n=0}^{\infty} \frac{\Gamma(n+1/2)}{\Gamma(1/2)} (-x^2)^{-n}.$$

6.17. At $x \to +\infty$

$$E_n \sim \frac{\exp(-x)}{x^n} \sum_{m=0}^{\infty} \frac{\Gamma(n+m)}{\Gamma(n)} (-x)^{-m}.$$

6.18. At $\alpha \to +\infty$

$$F(\alpha) \sim \sqrt{2\pi/3} \exp\left(-3\alpha^{2/3}/2\right) \left(1 + \mathcal{O}(\alpha^{-2/3})\right).$$

6.19. At $l \to +\infty$ and fixed $x > 1$

$$P_l(x) \sim \frac{\left(x + \sqrt{x^2 - 1}\right)^{l+\frac{1}{2}}}{\sqrt{2\pi l}\,(x^2 - 1)^{\frac{1}{4}}} \left(1 + \mathcal{O}(l^{-1})\right).$$

6.20. At $x \to \infty$ and fixed n

$$I_n(x) \sim \frac{e^x}{\sqrt{2\pi x}} \left(1 + \mathcal{O}(x^{-1})\right).$$

6.21. At $\nu \to \infty$ and $x > 0$

$$K_\nu(x) \sim \sqrt{\frac{\pi}{2\nu}} \left(\frac{2\nu}{xe}\right)^\nu \left(1 + \mathcal{O}(\nu^{-1})\right).$$

6.22. At $\nu \to \infty$ and fixed $\alpha > 0$

$$K_\nu\left(\frac{\nu}{\sinh \alpha}\right) \sim \sqrt{\frac{\pi \tanh \alpha}{2\nu}} \exp(\nu(\alpha - \coth \alpha)) \left(1 + \mathcal{O}(\nu^{-1})\right).$$

6.23. At $x \to +\infty$

$$\Phi(x) \sim \mathrm{i}\frac{\exp(\mathrm{i}x^2)}{2x}\left(1 + \mathcal{O}(x^{-2})\right).$$

6.24. At $x \to -\infty$

$$\mathrm{Ai}\,(x) \sim \frac{1}{|x|^{1/4}\sqrt{\pi}}\cos\left(\frac{2}{3}|x|^{3/2} - \frac{\pi}{4}\right)\left(1 + \mathcal{O}(|x|^{-3/2})\right).$$

6.25. At $x \to +\infty$ and fixed n

$$J_n(x) \sim \sqrt{\frac{2}{\pi x}}\cos\left(x - \pi n/2 - \pi/4\right).\left(1 + \mathcal{O}(x^{-1})\right).$$

6.26. At $n \to +\infty$

$$J_n(n) \sim \frac{\Gamma(1/3)}{2\pi\sqrt{3}}\left(\frac{6}{n}\right)^{1/3}\left(1 + \mathcal{O}(n^{-1/3})\right).$$

6.27. At $n \to +\infty$ and fixed β

$$J_n(n/\cos\beta) \sim \sqrt{\frac{2}{\pi n \tan\beta}}\cos\left\{n\left(\beta - \tan\beta\right) + \pi/4\right\}\left(1 + \mathcal{O}(n^{-1})\right).$$

6.28. The convergency sectors in the complex plane of the variable $x = |x|\mathrm{e}^{\mathrm{i}\theta}$ are:

$$\begin{aligned} F_1 &: \quad \theta \in [-\pi/4, \pi/4] \cup [3\pi/4, 5\pi/4]\,; \\ F_2 &: \quad \theta \in [0, \pi/3] \cup [2\pi/3, \pi] \cup [4\pi/3, 5\pi/3]\,; \\ F_3 &: \quad q\frac{2\pi}{n} + \frac{\pi}{2n} < \theta < q\frac{2\pi}{n} + \frac{3\pi}{2n}, \qquad q \text{ integer} \end{aligned}$$

6.29. **Hint.** Write the complex variable as $z = x + \mathrm{i}y$ and derive the equations for the curves in terms of (x, y) from $\mathrm{Re}\,w(z) = \text{const}$ or $\mathrm{Im}\,w(z) = \text{const}$. Pay attention to zeros and singularities of the functions $w(z)$.

6.30. All the functions have one stationary point $z_0 = 0$. The functions $\mathrm{Re}\,w_1(z)$ and $\mathrm{Re}\,w_3(z)$ grow lengthwise the real axis and decrease lengthwise the imaginary axis away from the saddle point. The function $\mathrm{Re}\,w_2(z)$ grows along the directions $\theta = 0,\ 2\pi/3,\ 4\pi/3$, and decreases along the directions $\theta = \pm\pi/3,\ \pi$ from the stationary point. Here θ is the argument of z: $z = |z|\mathrm{e}^{\mathrm{i}\theta}$.

6.31. Hint. Change the variable in the Airy function integral representation: $t \to t/\omega$ for the second term in the identity and $t \to t\omega$ for the third term. Convert sum of integrals to one contour integral (see Fig. 6.3).

6.32. At $x \to +\infty$

$$F(x) \sim -\frac{1}{2}\exp(1/3 - x) + \mathcal{O}\left(\exp(-2/3x^{3/2})\right).$$

6.33. At $l \to \infty$ and fixed θ

$$P_l(\cos\theta) \sim \sqrt{\frac{2}{\pi l \sin\theta}}\sin\left[\theta(l+1/2)+\pi/4\right]\left(1+\mathcal{O}(l^{-1})\right);$$
$$\theta l \gg 1,\ (\pi-\theta)l \gg 1.$$

6.34. The averaged oscillator amplitude decreases in time according to the law

$$J(t) = J(0)\exp(-\gamma t).$$

6.35. The averaged oscillator amplitude decreases in time according to the law

$$J(t) = J(0) - \frac{4\gamma t}{\pi\omega_0},\ \text{at}\ 0 < t < \frac{\pi\omega_0 J(0)}{4\gamma}.$$

6.36. The oscillation frequency of the pendulum is

$$\omega_{NL} = \sqrt{\frac{g}{l}}\left(1 - \frac{a^2}{16}\right),$$

where a is the amplitude.

6.37. The equation for the averaged oscillations amplitude has the form

$$\dot{J} = \frac{\epsilon J}{2}\left(1 - \frac{4J}{3\pi}\right).$$

For $\epsilon > 0$ it has one stable stationary point $J = 3\pi/4$.

Chapter 7

Green's Functions Method

7.1 Green's Functions

1. The method of Green's functions allows us to solve inhomogeneous linear differential equations with arbitrary right-hand sides. *Green's function of the first kind* $G(x, x')$, $x, x' \in \mathcal{D} \subset \mathbb{R}^n$ for a boundary value problem

$$\mathcal{L}u = f, \quad \mathcal{B}u|_{x\in S} = 0, \tag{7.1}$$

where $\mathcal{L}$ and $\mathcal{B}$ are some linear operator, satisfies the equation

$$\mathcal{L}G(x, x') = \delta(x - x') \tag{7.2}$$

with boundary condition $\mathcal{B}G(x, x')|_{x\in S} = 0$. The domain $\mathcal{D}$ and its boundary S are schematically depicted in Fig. 7.1. The solution of problem (7.1) is given by the *Duhamel integral*

$$u(x) = \int_{\mathcal{D}} G(x, x') f(x')\, dx'.$$

Expanding the operator $\mathcal{L}$ together with the inverse operator $\mathcal{L}^{-1}$ in terms of projectors to the subspaces of eigenfunctions $|n\rangle$

$$\mathcal{L} = \sum_n \lambda_n |n\rangle \langle n|, \quad \mathcal{L}^{-1} = \sum_n \frac{1}{\lambda_n} |n\rangle \langle n|, \quad \lambda_n \neq 0 \tag{7.3}$$

Mathematical Methods of Physics: Problems with Solutions
Igor V. Kolokolov, Evgeny A. Kuznetsov, Alexander I. Milstein, Evgeny V. Podivilov, Alexander I. Chernykh, David A. Shapiro, and Elena G. Shapiro

ISBN 978-981-5129-21-2 (Hardcover), 978-1-003-51360-5 (eBook)
www.jennystanford.com

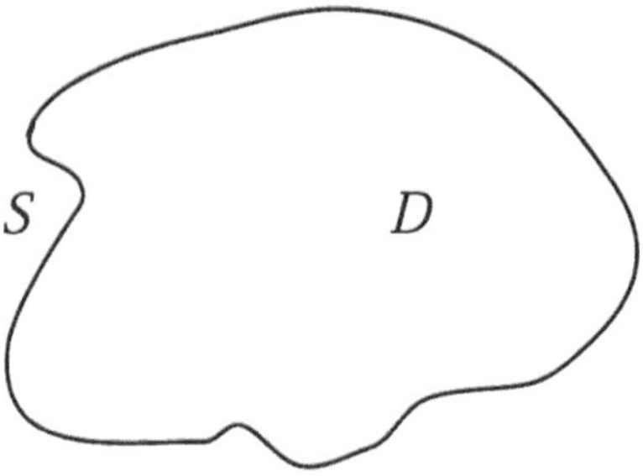

Figure 7.1 Domain of the boundary value problem.

we see that Green's function is the integral kernel of the inverse operator $G(x, x') = \langle x|\mathcal{L}^{-1}|x'\rangle$. Equation (7.2) is a manifestation of this fact. Green's function exists and it is unique if the spectral problem

$$\mathcal{L}\,|n\rangle = \lambda_n\,|n\rangle\,, \quad \mathcal{B}\,|n\rangle\,|_{x\in S} = 0 \tag{7.4}$$

has no zero eigenvalue. We can see from the expansion (7.3) that Green's function of a self-conjugate operator ($\mathcal{L} = \mathcal{L}^\dagger$, see Chapter 1) obeys to the *reciprocity principle*

$$G(x, x') = G^*(x', x).$$

To find Green's function of a differential equation, follow the following rules. Let $\mathcal{L}$ be an ordinary differential operator of N-th order. Integrate (7.2) over an infinitesimal neighborhood of the point x' and find the jump of the $(N-1)$-th derivative of G at the point $x = x'$. It remains to perform three steps:

(1) Solve the homogeneous equation.
(2) Write the solution in the domains $x < x'$ and $x > x'$ as two different linear combinations of solutions of the homogeneous equation containing $2N$ unknown coefficients.
(3) Find these coefficients using N boundary conditions, $N-1$ continuity conditions for derivatives of orders $0, 1, \ldots, N-2$ and one condition for the jump of the derivative of order $N-1$ at $x = x'$.

Green's function is represented, generally speaking, by different formulas for $x < x'$ and $x > x'$. If Green's function is symmetric, i.e., $G(x, x') = G(x', x)$, these formulas differ only by permutation x and x'. To shorten the notation, instead of x, x' the notation $x_< = \min(x, x')$, $x_> = \max(x, x')$ will be used.

2. If problem (7.4) has non-trivial solutions $|i\rangle$, $i = 1, \ldots, k$ with $\lambda_i = 0$ (so called *zero modes*), then the inhomogeneous problem (7.1) is solvable when its right-hand side f is orthogonal to zero modes of the conjugated problem

$$\mathcal{L}^\dagger v = 0, \quad \mathcal{B}^\dagger v\big|_{x\in S} = 0. \tag{7.5}$$

Here $\mathcal{L}^\dagger$ is the operator conjugated to $\mathcal{L}$: $(v, \mathcal{L}u) = (\mathcal{L}^\dagger v, u)$ and $\mathcal{B}^\dagger$ is the conjugate operator defining boundary conditions.

If our inhomogeneous problem is solvable then we can use *generalized (modified)* Green's function which obey the equation

$$\mathcal{L}G(x, x') = \delta(x - x') - \sum_{i=1}^{k} u_i(x)v_i^*(x'), \tag{7.6}$$

instead of (7.2). Here zero modes u_i and v_i of direct and conjugate problems are mutually orthogonal and numbered according to the condition

$$\int_{\mathcal{D}} v_i^*(x)u_j(x)\, dx = \delta_{ij}, \quad i, j = 1, \ldots, k.$$

The generalized Green's function is defined uniquely if you require its orthogonality to zero modes of the homogeneous conjugate problem (7.5):

$$\int_{\mathcal{D}} v_i^*(x)G(x, x')\, dx = 0, \quad i = 1, \ldots, k. \tag{7.7}$$

The expansion in projectors in subspace orthogonal to the zero modes is written as $\mathcal{L}^{-1} = \sum_n' \frac{1}{\lambda_n} |n\rangle \langle n|$, where $\sum_n' = \sum_{n=k+1}^{\infty}$ denotes the summation over non-zero modes.

To find the generalized Green's function of an ordinary differential equation with zero modes, you must first normalize them. In steps 1 and 2, a particular solution of the Eq. (7.6) without δ-function should be added to the solution of the homogeneous equation. In step 3, k conditions have been missed to find the unknown coefficients, since each zero mode simultaneously satisfies two boundary conditions. Therefore, it is necessary to add k orthogonality requirements (7.7). The solution is written as

$$u(x) = \int_{\mathcal{D}} G(x, x')f(x')\, dx' + \sum_{i=1}^{k} c_i u_i(x),$$

where c_i are arbitrary coefficients.

3. When, instead of (7.1), we should investigate a problem with inhomogeneous boundary conditions

$$\mathcal{L}u = 0, \quad \mathcal{B}u|_{x \in S} = g, \tag{7.8}$$

then we need to find *Green's function of the second kind* $G_S(x, x')$. The solution of problem (7.8) is written as an integral over the boundary S of the domain $\mathcal{D}$:

$$u(x) = \int_S G_s(x, x')g(x')\,dx', \quad x \in \mathcal{D}.$$

The solution of a linear problem with a non-zero right-hand side and non-zero boundary conditions

$$\mathcal{L}u = f, \quad \mathcal{B}u|_{x \in S} = g$$

we can search as the sum of solutions to two problems (7.1) and (7.8). As a result, we arrive at the sum of the volume integral in $\mathcal{D}$ with the function f and the surface integral in S with the function g.

Green's functions of the first and second kinds are related. The linkage between them for a particular equation can be found using the corresponding *Green's formula*, which is derived considering the difference of the scalar products $(u, \mathcal{L}v) - (\mathcal{L}^\dagger u, v)$ and reducing it to a surface integral.

4. We call *fundamental solution* any solution of Eq. (7.2) that does not necessarily satisfy the boundary conditions. The fundamental solution is determined up to any solution of the homogeneous equation. The behavior near singularity of the fundamental solution of the Poisson equation

$$\triangle g(\mathbf{r}, \mathbf{r}') = \delta(\mathbf{r} - \mathbf{r}')$$

can be found by integrating over $\mathbf{r}$ near the point $\mathbf{r} = \mathbf{r}'$. For the dimension $n = 1, 2, 3$, we get

Dimension	$g(\mathbf{r}, \mathbf{r}')$
1	$\frac{\lvert x - x'\rvert}{2}$
2	$\frac{\ln \lvert\mathbf{r} - \mathbf{r}'\rvert}{2\pi}$
3	$-\frac{1}{4\pi \lvert\mathbf{r} - \mathbf{r}'\rvert}$

Green's function of the first kind $G(x, x')$ can be constructed from the fundamental solution $g(x, x')$, if we add a linear combination of solutions of homogeneous equation without singularities at $\mathbf{r} = \mathbf{r}'$, and require the fulfillment of the boundary conditions.

It is possible to find this combination if the domain $\mathcal{D}$ has some symmetries. In a 3-dimensional case, the image method sometimes helps in many interesting cases, and in a two-dimensional case, the method of conformal transformations is useful in a number of interesting cases. The application of the image method is based on the inversion transformation $r' \to R^2/r'$ of the Laplace equation (see problem **7.5**) with respect of a sphere of radius R if $\mathcal{D}$ is a ball of radius R, or reflection transformation about the plane, if $\mathcal{D}$ is half-space. In a two-dimensional case, the Laplace equation is covariant with respect to conformal transformations. Conformal transformations are useful if they allow us to map the domain to a simpler one for which Green's function is easier to build.

For the *elliptic* Helmholtz operator $\mathcal{L} = \Delta - k^2$ or the Laplace operator $\mathcal{L} = \Delta$ the difference of volume integrals is expressed via integral over the surface by the following Green's formula:

$$(v, \mathcal{L}u) - (\mathcal{L}^{\dagger}v, u) = -\int_S \left(v^* \frac{\partial u}{\partial n} - \frac{\partial v^*}{\partial n} u \right) dS, \tag{7.9}$$

where $\partial/\partial n$ denotes the derivative along the inner normal. The formula is derived from the identity $v\Delta u - u\Delta v = \operatorname{div}(v\nabla u - u\nabla v)$ with transformation of the volume integral into the surface one.

Let us take as $v(x)$ the function $G(x, x')$ which obeys Eq. (7.2) and as $u(x)$ solution of problem (7.8) and then use Green's formula (7.9). Now let's change notation $x \leftrightarrow x'$, apply the reciprocity principle and get

$$u(x) = \int_S \left(G(x, x') \frac{\partial u(x')}{\partial n'} - \frac{\partial G(x, x')}{\partial n'} u(x') \right) dx'. \tag{7.10}$$

Hence Green's function of the second kind is found for different boundary conditions. For the *Dirichlet problem* $u(x)|_{x \in S} = g(x)$ $(\mathcal{B} = 1)$ Green's function of the first kind satisfies the boundary condition

$$G(x, x')|_{x \in S} = 0.$$

Green's function of the second kind (*double layer* potential) is obtained from the formula

$$G_s(x, x') = -\frac{\partial}{\partial n'} G(x, x')|_{x' \in S}.$$

For the *Neumann problem* $\partial u(x)/\partial n|_{x\in S} = g(x)$ ($\mathcal{B} = \partial/\partial n$), Green's function of the first kind obeys the boundary conditions

$$\partial G(x, x')/\partial n\big|_{x\in S} = 0.$$

Green's function of the second kind (*simple layer* potential) is given by the formula

$$G_s(x, x') = G(x, x')|_{x'\in S}.$$

5. For a *parabolic* operator, for example, thermal conductivity operator $\mathcal{L} = \partial/\partial t - \Delta$, two Green's functions can be introduced. In domain infinite in the coordinate space, Green's function of the first kind decreases at infinity and satisfies the equation

$$\mathcal{L}G(\mathbf{r}, t; \mathbf{r}', t') = \delta(\mathbf{r} - \mathbf{r}')\delta(t - t'),$$
$$G(\mathbf{r}, t; \mathbf{r}', t') \to 0 \quad \text{at} \quad |\mathbf{r} - \mathbf{r}'| \to \infty.$$

The function G allows to solve the problem

$$\mathcal{L}u = f(\mathbf{r}, t), \quad u(\mathbf{r}, t) \to 0 \quad \text{at} \quad t \to -\infty.$$

The solution of the Cauchy problem with the initial conditions

$$\mathcal{L}u = 0, \quad u(\mathbf{r}, 0) = \phi(\mathbf{r})$$

is expressed in terms of the Green's function of the second kind G_s, which starts from δ-function at the initial moment

$$\mathcal{L}G_s(\mathbf{r}, \mathbf{r}', t) = 0, \quad G_s(\mathbf{r}, \mathbf{r}', t) \to \delta(\mathbf{r} - \mathbf{r}') \quad \text{at} \quad t \to +0.$$

6. For the operator $\mathcal{L}$ of the *hyperbolic* type, for example, $\mathcal{L} = \Box$, there are several different Green's functions. We will only use *retarded Green's function* determined from the solution of the wave equation

$$\Box G = \delta(\mathbf{r} - \mathbf{r}')\delta(t - t'), \quad \Box \equiv \frac{1}{c^2}\frac{\partial^2}{\partial t^2} - \Delta. \tag{7.11}$$

It is equal to zero at $t < t'$ and goes to zero together with its derivatives at spatial infinity:

$$G(\mathbf{r}, t; \mathbf{r}', t') = 0,\ t < t'; \quad G(\mathbf{r}, t; \mathbf{r}', t') \to 0,\ |\mathbf{r} - \mathbf{r}'| \to \infty.$$

If the coefficients of the differential equation are constant, then the Green's function can be found using the Fourier transform. When performing the inverse transformation, a difficulty may arise if the pole of Green's function in the ω, $\mathbf{k}$-representation lies on the integration contour. The rules for bypassing the poles are fixed by physical considerations. For the Cauchy problem with initial conditions, instead of the Fourier transform, the Laplace transform can be used.

7.2 Continuous Spectrum

We will consider mainly the differential operators acting in $L^2(\Omega)$ which is the space of quadratically integrable functions of $x \in \Omega$. The value of λ is called the eigenvalue of the operator $\mathcal{A}$, if there is a solution $\psi_\lambda(x)$ of the equation

$$\mathcal{A}\psi_\lambda(x) = \lambda\psi_\lambda(x),$$

belonging to $L^2(\Omega)$. This solution is called an eigenfunction. It is said that λ is eigenvalue of a discrete spectrum and that all such λ form a discrete spectrum σ_p. For a self-conjugate operator $\mathcal{A} = \mathcal{A}^\dagger$ acting in $L^2(\Omega)$, where $\Omega = \mathcal{D}$ is some finite region of space, the statement holds: the eigenvalues of $\mathcal{A}$ form a discrete set $\sigma_p = \{\lambda_n\}$, $n = 0, 1, \ldots$ and the eigenfunctions corresponding to different eigenvalues are orthogonal.

For most differential operators used in physics (for example, Sturm–Liouville and Laplace operators), the set of eigenfunctions forms a complete set in $L^2(\Omega)$. Hence, we can construct an orthonormal basis by selecting and normalizing suitable linear combinations in degenerate cases. So such the operator $\mathcal{A}$ can be represented in the canonical form (7.3)

$$\mathcal{A} = \sum_n \lambda_n |n\rangle \langle n| .$$

If $A(x, x')$ is the kernel of integral operator

$$(\mathcal{A}\psi)(x) = \int_\Omega A(x, x')\psi(x')\, dx',$$

it can be expanded over its eigenfunctions as

$$A(x, x') = \langle x\, |\mathcal{A}|\, x' \rangle = \sum_n \lambda_n \psi_n(x)\psi_n^*(x'); \tag{7.12}$$

$$\psi_n(x) = \langle x|n\rangle , \quad \psi_n^*(x') = \langle n|x'\rangle .$$

This representation of the operator is called *spectral decomposition*.

If Ω is a noncompact domain then (7.12) may be incorrect and should be modified, since the operator $\mathcal{A}$ can have a *continuous spectrum*.

Let $\mathcal{A}$ still be self-conjugate. We say that the interval (a, b) of the real axis belongs to *continuous spectrum* σ_c, if for all $\lambda \in \sigma_c = (a, b)$ there are solutions $\psi_\lambda(x)$ of the equation

$$\mathcal{A}\psi_\lambda(x) = \lambda\psi_\lambda(x),$$

not belonging to $L^2(\Omega)$, but such that any superposition of them of the form

$$\Psi(x) = \int_a^b a(\lambda)\psi_\lambda(x)\,d\lambda, \quad a(\lambda) \in L^2(a, b)$$

already lies in $L^2(\Omega)$. Speaking in the language of quantum mechanics, from the wave functions of a continuous spectrum we can construct normalized wave packets that are arbitrarily close to any given $\psi_\lambda(x)$ in an arbitrarily large area of space. For example, the function

$$a(\lambda) = \frac{\varepsilon}{(\lambda - \lambda_0)^2 + \varepsilon^2}$$

with a small but finite ε has a finite norm, and the normalized packet $\Psi(x)$ in a region with a linear size $\sim 1/\varepsilon$ differs only slightly ($\sim \varepsilon$) from $\psi_{\lambda_0}(x)$.

The eigenfunctions of the continuous spectrum of the self-conjugate operator $\mathcal{A}$ are mutually orthogonal:

$$(\psi_\lambda, \psi_{\lambda'}) = \int_\Omega dx\, \psi_\lambda^*(x)\psi_{\lambda'}(x) = 0, \quad \lambda \neq \lambda'$$

and together with the functions of the discrete spectrum form a complete set: any $f \in L^2(\Omega)$ can be represented as a linear superposition

$$f(x) = \sum_n a_n\psi_n(x) + \int_{\sigma_c} d\lambda\, a(\lambda)\psi_\lambda(x). \tag{7.13}$$

For continuous spectrum functions, it is customary to choose the δ-function normalization

$$\int_\Omega dx\, \psi_\lambda^*(x)\psi_{\lambda_0}(x) = \delta(\lambda - \lambda_0). \tag{7.14}$$

Multiplying both parts of (7.13) by $\psi_{\lambda_0}(x)$ and integrating we find the coefficients $a(\lambda)$

$$a(\lambda) = \int_\Omega f(x)\psi_\lambda^*(x)\,dx. \tag{7.15}$$

Note that the normalization change, i.e., the appearance of the coefficient $C(\lambda)$ at δ-function in (7.14) will cause the coefficient $1/C(\lambda)$ to appear on the right side of (7.15).

Let us assume that the normalization (7.14) is selected. Then the spectral decomposition of the kernel $A(x, x')$ takes the form

$$A(x, x') = \sum_n \lambda_n \psi_n(x)\psi_n^*(x') + \int_{\sigma_c} d\lambda\, \lambda \psi_\lambda(x)\psi_\lambda^*(x'), \tag{7.16}$$

where summation is performed over the discrete spectrum σ_p and the integration goes over a continuous basis. The formula (7.16) implies that the eigenvalues of the continuous spectrum of the operator $\mathcal{A}$ are non-degenerate. If there is a degeneracy, then in each eigen-subspace with eigenvalue λ we can choose an orthonormal in the sense of (7.14) basis of $\psi_\lambda^{(j)}(x)$. The spectral decomposition in this case is supplemented by summing over all eigenfunctions that belong to this eigenvalue λ

$$A(x, x') = \sum_n \lambda_n \psi_n(x)\psi_n^*(x') + \int_{\sigma_c} d\lambda\, \lambda \sum_j \psi_\lambda^{(j)}(x)\psi_\lambda^{(j)*}(x'). \tag{7.17}$$

The possibility of degeneracy of the eigenvalues of the discrete spectrum is taken into account in (7.16), (7.17) by the fact that some λ_n in the first term can coincide.

7.3 Resolvent

Resolvent $\hat{R}_z$ of a given self-conjugate operator $\hat{H}$ is the following operator that depends on the complex variable z as a parameter

$$\hat{R}_z = (z - \hat{H})^{-1}.$$

From the definition follows the equation for the integral kernel $R_z(x, x')$ of the operator $\hat{R}_z$

$$(z - \hat{H})R_z(x, x') = \delta(x - x').$$

The eigenfunctions $\psi_\lambda(x)$ of $\hat{H}$ and $\hat{R}_z$ are obviously the same, and the eigenvalues are trivially recalculated. As a result, we obtain the spectral decomposition for $R_z(x, x')$:

$$R_z(x, x') = \sum_n \frac{1}{z - \lambda_n}\psi_n(x)\psi_n^*(x') + \int_{\sigma_c} \frac{d\lambda}{z - \lambda} \sum_j \psi_\lambda^{(j)}(x)\psi_\lambda^{(j)*}(x'). \tag{7.18}$$

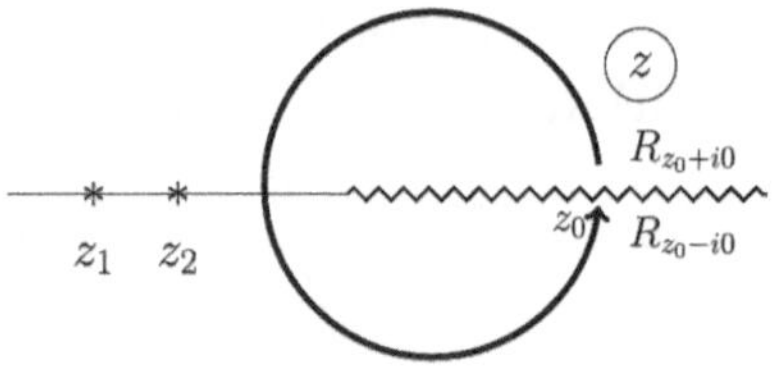

Figure 7.2 The complex plane of the spectral parameter z with poles at the points of the discrete spectrum and the cut corresponding to the continuous spectrum.

When $z = 0$, the resolvent coincides with (minus) Green's function. It is seen that the resolvent $R_z(x, x')$ contains all the information about the spectrum of the operator $\hat{H}$. As an analytical function of the variable $z \in \mathbb{C}$ resolvent is defined in z- plane with the cut along the interval of the real axis corresponding to the continuous spectrum and has poles corresponding to the discrete spectrum of the operator $\hat{H}$ (Fig. 7.2). The residue in the pole $z = \lambda_n$ of the resolvent $R_z(x, x')$, as follows from (7.18), is

$$\operatorname*{Res}_{z=\lambda_n} R_z = \sum_j \psi_{nj}(x)\psi^*_{nj}(x'),$$

where $\psi_{nj}(x)$ are eigenfunctions: $\hat{H}\,\psi_{nj} = \lambda_n \psi_{nj}$, $j = 1, \ldots, \kappa_n$ and κ_n is the multiplicity of degenerate eigenvalues λ_n.

The jump of the resolvent on the cut can be found using the formula

$$\lim_{\epsilon \to +0} \left(\frac{1}{z + \mathrm{i}\epsilon} - \frac{1}{z - \mathrm{i}\epsilon} \right) = -2\pi \mathrm{i}\delta(z)$$

and it is equal to

$$R_z|_{z=\lambda_0+\mathrm{i}0} - R_z|_{z=\lambda_0-\mathrm{i}0} = -2\pi\mathrm{i} \sum_j \psi^{(j)}_{\lambda_0}(x)\psi^{(j)}_{\lambda_0}(x'). \tag{7.19}$$

This relation allows us to find the eigenfunctions of the continuous spectrum normalized according to (7.14) with an accuracy up to a unitary rotation in their own subspace. This rotation leaves the following bilinear form invariant

$$\sum_j \psi^{(j)}_{\lambda_0}(x)\psi^{(j)}_{\lambda_0}(x').$$

Thus, knowing the properties of the resolvent as a function of its complex parameter z is equivalent to knowing the eigenvalues and eigenfunctions of both the discrete and continuous spectrum.

The method of Green's functions is discussed in books (Duffy, 2015; Mathews and Walker, 1970; Roach, 1992; Stakgold and Holst, 2011). The concept of resolvent and its relation to continuous spectrum functions are described in (Richtmyer, 1978). Conformal transformations important for two-dimensional harmonic functions are explained in textbooks on complex variables theory (Ablowitz *et al.*, 2003; Ahlfors, 1953; Brown and Churchill, 2009).

Worked Examples

7.1. Find Green's function and write out the solution of the inhomogeneous equation $u'' = f(x)$, if $u(0) = u(1) = 0$, $x \in [0, 1]$.

Solution. The solution of a homogeneous equation is a linear function. We immediately represent Green's function in the form that satisfies the boundary conditions

$$G(x, x') = \begin{cases} Ax, & \text{if } x < x'; \\ B(x-1), & \text{if } x > x'. \end{cases}$$

Here $0 \le x \le 1$, $0 < x' < 1$. The continuity condition of the function and the unit jump of the derivative give a system of two equations for the coefficients A, B

$$Ax' - B(x'-1) = 0; \quad B - A = 1.$$

The answer is conveniently expressed in terms of the variables $x_>$, $x_<$: $G(x, x') = x_< (x_> - 1)$.

7.2. Prove that Green's function of the equations $\mathcal{L}_r u(r) = f(r)$ with the operator

$$\mathcal{L}_r = \frac{d^2}{dr^2} + k^2(r)$$

and boundary conditions $u(0) = u(1) = 0$ can be represented in the form

$$G(r, r') = \frac{\chi_1(r_<)\chi_2(r_>)}{W(r')}, \quad W(r) = \begin{vmatrix} \chi_1(r) & \chi_2(r) \\ \chi_1'(r) & \chi_2'(r) \end{vmatrix}, \tag{7.20}$$

if there are no zero modes. Here the function $\chi_i(r)$ are lineary independent solutions of the homogeneous equation

$$\mathcal{L}_r \chi_i(r) = 0, \quad i = 1, 2, \quad \chi_1(0) = \chi_2(1) = 0.$$

Solution. Since χ_1 satisfies the left boundary condition, and χ_2 the right one, we can search for Green's function in the form

$$G(r, r') = \begin{cases} A\chi_1(r), & \text{if } r < r'; \\ B\chi_2(r), & \text{if } r > r'. \end{cases}$$

The boundary conditions are fulfilled automatically, and the continuity for $r = r'$ give a system of equations for the coefficients A, B

$$A\chi_1(r') - B\chi_2(r') = 0, \quad B\chi_2'(r') - A\chi_1'(r') = 1.$$

The absence of zero modes means that $\chi_1 \neq \chi_2$. The determinant of the system coincides with the Wronskian (7.20) of the fundamental system of solutions at the point $r = r'$, and is therefore nonzero. In this case the Wronskian $W(r')$ is independent on r'.

If the boundary conditions are not separated into right one and left one, then the formula (7.20) no longer gives Green's function $G(r, r')$, but only the fundamental solution $g(r, r')$. The fundamental solution can be turned into a Green's function by adding a linear combination of solutions to a homogeneous equation, and the boundary conditions allow us to determine the coefficients.

Here is the formula for Green's function when zero modes are absent. Let us assume that instead of zero conditions for the function $u(r)$ the following homogeneous boundary conditions of the general form are given: $\mathcal{B}_1 u = 0$, $\mathcal{B}_2 u = 0$, where $\mathcal{B}_{1,2}$ are boundary condition operators, which are some linear combinations of the values of the function and its first derivative on the left and right boundaries. Green's function is constructed with help of the fundamental solution using the formula

$$G(r, r') = \frac{Z(r, r')}{\Delta}; \ \Delta = \begin{vmatrix} \mathcal{B}_1\chi_1 & \mathcal{B}_1\chi_2 \\ \mathcal{B}_2\chi_1 & \mathcal{B}_2\chi_2 \end{vmatrix}, \ Z = \begin{vmatrix} g(r, r') & \chi_1(r) & \chi_2(r) \\ \mathcal{B}_1 g & \mathcal{B}_1\chi_1 & \mathcal{B}_1\chi_2 \\ \mathcal{B}_2 g & \mathcal{B}_2\chi_1 & \mathcal{B}_2\chi_2 \end{vmatrix}.$$

Indeed, the formula gives the sum of the fundamental solution $g(r, r')$ and a linear combination of functions $\chi_{1,2}(r)$ that have no singularities at $r = r'$. After action of the operators $\mathcal{B}_1$ or $\mathcal{B}_2$ on the determinant Z we get the determinant with a pair of coinciding rows, thus $\mathcal{B}_{1,2} Z = 0$. Hence $G(r, r')$ obeys equation and boundary conditions. [1]

[1] How to generalize this formula to the N-th order equation and to the operator with zero modes is discussed in the reference book (Kamke, 1969).

7.3. Find the generalized Green's function of the homogeneous boundary value problem $L = d^2/dx^2, u'(0) = u'(1) = 0$.

Solution. The zero mode in this case is a constant solution, if it is normalized, then $u_0(x) = 1$. Therefore, the Green's function satisfies the equation

$$G''(x, x') = \delta(x - x') - 1. \qquad (7.21)$$

Tha solution of the equation without δ-function is $-x^2/2$, thus we look for G in the form

$$G(x, x') = -\frac{x^2}{2} + \begin{cases} Ax + B, & \text{if } x < x'; \\ Cx + D, & \text{if } x > x'. \end{cases}$$

The boundary conditions allow us to find two coefficients $A = 0, C = 1$. Matching at $x = x'$ gives only one condition $B = x' + D$. The missing condition follows from the requirement of orthogonality to the zero mode

$$\int_0^1 dx\, G(x, x') = -\frac{1}{2}\int_0^1 x^2\, dx + \int_0^{x'} B\, dx + \int_{x'}^1 (x + D)\, dx = 0,$$

which results in $B = x' - x'^2/2 - 1/3$, $D = -x'^2/2 - 1/3$ and

$$G(x, x') = -\frac{1}{3} - \frac{x_>^2 + x_<^2}{2} + x_>.$$

7.4. Under what condition the inhomogeneous problem $u'' = f(x), u'(0) = a, u'(1) = b$ is solvable? Write out the solution.

Solution. Multiply the equation by the zero mode $u_0(x) \equiv 1$ and integrate from 0 to 1. We obtain the solvability condition

$$b - a - \int_0^1 f(x)\, dx = 0, \qquad (7.22)$$

that is, the condition of orthogonality to the zero mode of the conjugate homogeneous problem.

One can clearly understand the algebraic meaning of the condition if, instead of the differential one, one considers the difference equation, as if preparing a discrete model for a numerical solution. To do this, we approximate the first derivative by the difference scheme $u'(x) \approx [u(x + h) - u(x)]/h$, and the second derivative by the scheme

$$u''(x) \approx [u(x - h) - 2u(x) + u(x + h)]/h^2, \quad h = 1/N.$$

Then our problem takes the $(N+1)\times(N+2)$ extended matrix form:

$$A = \left(\begin{array}{cccccccc|l} -1 & 1 & 0 & 0 & 0\ldots0 & 0 & 0 & ah \\ 1 & -2 & 1 & 0 & 0\ldots0 & 0 & 0 & f(h)h^2 \\ 0 & 1 & -2 & 1 & 0\ldots0 & 0 & 0 & f(2h)h^2 \\ 0 & 0 & 1 & -2 & 1\ldots0 & 0 & 0 & f(3h)h^2 \\ . & . & . & . & . \;\; . \;\; . & . & . & \ldots \\ 0 & 0 & 0 & 0 & 0\ldots1 & -2 & 1 & f(1-h)h^2 \\ 0 & 0 & 0 & 0 & 0\ldots0 & 1 & -1 & -bh \end{array}\right).$$

The first and last lines correspond to the boundary conditions, and the rest correspond to the differential equation. A vertical line separates the matrix of the system from the column of the right parts. As it is easy to see, the sum of all the rows of the matrix of the system is zero, which means that the determinant vanishes and zero is an eigenvalue. The eigenvector corresponding to the zero eigenvalue is a discrete analog of the zero mode. For solvability, it is necessary that the sum of the elements of the last column also vanishes

$$h(a-b)+h^2\sum_{k=1}^{N-1} f(kh)=0.$$

Then the rank of the extended matrix of the system N will coincide with the rank of the matrix of the system. Hence, for $N\to\infty$ the condition (7.22) is reproduced.

If the conditions were imposed not on the derivative, but on the function $u(0)=a$, $u(1)=b$, then in the first line would disappear the second unit and, in the last line would disappear the penultimate unit. Then the matrix of the system would be non-degenerate, the zero modes would disappear, and the boundary value problem would become solvable for arbitrary a, b, $f(x)$.

To write out a solution to an inhomogeneous problem, you need to find some simple function that satisfies the boundary conditions $U'(0)=a$, $U'(1)=b$. Let's choose, for example, $U(x)=ax+(b-a)x^2/2$. We will look for a solution in the form $u(x)=v(x)+U(x)$. Then the function $v(x)$ satisfies the equation $v''=f(x)+a-b$ and zero boundary conditions on the derivative $v'(0)=v'(1)=0$. The problem is reduced to the previous one, and the solution is written as

$$u(x)=\int_0^1 G(x,x')f(x')\,dx'+ax+\frac{(b-a)x^2}{2}+C,$$

where an arbitrary constant C is coefficient at zero mode. We did not write $a - b$ under the sign of the integral, using the orthogonality condition of the modified Green function and the zero mode.

If we had chosen another function $U(x)$, the solution would have been

$$u(x) = \int_0^1 G(x, x') \left[f(x') - U''(x') \right] dx' + U(x) + C, \qquad (7.23)$$

which is the same as the previous one up to zero mode. This can be shown by integrating in parts. Let's check that (7.23) is the solution to the problem. To do this, we differentiate (7.23) twice and use the equation for Green's function (7.21). Then we get

$$u''(x) = f(x) - \int_0^1 \left[f(x') - U''(x') \right] dx'.$$

From the solvability condition (7.22) it follows that the integral on the right side of this expression is zero.

7.5. Show that if $\phi(\mathbf{r})$ is a solution of the three-dimensional Laplace equation, then $\psi(\mathbf{r}) = \phi(\mathbf{r}/r^2)/r$ is also a solution.

Solution. In spherical coordinates

$$\Delta_\mathbf{r} \psi(\mathbf{r}) = \left[\frac{1}{r} \frac{\partial^2}{\partial r^2} r + \frac{\Delta_\Omega}{r^2} \right] \psi(\mathbf{r}) = \frac{1}{r} \left[\frac{\partial^2}{\partial r^2} + \frac{\Delta_\Omega}{r^2} \right] \phi \left(\frac{\mathbf{r}}{r^2} \right),$$

Changing the variable $\boldsymbol{\xi} = \mathbf{r}/r^2$ we obtain $\xi = 1/r$,

$$\frac{\partial^2}{\partial r^2} = \xi^4 \frac{\partial^2}{\partial \xi^2} + 2\xi^3 \frac{\partial}{\partial \xi},$$

$\phi(\mathbf{r}/r^2) = \phi(1/r, \theta, \varphi) = \phi(\xi, \theta, \phi)$. Then

$$\Delta_\mathbf{r} \psi(\mathbf{r}) = \xi^5 \left[\frac{\partial^2}{\partial \xi^2} + \frac{2}{\xi} \frac{\partial}{\partial \xi} + \frac{\Delta_\Omega}{\xi^2} \right] \phi(\xi, \theta, \varphi) = \xi^5 \, \Delta_\xi \, \phi(\boldsymbol{\xi}) = 0.$$

7.6. Find Green's function of the two-dimensional Poisson equation $\Delta u = f(\mathbf{r})$ in a disk of radius R, $u|_{r=R} = 0$. Get the solution of the Dirichlet problem for the Laplace equation $\Delta u = 0$ in a disk with a boundary condition $u|_{r=R} = A(\varphi)$, where φ is the angle. Express the solution as a contour integral in the plane of a complex variable.

Solution. We find Green's function of the first kind using the image method by placing a charged "thread" of the opposite sign at the inversion point $r'_* = R^2/r'$ and adding a suitable constant. Then

$$G(r,\varphi;r'\varphi') = \frac{1}{4\pi}\ln\frac{r^2 - 2rr'\cos\psi + r'^2}{r^2r'^2/R^2 - 2rr'\cos\psi + R^2}; \quad \psi = \varphi - \varphi',$$

so $G|_{r=R} = 0$ regardless of r', φ'. The function of the second kind is found by differentiating

$$G_s(r,\varphi,\varphi') = \left.\frac{\partial G}{\partial r'}\right|_{r'=R} = \frac{1}{2\pi R}\frac{R^2 - r^2}{R^2 - 2Rr\cos\psi + r^2}.$$

The solution to the Dirichlet problem is given by the *Poisson formula*

$$u(r,\varphi) = \frac{1}{2\pi}\int_0^{2\pi}\frac{R^2 - r^2}{R^2 + r^2 - 2Rr\cos(\varphi - \varphi')}A(\varphi')\,d\varphi'. \tag{7.24}$$

The problem "Find an analytic function in a disk whose real part takes the value $f(z)$ at the boundary of the disk" is solved using the *Schwartz formula*

$$w(z) = \frac{1}{2\pi\mathrm{i}}\oint\frac{d\zeta}{\zeta}\frac{\zeta + z}{\zeta - z}f(\zeta) + \mathrm{i}C, \tag{7.25}$$

the real part of which is reduced to (7.24) by replacing $z = re^{\mathrm{i}\varphi}$, $\zeta = Re^{\mathrm{i}\varphi'}$. Then $|\zeta - z|^2 = R^2 + r^2 - 2Rr\cos\psi$ gives denominator of the Green's function and $(|\zeta|^2 - |z|^2)/|\zeta - z|^2 = \mathrm{Re}\,(\zeta + z)/(\zeta - z)$ gives the Green's function G_s itself. The harmonic function $u(z)$ is the real part of the analytic function $w(z)$.

7.7. A pipe of radius R and infinite length is placed in the ground to a depth of h and maintained at a constant temperature T_0. Find the temperature distribution if on the earth's surface $T = 0$.

Solution. The temperature distribution obeys the Laplace equation $\Delta T = 0$. Since T does not depend on the coordinate along the pipe, the area in which the Laplace equation is to be solved is two-dimensional (Fig. 7.3). Let's use the method of conformal transformations. With the help of the linear-fractional function

$$\zeta = \frac{z - c}{z + c}, \quad z = x + \mathrm{i}y \tag{7.26}$$

you can map the area in question to a ring. The line $z = \mathrm{i}y$ will move to the unit circle $|\zeta|^2 = 1$ if c is a real number. The surface of the pipe will also become a circle $|\zeta|^2 = a^2$ of radius $a = \sqrt{(h-c)/(h+c)} < 1$, if

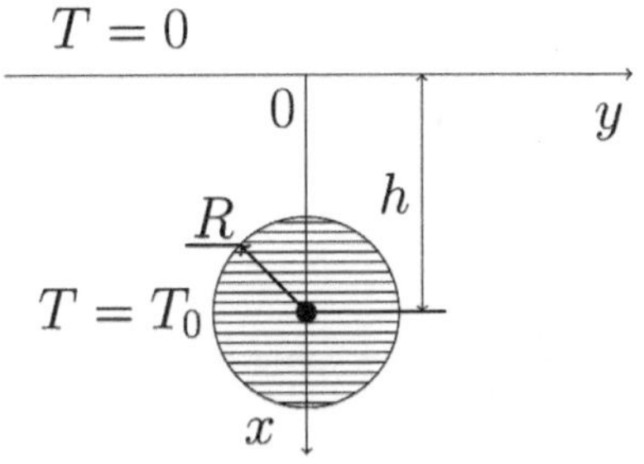

Figure 7.3 Two-dimensional domain to problem **7.7**.

$c = \sqrt{h^2 - R^2}$. This can be verified by substituting $z = h + Re^{i\varphi}$ into the fractional-linear transformation (7.26). The two-dimensional Laplace equation is covariant with respect to conformal transformations, so in the new variables the temperature also satisfies the Laplace equation. The conditions at the ring boundaries will remain the same

$$T\,|_{|\zeta|=1} = 0, \qquad T\,|_{|\zeta|=a} = T_0.$$

Since the boundary conditions do not depend on the angle φ in polar coordinates, the problem can be reduced to one-dimensional one. The two-dimensional Laplace equation has two angle-independent solutions: $\ln|\zeta|$ and 1. Their combination satisfying the boundary conditions has the form

$$T(\zeta) = T_0 \frac{\ln|\zeta|}{\ln a}.$$

We find the solution to the problem by returning to the initial variables $\zeta = (x + iy - c)/(x + iy + c)$:

$$T(x, y) = \frac{T_0}{2} \frac{\ln \frac{(x+c)^2+y^2}{(x-c)^2+y^2}}{\ln \frac{h+c}{R}}.$$

7.8. Find Green's function of the three-dimensional Helmholtz equation

$$\Delta u + k_0^2 u = f(\mathbf{r}),$$

satisfying the Sommerfeld radiation condition (there is no converging wave coming from infinity) $u|_{r\to\infty} \sim e^{ik_0 r}/r$.

Solution. The coefficients of the operator do not depend on $\mathbf{r}$, so the equation

$$(\Delta + k_0^2)G(\mathbf{r}, \mathbf{r}') = \delta(\mathbf{r} - \mathbf{r}')$$

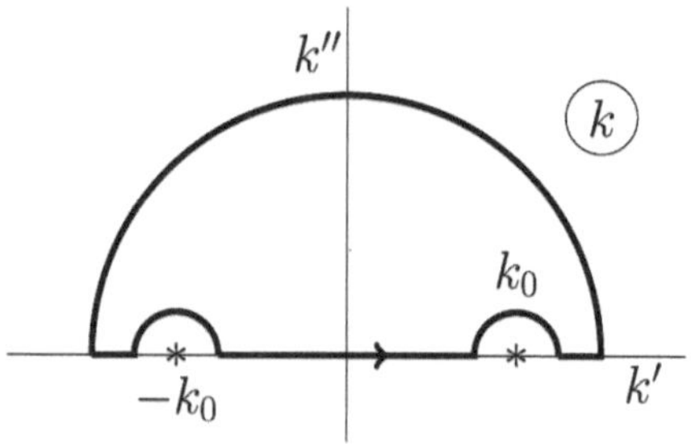

Figure 7.4 The integration contour for the Helmholtz equation.

equation can be reduced to an algebraic one by using the Fourier transform on the variable $\mathbf{r} - \mathbf{r}'$:

$$(k_0^2 - k^2)G_{\mathbf{k}} = 1.$$

Green's function $G(\mathbf{r}, \mathbf{r}')$ depends only on the difference $\mathbf{r} - \mathbf{r}'$, so it is equal to $G(\mathbf{r} - \mathbf{r}', 0)$ and can be found using the inverse transformation:

$$G(\mathbf{r}, 0) = \int \frac{d\mathbf{k}}{(2\pi)^3} \frac{\mathrm{e}^{i\mathbf{kr}}}{k_0^2 - k^2}.$$

It is convenient to perform the integration in spherical coordinates directing the z axis along the vector $\mathbf{r}$. Integration with respect to φ gives 2π and after integrating over θ a simple integral remains

$$G(\mathbf{r}, 0) = \frac{1}{ir} \frac{1}{(2\pi)^2} \int_{-\infty}^{\infty} \frac{k\mathrm{e}^{ikr}\, dk}{k_0^2 - k^2}.$$

The poles of the integrand lie on the integration contour, so you need to choose a rule for bypassing them based on the boundary conditions. To obtain a function that satisfies the radiation condition, we must leave the divergent wave and discard the convergent one. Since r is a positive value, the contour should be closed from above. So that the integral reduces to a residue at the point $k = k_0$, select the contour shown in fig. 7.4.

$$G^{(+)}(\mathbf{r}, \mathbf{r}') = -\frac{\mathrm{e}^{ik_0|\mathbf{r}-\mathbf{r}'|}}{4\pi|\mathbf{r} - \mathbf{r}'|}.$$

Such a Green function in the form of a divergent spherical wave is used, for example, in problems of diffraction and scattering theory (see problem **7.39**).

7.9. Find Green's functions of the three-dimensional heat equation $u_t = \Delta u$ and the nonstationary free Schrödinger equation $\mathrm{i}u_t = -\Delta u/2$.

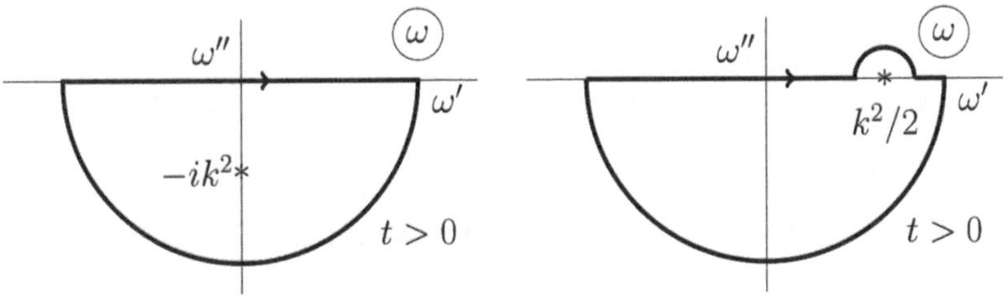

Figure 7.5 Integration contours for the heat conduction equation (left) and Schrödinger equation (right).

Solution. Since the coefficients of the equations are constant, the Green's function can only depend on the time and coordinate differences. Then the equations for Green's functions can be written as

$$G_t - \triangle G = \delta(\mathbf{r})\delta(t), \quad \mathrm{i}G_t + \frac{1}{2} \triangle G = \delta(\mathbf{r})\delta(t),$$

and in the final expression return the original arguments, i.e., perform a replacement $\mathbf{r} \to \mathbf{r} - \mathbf{r}'$, $t \to t - t'$.

The Fourier transform

$$u_{\mathbf{k}\omega} = \int u(\mathbf{r}, t)\mathrm{e}^{\mathrm{i}\omega t - \mathrm{i}\mathbf{k}\mathbf{r}}\, d\mathbf{r}dt, \quad u(\mathbf{r}, t) = \int u_{\mathbf{k}\omega}\mathrm{e}^{-\mathrm{i}\omega t + \mathrm{i}\mathbf{k}\mathbf{r}} \frac{d\mathbf{k}d\omega}{(2\pi)^4}$$

reduces the equations to the form

$$\left(-\mathrm{i}\omega + k^2\right) G_{\mathbf{k}\omega} = 1, \quad \left(\omega - k^2/2\right) G_{\mathbf{k}\omega} = 1.$$

The inverse Fourier transform for the heat equation can be performed by integrating in the plane of the complex variable ω. For $t < 0$, the contour should be closed in the upper half-plane, so that the integral over the infinitely distant semicircle disappears. The only pole of the integrand is located outside the area covered by the contour, so at $t < 0$, Green's function vanishes. At $t > 0$, we close the contour in the lower half-plane (Fig. 7.5, left contour). The integral is given by the residue

$$G_{\mathbf{k}}(t) = \int \frac{d\omega}{2\pi} \frac{\mathrm{e}^{-\mathrm{i}\omega t}}{-\mathrm{i}\omega + k^2} = -\mathrm{i} \operatorname*{Res}_{\omega=-\mathrm{i}k^2} \frac{\mathrm{e}^{-\mathrm{i}\omega t}}{-\mathrm{i}\omega + k^2} = \mathrm{e}^{-k^2 t}.$$

The transformation by $\mathbf{k}$ reduces to the calculation of the Gaussian integral and gives

$$G(\mathbf{r}, t; \mathbf{r}', 0) = \frac{\theta(t)}{(4\pi t)^{3/2}} \exp\left[-(\mathbf{r} - \mathbf{r}')^2/4t\right].$$

In the case of the Schrödinger equation, the pole falls on the real axis, so for the convergence of the integral the integration contour should be shifted from the real axis, i.e., the pole bypass rule should be chosen. An additional consideration that allows you to choose from two possibilities is the causality principle which is the condition for Green's function to vanish when $t < 0$. Then when $t > 0$ the contour is selected according to Fig. 7.5, and the integral is equal to

$$G_{\mathbf{k}}(t) = -\mathrm{i} \operatorname*{Res}_{\omega=k^2/2} \frac{\mathrm{e}^{-\mathrm{i}\omega t}}{\omega - k^2/2} = -\mathrm{i}\mathrm{e}^{-\mathrm{i}k^2 t/2}.$$

After the inverse Fourier transform in $\mathbf{k}$ we get the retarded Green function:

$$G(\mathbf{r}, t; \mathbf{r}', 0) = \frac{\theta(t)}{(2\pi \mathrm{i}t)^{3/2}} \exp\left[\mathrm{i}(\mathbf{r} - \mathbf{r}')^2/2t\right], \tag{7.27}$$

which describes the spreading of a wave packet localized at the initial moment. The retarded Green's function allows us to solve the Cauchy problem with initial conditions. This Green function can be obtained also by using the Laplace transform which avoids the problem of pole bypass:

$$\left(\mathrm{i}p + k^2/2\right) G_{p\mathbf{k}} = 1, \quad G_{p\mathbf{k}} = \int_0^\infty G_{\mathbf{k}}(t)\mathrm{e}^{-pt}\, dt.$$

The inverse Laplace transform is defined by the contour integral

$$G_{\mathbf{k}}(t) = \frac{1}{2\pi \mathrm{i}} \oint G_{p\mathbf{k}} \mathrm{e}^{pt}\, dp.$$

which fixes unambiguously that at $t > 0$ all the poles in the variable p must be bypassed from the right. As a result, we arrive at the expression (7.27).

Some physical problems require a function that vanishes when $t > 0$. To get it, the pole must be bypassed from below and the resulting *advanced* Green function is $G^{(-)} \propto \theta(-t)$.

7.10. Find Green's function $G(x, x', t)$ of a one-dimensional Fokker–Planck equation:

$$\frac{\partial u}{\partial t} = \frac{\partial}{\partial x}(xu) + \frac{1}{2}\frac{\partial^2 u}{\partial x^2}, \quad u(x, 0) = g(x).$$

Solution. The solution of the Cauchy problem

$$G_t = G + xG_x + \frac{1}{2}G_{xx}, \quad G(x, x', 0) = \delta(x - x') \tag{7.28}$$

for the retarded Green function G can be found with the help of the Gaussian ansatz:

$$G = C(t)\exp\left[-\frac{(x-a(t))^2}{2D(t)}\right]. \tag{7.29}$$

Fokker–Planck equations conserve the integral over space which is simply the number of particles:

$$\int G(x, x', t)\,dx = 1,$$

$C(t) = (2\pi D(t))^{-1/2}$. Substituting (7.29) into (7.28) and equating the coefficients at x, x^2, we arrive at ordinary differential equations for the mean value $a(t)$ and the dispersion $D(t)$:

$$\dot{a} = -a, \quad \dot{D} = 1 - 2D.$$

The initial conditions $a(0) = x'$, $D(0) = 0$ correspond to the δ-function form of $G(x, x', t)$ at $t \to +0$. Finally we obtain:

$$G(x, x', t) = \frac{1}{\sqrt{\pi}}\frac{\exp\left[-\left(x - x'e^{-t}\right)^2/\left(1 - e^{-2t}\right)\right]}{\sqrt{1 - e^{-2t}}}.$$

At small times $t \ll 1$ this expression coincides with Green's function of the diffusion equation:

$$G(x, x', t) = \frac{\exp\left[-(x - x')^2/2t\right]}{\sqrt{2\pi t}}.$$

At large times $t \gg 1$ it does not depend on the variable x':

$$G(x, x', t) = \frac{\exp\left(-x^2\right)}{\sqrt{\pi}}.$$

The Fokker–Planck equation describes a random walk of particles and this kind of asymptotics at large times means that after several collisions the Brownian particle "forgets" the initial conditions.

This problem can also be solved using the Fourier transform in the coordinate x which lowers the order of the equation from the second to the first one. The resulting first order partial differential equation can be solved by the method of characteristics (see Chapter 2).

7.11. Find Green's function of a one-dimensional wave equation

$$\frac{1}{c^2}u_{tt} - u_{xx} = f(x, t), \quad u(x, t) \to 0 \quad \text{at} \quad t \to -\infty.$$

Solution. After the Fourier transform in coordinate and time, we get Green's function in the ω, **k**-representation

$$G_{k\omega} = \frac{1}{k^2 - \omega^2/c^2}.$$

Let us perform the inverse Fourier transform in the variable ω bypassing the poles according to the causality condition:

$$G_k(t) = \theta(t)\frac{\mathrm{i}c}{2k}\left(\mathrm{e}^{-ikct} - \mathrm{e}^{ikct}\right).$$

The inverse transformation in k gives:

$$G(x, t; 0, 0) = \frac{c}{2}\left[\theta(x + ct) - \theta(x - ct)\right]\theta(t). \tag{7.30}$$

This means that the local perturbation at the point x', which acted at the time t', has an effect only within the interval $x' - c(t - t') < x < x' + c(t - t')$ due to the finite velocity c of perturbations propagation (*influence zone*).

7.12. Show that the solution of the Cauchy problem

$$\Box u = 0, \qquad u(\mathbf{r}, 0) = \varphi(\mathbf{r}), \quad u_t(\mathbf{r}, 0) = \psi(\mathbf{r}) \tag{7.31}$$

for the wave equation can be expressed in terms of the corresponding Green function by the following formula:

$$u(\mathbf{r}, t) = \frac{1}{c^2}\int\left[\frac{\partial}{\partial t}G(\mathbf{r}, t; \mathbf{r}', 0)\varphi(\mathbf{r}') + G(\mathbf{r}, t; \mathbf{r}', 0)\psi(\mathbf{r}')\right] d\mathbf{r}'.$$

Solution. The wave equation has a second order in time, so the corresponding Cauchy problem contains initial conditions both for the function and the first derivative. Hence, there are a pair of Green's functions of the second kind $G_s^{(1)}$ and $G_s^{(2)}$, and the solution is given by the formula

$$u(\mathbf{r}, t) = \int G_s^{(1)}(\mathbf{r}, \mathbf{r}', t)\varphi(\mathbf{r}')\,d\mathbf{r}' + \int G_s^{(2)}(\mathbf{r}, \mathbf{r}', t)\psi(\mathbf{r}')\,d\mathbf{r}'. \tag{7.32}$$

Let us find the connection of the functions $G_s^{(1)}$, $G_s^{(2)}$ with Green's function of the first kind obeying the equation

$$\Box G(\mathbf{r}, t; \mathbf{r}', t') = \delta(\mathbf{r} - \mathbf{r}')\delta(t - t') \tag{7.33}$$

and diminishing together with its first derivatives at $t \to \infty$ and $r \to \infty$. Multiplying Eq. (7.33) by $u(\mathbf{r}, t)$ and subtracting then (7.31) multiplied by $G(\mathbf{r}, t; \mathbf{r}', t')$ we get after integration over $\mathbf{r}$, t for $t' > 0$:

$$u(\mathbf{r}', t') = \int d\mathbf{r}\,dt\,(u\Box G - G\Box u) =$$

$$= \int d\mathbf{r}\int dt\,\frac{1}{c^2}\frac{\partial}{\partial t}\left(u\frac{\partial G}{\partial t} - G\frac{\partial u}{\partial t}\right) - \int dt\int d\mathbf{r}\,\mathrm{div}\,(u\nabla G - G\nabla u).$$

The integral of the divergence is transformed into an integral over an infinitely distant surface and vanishes. In the first term the integral over t is determined by the contribution at $t = 0$:

$$u(\mathbf{r}', t') = \frac{1}{c^2} \int d\mathbf{r} \left[-\frac{\partial G}{\partial t} \varphi(\mathbf{r}) + G\psi(\mathbf{r}) \right] \Bigg|_{t=0} .$$

Changing the definition of variables $\mathbf{r} \leftrightarrow \mathbf{r}'$, $t \leftrightarrow t'$ we obtain:

$$u(\mathbf{r}, t) = \frac{1}{c^2} \int d\mathbf{r}' \left[\frac{\partial G}{\partial t}(\mathbf{r}, t; \mathbf{r}', 0)\varphi(\mathbf{r}') + G(\mathbf{r}, t; \mathbf{r}', 0)\psi(\mathbf{r}') \right] .$$

Comparing this result with (7.32) we obtain the desired relation:

$$G_s^{(1)}(\mathbf{r}, \mathbf{r}', t) = \frac{1}{c^2} \frac{\partial}{\partial t} G(\mathbf{r}, t; \mathbf{r}', 0), \quad G_s^{(2)}(\mathbf{r}, \mathbf{r}', t) = \frac{1}{c^2} G(\mathbf{r}, t; \mathbf{r}', 0).$$

7.13. Knowing Green's function of a one-dimensional wave equation (see problem **7.11**), solve the Cauchy problem

$$\frac{1}{c^2} u_{tt} - u_{xx} = 0; \quad u(x, 0) = \varphi(x), u_t(x, 0) = \psi(x).$$

Solution. The solution of the Cauchy problem can be written as the sum of two integrals (problem **7.12**)

$$u(x, t) = \int_{-\infty}^{\infty} G_s^{(1)}(x, x', t)\varphi(x')\, dx' + \int_{-\infty}^{\infty} G_s^{(2)}(x, x', t)\psi(x')\, dx'. \qquad (7.34)$$

Using the expression (7.30) for Green's function of a one-dimensional equation, we get

$$G_s^{(1)}(x, x', t) = \frac{1}{c^2} \frac{\partial}{\partial t} G(x, t; x', 0) = \frac{1}{2} \left[\delta(x - x' + ct) + \delta(x - x' - ct) \right],$$
$$G_s^{(2)}(x, x', t) = \frac{1}{c^2} G(x, t; x', 0) = \frac{1}{2} \left[\theta(x - x' + ct) - \theta(x - x' - ct) \right].$$

Hence follows the *Dalembert formula*

$$u(x, t) = \frac{1}{2} \left[\varphi(x + ct) + \varphi(x - ct) \right] + \frac{1}{2c} \int_{x-ct}^{x+ct} \psi(x')\, dx'. \qquad (7.35)$$

The solution at the point x is determined by the values of the functions φ at the points $x' = x \pm ct$ and ψ on the interval $x' \in (x - ct, x + ct)$, which is called *dependency domain*. **7.14**. Find Green's functions of the second kind and write out the solution of the Cauchy problem for a three-dimensional wave equation. Show that the solution of $u(\mathbf{r}, t)$ at the point $\mathbf{r}$ is completely determined by the values of the functions

$$\phi(\mathbf{r}) = u(\mathbf{r}, 0), \quad \psi(\mathbf{r}) = u_t(\mathbf{r}, 0)$$

and the normal derivative $\partial\phi/\partial n$ on the sphere of the radius ct with the center in the point $\mathbf{r}$ (*the Huygens principle*).

Solution. Using Green's function of the first kind (problem **7.48**) and the result of problem **7.12**, we find Green's functions of the second kind

$$G_s^{(1)} = -\frac{\delta'(\rho - ct)}{4\pi\rho}, \quad G_s^{(2)} = \frac{\delta(\rho - ct)}{4\pi\rho c},$$

where $\rho = |\mathbf{r} - \mathbf{r}'|$. The solution of the Cauchy problem for the wave equation

$$u(\mathbf{r}, t) = \frac{1}{4\pi}\int\left[-\frac{\delta'(\rho - ct)}{\rho}\varphi(\mathbf{r}') + \frac{\delta(\rho - ct)}{\rho c}\psi(\mathbf{r}')\right] d\mathbf{r}'$$

can be rewritten through the functional M acting on the functions of three variables $\mu(\mathbf{r})$ which is simply the averaging over a unit sphere

$$M[\mu(\mathbf{r})] = \frac{1}{4\pi}\int\limits_{|\boldsymbol{\xi}|=1} \mu(\mathbf{r} + ct\boldsymbol{\xi})\, d^2 S_{\xi}.$$

It turns out the *Kirchhoff formula*

$$u(\mathbf{r}, t) = \frac{\partial}{\partial t}(tM[\varphi]) + tM[\psi].$$

7.15. Find Green's function of the Helmholtz equation decreasing at infinity

$$\left(\triangle_D - q^2\right) G(x, x') = \delta(x - x'), \tag{7.36}$$

where $\triangle_D$ is the D- dimensional Laplacian

Solution. Using the Fourier transform , we find

$$G(x, 0) = -\int \frac{d^D k}{(2\pi)^D}\frac{e^{ikx}}{k^2 + q^2}.$$

The denominator can be written as

$$\frac{1}{k^2 + q^2} = \int_0^{\infty} dt\, e^{-(k^2+q^2)t}$$

Changing the integration order we compute first the Gaussian D-dimensional integral with the use the formula

$$\int d^D y \exp\left[-\frac{1}{2}(y, Ay) + i(b, y)\right] = \frac{(2\pi)^{D/2}}{(\det A)^{1/2}} \exp\left[-\frac{1}{2}(b, A^{-1}b)\right],$$

where A is the $D \times D$ matrix and b is D-dimensional vector. As a result, we get

$$\int d^D k\, e^{-k^2 t + ikx} = \left(\frac{\pi}{t}\right)^{D/2} e^{-x^2/4t}.$$

It remains to calculate the integral over t, which reduces to the integral representation of the MacDonald function (A.31)

$$G(x,0) = -\frac{1}{(4\pi)^{D/2}} \int_0^\infty ds\, s^{D/2-2} e^{-x^2 s/4 - q^2/s} = -\frac{q^\alpha}{(2\pi)^{D/2}} \frac{K_\alpha(q|x|)}{|x|^\alpha},$$

where $\alpha = D/2 - 1$, $s = 1/t$. Hence

$$G(x,x') = -\frac{q^\alpha}{(2\pi)^{D/2}} \frac{K_\alpha(q|x-x'|)}{|x-x'|^\alpha}.$$

Let's consider special cases. If $D = 2$, then $\alpha = 0$, and the behavior of the MacDonald function of the zero order for small argument values is logarithmic (A.33):

$$G(r,0) \sim \frac{1}{2\pi} \ln r,$$

For $D > 2$ we get at $q|x| \ll 1$

$$G(r,0) \sim -\frac{\Gamma(\alpha)}{4\pi^{D/2} r^{D-2}}.$$

In particular

$$G(r,0) \sim \begin{cases} -\frac{1}{4\pi r}, & D = 3; \\ -\frac{1}{4\pi^2 r^2}, & D = 4. \end{cases}$$

At odd dimensions $D = 3, 5, \ldots$ the function K_α is expressed through elementary functions. Let's take a closer look at a special case $D = 4$: $x = (x_0, x_1, x_2, x_3)$. If we substitute $x_0 \to it$ (that is we rotate on the angle $\pi/2$ in the x_0 complex plane) and instead of q we write m, then Eq. (7.36) takes the form of the Klein–Gordon–Fock equation

$$\left(\Box + m^2\right) G(x,0) = -\delta(x),$$

which describes the motion of a relativistic spinless particle of mass m.

The so-called *Feynman* Green function has the form

$$G(x,0) = -\frac{m K_1(mz)}{4\pi^2 z}, \quad z = \sqrt{r^2 - t^2},$$

where $r^2 = x_1^2 + x_2^2 + x_3^2$, $c = 1$. When the interval is space-like one ($r^2 > t^2$), then the variable z is a real number and the function decays exponentially in

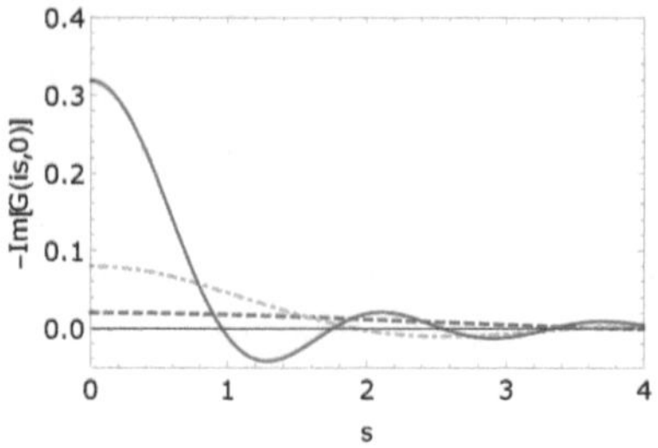

Figure 7.6 The imaginary part of the Feynman Green function as a function of $s = \sqrt{t^2 - r^2}$ time-like interval for three different masses: 1—m, 2—$2m$, 3—$4m$.

the domain $r > t$: $K_1(mz) \simeq \sqrt{\pi/2mz}\mathrm{e}^{-mz}$. If the interval is time-like one ($r^2 < t^2$), then the variable $z \simeq \mathrm{i}\sqrt{t^2 - r^2}$ is imaginary and the function G oscillates (see pic. 7.6). In the non-relativistic limit $r^2 \ll t^2$, $mt \gg 1$ we obtain

$$G(x, 0) \simeq \frac{m}{2(2\pi \mathrm{i}t)^{3/2}}\mathrm{e}^{-\mathrm{i}m(t - r^2/2t)},$$

With accuracy up to a constant factor, we arrive at Green's function of the Schrödinger equation (7.27).

7.16. Find the continuous spectrum of the operator $A = -\mathrm{i}d/dx$, $x \in \mathbb{R}$.

Solution. To show that the eigenfunction $u(x) = \exp(\mathrm{i}\lambda x)$ of the operator $A = -\mathrm{i}d/dx$ belongs to a continuous spectrum, let's make sure that $Au = \lambda u$. It remains to construct an approximate normalized eigenfunction $\tilde{u}$ and make sure that the norm $\|A\tilde{u} - \lambda\tilde{u}\|$ tends to zero. Let's choose an approximate function in the form of a wave packet, for example $\tilde{u}(x) = \alpha(x)\exp(\mathrm{i}\lambda x)$, $\alpha(x) = \varepsilon^{1/4}\pi^{-1/4}\exp(-\varepsilon x^2/2)$, where $\|\alpha\| = 1$. Find

$$\|A\tilde{u} - \lambda\tilde{u}\|^2 = \int_{-\infty}^{\infty} \alpha'^2(x)\, dx = \varepsilon/2 \to 0, \ \ \varepsilon \to 0.$$

Thus, the continuous spectrum coincides with the entire real axis $\mathbb{R}$.

7.17. Find the resolvent of the operator $-d^2/dx^2$.

Solution. The equation

$$\left(z + \frac{d^2}{dx^2}\right) R_z(x, x') = \delta(x - x')$$

with zero boundary conditions at infinity has the solution

$$R_z(x, x') = \frac{1}{2i\sqrt{z}} e^{i\sqrt{z}|x-x'|}.$$

Here the main branch of the function $\sqrt{z}$: $\operatorname{Im}\sqrt{z} > 0$ is assumed. The resolvent is analytical in the whole z-planes with the cut along the real positive semiaxis $\mathbb{R}_+ = \{z \in \mathbb{R},\ z > 0\}$. The jump on the cut is determined by the values of the square root of $z = k^2 \pm i\epsilon$ on the edges of the cut

$$k > 0, \quad \lim_{\epsilon \to +0} \sqrt{k^2 + i\epsilon} = k, \quad \lim_{\epsilon \to +0} \sqrt{k^2 - i\epsilon} = -k$$

and it is equal to

$$R_{k^2+i0} - R_{k^2-i0} = \frac{1}{ik} \cos k(x - x') = -\frac{i}{k} (\cos kx \cos kx' + \sin kx \sin kx').$$

Comparing with (7.19), we get that the operator $-d^2/dx^2$ has only a continuous spectrum $z = k^2 > 0$, each eigenvalue is twice degenerate and a possible basis normalized according to (7.14) in a two-dimensional eigenspace corresponding to the eigenvalues of $z = k^2$ consists of a pair of functions

$$\psi_k^{(1)}(x) = \frac{\cos kx}{\sqrt{2\pi k}}, \quad \psi_k^{(2)}(x) = \frac{\sin kx}{\sqrt{2\pi k}}.$$

7.18. Find the resolvent of the radial Laplace operator

$$\mathcal{L}_r = -\frac{1}{r}\frac{\partial}{\partial r} r \frac{\partial}{\partial r}$$

in the functional space with the scalar product $(v, u) = \int r\, dr\, v^*(r) u(r)$.

Solution. The resolvent obeys the equation

$$\left(\frac{1}{r}\frac{d}{dr} r \frac{d}{dr} + z\right) g_z^0(r, r') = \frac{\delta(r - r')}{r},$$

from where, by integrating over rdr, it turns out that the jump of the derivative at $r = r'$ is equal to $1/r'$. Two linearly independent solutions are $\chi_1(r) = J_0(kr)$, satisfying the regularity condition for $kr \to 0$, and $\chi_2(r) = J_0(kr) + iY_0(kr)$ taking form of a divergent cylindrical wave at $kr \to \infty$, where $k = \sqrt{z}$.

The asymptotics of the second solution can be checked using the definition of the Neumann function

$$Y_0(x) = \lim_{\nu \to 0} \frac{J_\nu(x) \cos \pi\nu - J_{-\nu}(x)}{\sin \pi\nu}. \tag{7.37}$$

We arrive at

$$J_0(x) \sim \sqrt{\frac{2}{\pi x}} \cos\left(x - \frac{\pi}{4}\right), \quad Y_0(x) \sim \sqrt{\frac{2}{\pi x}} \sin\left(x - \frac{\pi}{4}\right),$$

$$\chi_2(r) \sim \sqrt{\frac{2}{\pi \mathrm{i} kr}} \mathrm{e}^{\mathrm{i}kr}.$$

The Wronskian can be found from the asymptotics for $r \to \infty$: $W(r') = 2\mathrm{i}/\pi r'$. Acting similarly to problem (**7.2**), we get the resolvent

$$g_z^0(r, r') = \frac{\pi}{2\mathrm{i}} J_0(kr_<) \left[J_0(kr_>) + \mathrm{i}Y_0(kr_>)\right].$$

The branch of $\sqrt{z}$ in the plane z, cut by $\mathbb{R}_+$, we choose from the condition $\sqrt{1} = +1$ on the upper edge of the cut. When passing through the cut, the even function $J_0(x)$ does not change, and the jump Y_0 is found from formula $J_\nu(-x) = \mathrm{e}^{\mathrm{i}\pi\nu} J_\nu(x)$ and definitions (A.21):

$$Y_0(+x) - Y_0(-x) = -2\mathrm{i}J_0(x).$$

Comparing the expression

$$g^0_{k^2+\mathrm{i}\epsilon} - g^0_{k^2-\mathrm{i}\epsilon} = -\mathrm{i}\pi J_0(kr) J_0(kr')$$

with (7.19) we find eigenfunctions of a continuous spectrum

$$\psi_k(r) = \frac{1}{\sqrt{2}} J_0(kr).$$

We emphasize that these functions are normalized by the condition

$$\int \psi_k^*(r) \psi_{k'}(r)\, r dr = \delta(k^2 - k'^2).$$

7.19. Find the resolvent of the operator $-\Delta$ in three dimensions in spherical coordinates and construct a spherically symmetric normalized eigenfunction.

Solution. The equation

$$(z + \Delta)\, R_z(\mathbf{r}, \mathbf{r}') = \delta(\mathbf{r} - \mathbf{r}')$$

is solved by the Fourier transform and it is equivalent to the Helmholtz equation with the boundary conditions $R_z \to 0$ at $\operatorname{Im}\sqrt{z} > 0$ and $r \to \infty$ (problem **7.8**). We get

$$R_z(\mathbf{r}, \mathbf{r}') = -\frac{\mathrm{e}^{\mathrm{i}\sqrt{z}|\mathbf{r}-\mathbf{r}'|}}{4\pi\, |\mathbf{r} - \mathbf{r}'|}.$$

The resolvent is defined in a plane with a cut along $\mathbb{R}_+$.

It is possible to expand the resolvent in a series in spherical harmonics

$$R_z(\mathbf{r}, \mathbf{r}') = \sum_{l,m} g^{lm}(r, r') Y_{lm}(\mathbf{n}) Y^*_{lm}(\mathbf{n}'),$$

where $\mathbf{n} = \mathbf{r}/r$, $\mathbf{n}' = \mathbf{r}'/r$ are unit vectors. We are only interested in the coefficient at $Y_{00}^2 = 1/4\pi$, which is found by integrating over all four angles in spherical coordinates $\mathbf{n} = (\theta, \varphi)$, $\mathbf{n}' = (\theta', \varphi')$. Let's move on to the integration by the angle ψ between the vectors $\mathbf{n}$ and $\mathbf{n}'$

$$g_z^{00}(r, r') = 2\pi \int_0^{\pi} d\psi \, \sin\psi \, R_z(|\mathbf{r} - \mathbf{r}'|),$$

and then to the variable $t = |\mathbf{r} - \mathbf{r}'|^2 = r^2 + r'^2 - 2rr' \cos\psi$

$$g^{00}(r, r') = -\frac{1}{4rr'} \int_{(r-r')^2}^{(r+r')^2} dt \, \frac{e^{i\sqrt{zt}}}{\sqrt{t}} =$$

$$= -\frac{e^{i\sqrt{z}(r+r')} - e^{i\sqrt{z}|r-r'|}}{2rr' i\sqrt{z}} = -\frac{e^{i\sqrt{z}r_>} \sin\sqrt{z}r_<}{rr'\sqrt{z}}.$$

The function g_z^{00} is analytic in the plane z cut along the positive real semiaxis $\mathbb{R}_+$. Jump on the cut $z = k^2$

$$g^{00}_{k^2+i0} - g^{00}_{k^2-i0} = -\frac{e^{ikr_>} \sin kr_< - e^{-ikr_>} \sin kr_<}{krr'} = -\frac{2i \sin kr \sin kr'}{krr'}$$

we equate to $-2\pi i \psi_k(r)\psi_k(r')$. This is how we can find spherically symmetric eigenfunctions of a continuous spectrum:

$$\psi_k(\mathbf{r}) = \frac{\sin kr}{\sqrt{\pi} kr} Y_{00}(\mathbf{n}).$$

We emphasize that these functions are normalized by the condition

$$\int \psi_k^*(\mathbf{r}) \psi_{k'}(\mathbf{r}) \, d\mathbf{r} = \delta(k^2 - k'^2),$$

where the integral is taken over the entire three-dimensional space.

7.20. Find the value of the resolvent $R_z(x, x')$ of the operator

$$\hat{H} = -\frac{d^2}{dx^2} + G\left[\delta(x + a) + \delta(x - a)\right],$$

at $x = 0$ and $x' = 0$. How does $R_z(0, 0)$ behave when $Ga \gg 1$?

Solution. The resolvent satisfies the equation

$$(z - \hat{H})R_z(x, x') = \delta(x - x')$$

and can be represented in the form (7.20)

$$R_z(x, x') = \frac{v(x_<)u(x_>)}{W},$$

where $u(x)$ and $v(x)$ are solutions of the homogeneous problem

$$(z - \hat{H})u(x) = (z - \hat{H})v(x) = 0$$

with the conditions $u(x) \to 0$ at $x \to +\infty$ and $v(x) \to 0$ at $x \to -\infty$, $W = v(x)u'(x) - u(x)v'(x)$ is the Wronskian of the solutions.

In the intervals $x < -a$, $x > a$ and $-a < x < a$, the homogeneous problems look simple:

$$\left(z + \frac{d^2}{dx^2}\right)u(x) = \left(z + \frac{d^2}{dx^2}\right)v(x) = 0$$

and their solutions are there linear superpositions of exponents $e^{\pm ix\sqrt{z}}$. As a function of the parameter z they are single-valued in the complex plane with a cut from $z = 0$ to an infinitely distant point. The asymptotic conditions for $x \to \pm\infty$ can be satisfied if the imaginary part of $\sqrt{z}$ has a fixed sign. This is so if the cut is drawn along $\mathbb{R}_+$.

Let's choose in the complex plane such a branch $\sqrt{z}$ that $\mathrm{Im}\sqrt{z} > 0$, then

$$u(x) = e^{ix\sqrt{z}}, \quad x > a;$$
$$v(x) = e^{-ix\sqrt{z}}, \; x < -a.$$

The values of $u(x)$ and $v(x)$ for $-a < x < a$ (coefficients in linear combinations of exponents $e^{\pm ix\sqrt{z}}$) are determined from the requirement of continuity of these functions for $x = \pm a$ and jumps of the derivative

$$u'(a + 0) - u'(a - 0) = Gu(a),$$
$$v'(-a + 0) - v'(-a - 0) = Gv(-a).$$

As a result, for $-a < x < a$ we get

$$u(x) = \left(1 + \frac{iG}{2\sqrt{z}}\right)e^{ix\sqrt{z}} - \frac{iG}{2\sqrt{z}}e^{-ix\sqrt{z}+2ia\sqrt{z}},$$
$$v(x) = \left(1 + \frac{iG}{2\sqrt{z}}\right)e^{-ix\sqrt{z}} - \frac{iG}{2\sqrt{z}}e^{ix\sqrt{z}+2ia\sqrt{z}},$$

and the Wronskian has the form:

$$W = 2i\sqrt{z}\left[1 + \frac{iG}{2\sqrt{z}}\left(1 - e^{2ia\sqrt{z}}\right)\right]\left[1 + \frac{iG}{2\sqrt{z}}\left(1 + e^{2ia\sqrt{z}}\right)\right].$$

It follows that the required value of the resolvent is

$$R_z(0,0) = -\frac{\mathrm{i}}{2\sqrt{z}} \frac{2\sqrt{z} + \mathrm{i}G\left(1 - \mathrm{e}^{2\mathrm{i}a\sqrt{z}}\right)}{2\sqrt{z} + \mathrm{i}G\left(1 + \mathrm{e}^{2\mathrm{i}a\sqrt{z}}\right)}.$$

The function $R_z(0, 0)$ has a branching point at $z = 0$ and its jump on the $\mathbb{R}_+$ cut determines the exact eigenfunctions of $\hat{H}$ belonging to a continuous spectrum. Now let $Ga \gg 1$, and $z \sim 1/a^2$. In the leading approximation in $(Ga)^{-1}$ the resolvent has the form

$$R_z(0,0) \approx -\frac{\mathrm{i}}{2\sqrt{z}} \frac{1 - \mathrm{e}^{2\mathrm{i}a\sqrt{z}}}{1 + \mathrm{e}^{2\mathrm{i}a\sqrt{z}}}.$$

This approximate expression has singularities in form of poles on the real axis in the points

$$\sqrt{z_n} = \frac{\pi(2n+1)}{2a}.$$

The next order in $(Ga)^{-1}$ gives values for poles with a nonzero imaginary part

$$\sqrt{z_n} \approx \frac{\pi(2n+1)}{2a} - \frac{\pi(2n+1)}{2Ga^2} + \frac{\pi(2n+1)}{2G^2a^3} - \mathrm{i}\frac{\pi^2(2n+1)^2}{4G^2a^3} + \ldots \tag{7.38}$$

We see that the sign of the imaginary part *does not correspond* to the selected branch of the function $\sqrt{z}$, and the resolvent for those z, for which we built it, has no poles. We can say that its poles lie on another sheet of the Riemann surface of the square root $\sqrt{z}$. Nevertheless, at $Ga \gg 1$, these "non-physical" poles lie close to the cut, that is, to our sheet of the Riemann surface $\sqrt{z}$, and this gives grounds for the pole approximation $R_z(x, x')$. Thus, for z from the upper half-plane the resolvent can be approximately represented as the sum of the pole contributions at the points (7.38).

The positions of such poles are usually interpreted as complex energies of decaying quasi-stationary states whose energies have a small imaginary part. Quasi-stationary states are used to describe the decay of radioactive nuclei and unstable particles (Perelomov and Zel'dovich, 1998). However, it is necessary to use the pole approximation with caution, because, strictly speaking, quasi-stationary states belong to a continuous spectrum. In particular, if we define the eigenfunctions $\tilde{\psi}_n(x)$ with complex energies via residues of the resolvent at such poles, then they will be exponentially growing at $|x| \to \infty$. However, this does not mean that they cannot be used in any tasks. In a restricted region of space, the wave function of the decaying state can be represented as a linear superposition of such non-normalized eigenfunctions with complex energies. At the same time, it is necessary that the size of this area be smaller than the characteristic inverse exponential growth increment of the function $\tilde{\psi}_n(x)$.

Exercises

7.21. Find Green's function of the operator $L = d/dx + 1$ acting on the space of functions $u(x),\ x \in [0, 1]$ with a periodic boundary condition $u(0) = u(1)$.

7.22. Find Green's function and write out the solution of the inhomogeneous equation $u'' = f(x)$ if:

(a) $u(-1) = u(1) = 0, \quad x \in [-1, 1]$;

(b) $u(0) = u'(1) = 0, \quad x \in [0, 1]$.

7.23. Find Green's function of the third-order equation $u''' = f(x)$ if $u(0) = u(1) = 0,\ u'(0) = u'(1)$.

7.24. Construct Green's function of the problems with boundary conditions:

(a) $dG/dx + kG = \delta(x - x'), G(x, x') = 0$ at $x < x'$;

(b) $d^2G/dx^2 + k^2G = \delta(x - x'), G(x, x') = 0$ at $x > x'$.

7.25. Construct a Green's function for the operator

$$L = \frac{d^2}{dx^2} - \frac{2}{x^2}, \quad u(0) = 0, \quad u(\infty) < \infty, \quad 0 \le x < \infty.$$

7.26. Find Green's function of the following boundary value problems for the string oscillation equation $u'' + k^2u = f(x),\ x \in [0, 1]$:

(a) $u(0) = u(1) = 0$;

(b) $u'(0) = u'(1) = 0$;

(c) $u(0) = u(1), \quad u'(0) = u'(1)$.

7.27. Find Green's function of the equation $u'' - k^2u = f(x)$ with boundary conditions:

(a) $u(0) = 0$, the function $u(x)$ is bounded at $x \to +\infty$ x belongs to the half-axis $x \ge 0$;

(b) $u(x) \to 0$ at $x \to \pm\infty$.

7.28. Using Green's function from problem **7.24**(b), get the solution of the equation

$$\psi'' + k^2\psi = f(x), \quad \mathrm{Im}\, k = 0$$

with the boundary condition $\psi \to e^{ikx}$ at $x \to +\infty$; $f(x) \to 0$ at $x \to \pm\infty$.

7.29. Find Green's function of the equation

$$u'' + \frac{1}{x}u' - \frac{m^2}{x^2}u = f(x), \quad x \in [0, 1]$$

with boundary conditions: $u(x)$ is bounded at $x \to 0$, $u(1) = 0$. Why the function turns out to be asymmetric: $G(x, x') \neq G(x', x)$?

7.30. Find the generalized Green's function for operators:
(a) $L = d^2/dx^2, u(0) = u(1), \; u'(0) = u'(1)$;
(b) $L = d^2/dx^2 + \pi^2, u(0) = u(1) = 0$;
(c) $L = d^2/dx^2 + \pi^2, u'(0) = u'(1) = 0$;
(d) $L = d^2/dx^2 - 1, u(0) = u'(0), \; u(1) = u'(1)$.

7.31. Express the solution of inhomogeneous equations with inhomogeneous boundary conditions in terms of the generalized Green's function:
(a) $Lu = f(x), u(0) = a, u(1) = b, L = d^2/dx^2 + \pi^2$;
(b) $Ly = f(x), u(0) - u'(0) = a, u(1) - u'(1) = b, L = d^2/dx^2 - 1$?
Under what conditions on $f(x)$ do the equations have solutions?

7.32. Find Green's function of the three-dimensional Poisson equation $\Delta u = f(x, y, z)$ in the half-space $z > 0$ with the boundary condition $u(x, y, 0) = 0$.

7.33. Find Green's function of the Poisson equation $\Delta u = f(x, y, z)$ in the domain $x > 0, \; y > 0$ with the boundary condition $u(0, y \geq 0, z) = u(x \geq 0, 0, z) = 0$.

7.34. Find Green's function of the Helmholtz equation

$$\Delta u - k^2 u = f(x, y, z)$$

in the half-space $z > 0$ with the boundary condition $u(x, y, 0) = 0$.

7.35. Find Green's function of the Poisson equation inside the hemisphere $x^2 + y^2 + z^2 < R^2, z > 0$ with zero boundary condition $G(x^2 + y^2 + z^2 = R^2) = G(z = 0) = 0$.

7.36. Show that the two-dimensional Laplace equation is covariant with respect to the conformal inversion transformation $r \to \xi = 1/r$.

7.37. Using Green's function from problem **7.6**, solve the Dirichlet problem $u|_S = \sin 2\varphi$ in the disk $r \leq R$.

7.38. Find a solution to the Neumann problem for the Poisson equation in the half-space $z > 0$:

$$\left.\frac{\partial}{\partial z}u(x, y, z)\right|_{z=0} = g(x, y).$$

7.39. Using Green's function of the three-dimensional Schrödinger equation

$$\frac{1}{2}\left(\Delta + k^2\right) G(\mathbf{r}, \mathbf{r}') = \delta(\mathbf{r} - \mathbf{r}'), \quad \lim_{r\to\infty} G(\mathbf{r}, \mathbf{r}') = 0$$

derive the integral equation for the wave function $\psi^{(+)}(\mathbf{r})$ with asymptotic behavior at infinity: a plane wave plus a divergent wave (the basic equation of the scattering theory). The potential is assumed to be decreasing $U(\mathbf{r}) \to 0$ at $r \to \infty$.

7.40. Prove that Green's function of the heat equation $u_t = u_{xx}$ allows us to obtain the evolution operator (Green's function of the second kind) by the formula

$$G_s(x - x', t) = G(x, t; x', 0).$$

7.41. Find Green's function $G_s(x - x', t)$ of the heat equation using the Fourier transform: in time, by coordinate, by both variables.

7.42. The initial temperature distribution is Gaussian:

$$u(x, 0) = \frac{\exp\left(-x^2/2a^2\right)}{a\sqrt{2\pi}}.$$

How will it evolve in time?

7.43. Find Green's function of the heat equation $u_t = u_{xx}$ with boundary conditions $u(t, 0) = 0$, $u(t, \infty) = 0$.

7.44. Find a solution of the Cauchy problem $u(x, 0) = x^5$, $u(0, t) = 0$ for the heat equation in the domain $x \geq 0, t \geq 0$.

7.45. Construct the temporal Green's function to the Cauchy problem for the Schrödinger equation for a one-dimensional harmonic oscillator

$$\mathrm{i}\frac{\partial \Psi}{\partial t} = \hat{H}\Psi, \quad \hat{H} = \frac{\hat{p}^2}{2} + \frac{\omega^2 x^2}{2}, \quad \hat{p} = -\mathrm{i}\frac{\partial}{\partial x}.$$

7.46. Find Green's function of two-dimensional wave equation with a source.

7.47. Find Green's functions of the second kind of two-dimensional wave equation.

7.48. Find the retarded Green function of the three-dimensional wave equation (7.11).

7.49. The point charge moves according to the law $\mathbf{r} = \mathbf{R}(t)$. Using Green's function, find the scalar potential in the entire space (Lienard–Wichert potential).

7.50. Show that the solution of the inhomogeneous wave equation $u_{tt} - u_{xx} = f(x, t)$ can be obtained from solving the Cauchy problem for the auxiliary function $v(x, t; \tau)$, $0 \le \tau \le t$ is a parameter:

$$v_{tt} - v_{xx} = 0 \qquad v(x, \tau; \tau) = 0, \quad \left.\frac{\partial v}{\partial t}(x, t; \tau)\right|_{t=\tau} = f(x, \tau),$$

according to the formula:

$$u(x, t) = \int_0^t v(x, t; \tau)\, d\tau.$$

7.51. Find a continuous spectrum of the following operators defined on the entire real axis:

(a) $A = x$;
(b) $A = -d^2/dx^2$;
(c) $A = -d^2/dx^2 + x^2$.

7.52. Find a continuous spectrum of an infinite matrix in which the elements above and below the main diagonal are equal to one, and the remaining elements are zero:

$$A = \begin{pmatrix} 0\,1\,0\,0 \ldots \\ 1\,0\,1\,0 \ldots \\ 0\,1\,0\,1 \ldots \\ 0\,0\,1\,0 \ldots \\ \ldots\ldots\ldots \end{pmatrix}.$$

Answers

7.21.

$$G(x, x') = \begin{cases} \dfrac{e^{-(x-x')}}{e-1}, & \text{if } x < x'; \\ \dfrac{e^{1-(x-x')}}{e-1}, & \text{if } x > x'. \end{cases}$$

7.22(a).

$$G(x, x') = \frac{(x_< + 1)\,(x_> - 1)}{2};$$

$$u(x) = \frac{x-1}{2}\int_0^x (x'+1)\,f(x')\,dx' + \frac{x+1}{2}\int_x^1 (x'-1)\,f(x')\,dx'$$

7.22(b).

$$G(x, x') = -\min(x, x') = -x_<, \quad u(x) = -\int_0^x x' f(x')\,dx' - x\int_x^1 f(x')\,dx'.$$

7.23.

$$G(x, x') = \frac{1}{2}x_<(x_> - 1)(x - x').$$

7.24(a).

$$G(x, x') = \theta(x - x')e^{-k(x-x')}.$$

7.24(b).

$$G(x, x') = -\theta(x' - x)\sin k(x - x')/k.$$

7.25.

$$G(x, x') = -x_<^2/3x_>.$$

7.26(a).

$$G(x, x') = \frac{\sin kx_< \sin k(x_> - 1)}{k \sin k}; \quad k \neq \pi n,\ n = 0, 1, \ldots.$$

7.26(b).

$$G(x, x') = \frac{\cos kx_< \cos k(x_> - 1)}{k \sin k}; \quad k \neq \pi n,\ n = 0, 1, \ldots.$$

7.26(c).

$$G(x, x') = \frac{\cos k\,(x_> - x_< - 1/2)}{2k \sin k/2}; \quad k \neq 2\pi n,\ n = 0, 1, \ldots .$$

7.27(a).

$$G(x, x') = -\frac{1}{k} \exp\left(-kx_>\right) \sinh kx_<, \quad k \neq 0.$$

7.27(b).

$$G(x, x') = -\frac{1}{2k} \exp\left(-k(x_> - x_<)\right) = -\frac{1}{2k} \exp\left(-k|x - x'|\right), \quad k \neq 0.$$

7.28.

$$\psi(x) = \mathrm{e}^{\mathrm{i}kx} - \frac{1}{k} \int_x^\infty \sin k(x - x') f(x')\, dx'.$$

7.29.

$$G(x, x') = \begin{cases} (x'/2m)x_<^m \left(x_>^m - x_>^{-m}\right), & m > 0, \\ x' \ln x_>, & m = 0. \end{cases}$$

7.30(a).

$$G(x, x') = -\frac{1}{2}(x_> - x_<)^2 + \frac{1}{2}(x_> - x_<) - \frac{1}{12}.$$

7.30(b).

$$G(x, x') = \frac{x_> \cos \pi x_> \sin \pi x_< + x_< \cos \pi x_< \sin \pi x_>}{\pi} - \\ - \frac{\sin \pi x_> \sin \pi x_<}{2\pi^2} - \frac{\sin \pi x_< \cos \pi x_>}{\pi}.$$

7.30(c).

$$G(x, x') = -\frac{x_> \sin \pi x_> \cos \pi x_< + x_< \sin \pi x_< \cos \pi x_>}{\pi} - \\ - \frac{\cos \pi x_> \cos \pi x_<}{2\pi^2} + \frac{\sin \pi x_< \cos \pi x_<}{\pi}.$$

7.30(d).

$$G(x, x') = \alpha \mathrm{e}^{x+x'} \left[2\alpha \mathrm{e}^2 - (x + x')\right] - \alpha \cosh(x - x') - \tfrac{1}{2}\mathrm{e}^{-|x-x'|}, \alpha = \tfrac{1}{\mathrm{e}^2 - 1}.$$

7.31(a).

$$u(x) = a(1-x) + bx + \int_0^1 G(x,x')\left[f(x') + \pi^2\left(a(1-x') + bx'\right)\right]dx' +$$

$$+C\sin\pi x, \quad \pi(a+b) = \int_0^1 \sin\pi x f(x)\,dx.$$

7.31.

$$u(x) = b(1+x) - ax + \int_0^1 G(x,x')\left[f(x') + ax' - b(1+x')\right]dx' +$$

$$+Ce^x, \quad a - be = \int_0^1 e^x f(x)\,dx.$$

7.32.

$$G(x,y,z;x',y',z') = F(z') - F(-z');$$

$$F(x,y,z) = -\frac{1}{4\pi\sqrt{(x-x')^2 + (y-y')^2 + (z-z')^2}}.$$

7.33.

$$G(x,y,z;x',y',z') = F(x',y') - F(-x',y') - F(x',-y') + F(-x',-y');$$

$$F(x',y') = -\frac{1}{4\pi\sqrt{(x-x')^2 + (y-y')^2 + (z-z')^2}}.$$

7.34.

$$G(x,y,z;x',y',z') = F(x,y,x;x',y',z') - F(x,y,x;x',y',-z');$$

$$F(x,y,z;x',y',z') = -\frac{\exp\left(\sqrt{(x-x')^2 + (y-y')^2 + (z-z')^2}\right)}{4\pi\sqrt{(x-x')^2 + (y-y')^2 + (z-z')^2}}.$$

7.35.

$$G(r,\theta,\varphi;r',\theta',\varphi') = F(r,r',\psi) - F(r,r',\tilde{\psi}) -$$

$$-\frac{R}{r'}F\left(r,\frac{R^2}{r'},\psi\right) + \frac{R}{r'}F\left(r,\frac{R^2}{r'},\tilde{\psi}\right);$$

$$F(r,r',\psi) = -\frac{1}{4\pi\sqrt{r^2 + r'^2 - 2rr'\cos\psi}};$$

$$\cos\psi = \cos\theta\cos\theta' + \sin\theta\sin\theta'\cos(\varphi - \varphi'),$$

$$\cos\tilde{\psi} = -\cos\theta\cos\theta' + \sin\theta\sin\theta'\cos(\varphi - \varphi').$$

7.36. $\triangle_{\mathbf{r}} = \xi^4 \triangle_{\xi}$.

7.37. $u(r, \varphi) = r^2 \sin 2\varphi / R^2$.

7.38.

$$u(x, y, z) = \frac{1}{2\pi} \iint \frac{g(x', y')\, dx'dy'}{\sqrt{(x - x')^2 + (y - y')^2 + z^2}}.$$

7.39.

$$\psi^{(+)}(\mathbf{r}) = \exp(\mathrm{i}\mathbf{kr}) - \frac{1}{2\pi} \int \frac{\exp(\mathrm{i}k|\mathbf{r} - \mathbf{r}'|)}{|\mathbf{r} - \mathbf{r}'|} U(\mathbf{r}')\psi^{(+)}(\mathbf{r}')\, d\mathbf{r}'.$$

7.40. Hint. The initial condition $u(x, 0) = \phi(x)$ can be moved to the right side of the equation, resulting in $u_t = u_{xx} + \phi(x)\delta(t)$.

7.41.

$$G(x, t; x', 0) = \frac{\theta(t)}{\sqrt{4\pi t}} \exp\left[-(x - x')^2/4t\right] \to \delta(x - x'), \quad t \to +0.$$

7.42.

$$\frac{1}{\sqrt{2\pi}} \frac{1}{\sqrt{a^2 + 2t}} \exp\left(-\frac{x^2}{2(a^2 + 2t)}\right).$$

When the Gaussian packet is spread out, the variance increases linearly with time $\sigma^2(t) = a^2 + 2t$.

7.43.

$$G(x, t; x', 0) = \frac{\theta(t)}{\sqrt{4\pi t}} \left[\mathrm{e}^{-\frac{(x-x')^2}{4t}} - \mathrm{e}^{-\frac{(x+x')^2}{4t}}\right].$$

7.44. $u(x, t) = x^5 + 20x^3 t + 60xt^2$.

7.45.

$$G(x, t; x', 0) = \frac{\theta(t)}{\sqrt{2\pi \mathrm{i}}} \sqrt{\cot \omega t} \exp\left[\mathrm{i}\frac{\omega(x \cos \omega t - x')^2}{\sin 2\omega t}\right].$$

7.46.

$$G(\mathbf{r}, t; \mathbf{r}', t') = \frac{c}{2\pi} \frac{\theta(c\tau - \rho)}{\sqrt{c^2\tau^2 - \rho^2}}, \quad \rho = |\mathbf{r} - \mathbf{r}'|, \tau = t - t'$$

7.47.

$$G_s^{(1)}(\mathbf{r}, \mathbf{r}', t) = \frac{c}{2\pi} \frac{\partial}{\partial t} \frac{1}{\sqrt{c^2t^2 - \rho^2}}, \quad G_s^{(2)}(\mathbf{r}, \mathbf{r}', t) = \frac{c}{2\pi} \frac{1}{\sqrt{c^2t^2 - \rho^2}},$$

The solution is given by the *Poisson formula*

$$u(\mathbf{r}, t) = \frac{1}{2\pi c}\frac{\partial}{\partial t}\int_D \frac{\varphi(\mathbf{r}')\, d^2r'}{\sqrt{c^2t^2 - (\mathbf{r}-\mathbf{r}')^2}} + \frac{1}{2\pi}\int_D \frac{\psi(\mathbf{r}')\, d^2r'}{\sqrt{c^2t^2 - (\mathbf{r}-\mathbf{r}')^2}}.$$

7.48.

$$G(\mathbf{r}, t; \mathbf{r}', t') = \frac{c}{4\pi\rho}\delta(\rho - c\tau), \quad \rho = |\mathbf{r}-\mathbf{r}'|, \ \tau = t - t'.$$

7.49.

$$\phi(\mathbf{r}, t) = \frac{c}{4\pi}\frac{1}{|\mathbf{r}-\mathbf{r}'| - \frac{1}{c}\dot{\mathbf{R}}(t')(\mathbf{r}-\mathbf{r}')}; \quad \mathbf{r}' = \mathbf{R}(t), \quad t' = t - \frac{1}{c}|\mathbf{r}-\mathbf{r}'|.$$

7.50. Hint. Using the d'Alembert formula (7.35) we find a solution to the auxiliary problem

$$v(x, t; \tau) = \frac{1}{2}\int_{x-(t-\tau)}^{x+(t-\tau)} f(x', \tau)\, dx'.$$

Differentiating by t and x the function $u(x, t)$, show that it satisfies the inhomogeneous wave equation.

7.51(a). The continuous spectrum coincides with the entire real axis.

7.51(b). The continuous spectrum is $\lambda \geq 0$.

7.51(c). There is no continuous spectrum. The eigenfunctions of the discrete spectrum are Hermite polynomials and form a complete set.

7.52. The continuous spectrum forms the segment $-2 \leq \lambda \leq 2$.

Chapter 8

Integral Equations

8.1 Fredholm Equations

Equations that include the unknown function under the integral sign are called integral equations. As shown in the previous chapter, the linear equation

$$\hat{L}\psi(x) = U(x)\psi(x) + g(x)$$

with differential operator $\hat{L}$ acting at the definite class of functions $\psi(x)$, $x \in \Omega$ and linear boundary conditions can be reduced to the integral equation

$$\psi(x) = \int_{\Omega} G(x, x')U(x')\psi(x')dx' + f(x). \tag{8.1}$$

Here $G(x, x')$ is the Green function of operator $\hat{L}$,

$$f(x) = \int_{\Omega} G(x, x')g(x')dx' + r(x),$$

$r(x)$ is the contribution of boundary conditions.

The type with integration over a fixed interval is called a *Fredholm* equation

$$\psi(x) = \lambda \int_{\Omega} K(x, x')\psi(x')dx' + hf(x), \tag{8.2}$$

Mathematical Methods of Physics: Problems with Solutions
Igor V. Kolokolov, Evgeny A. Kuznetsov, Alexander I. Milstein, Evgeny V. Podivilov, Alexander I. Chernykh, David A. Shapiro, and Elena G. Shapiro

ISBN 978-981-5129-21-2 (Hardcover), 978-1-003-51360-5 (eBook)
www.jennystanford.com

while if the upper limit is x it is a *Volterra* equation:

$$\psi(x) = \lambda \int_{x' \leqslant x} K(x, x')\psi(x')dx' + hf(x) \tag{8.3}$$

Here λ is the spectral parameter. When coefficient $h = 0, 1$ the equation is of the first or second kind, respectively. Eq. (8.2) or (8.3) can be written also in bra-ket notation:

$$(1 - \lambda\hat{K})|\psi\rangle = |f\rangle. \tag{8.4}$$

Here $\hat{K}$ is the operator with kernel $K(x, x')$.

Eq. (8.4) is solvable if the inverse operator for $(1 - \lambda\hat{K})^{-1}$, the resolvent, exists. In this case, the solution to (8.4) is

$$|\psi\rangle = \hat{R}|f\rangle, \quad \hat{R} = (1 - \lambda\hat{K})^{-1}. \tag{8.5}$$

Explicit expression for the resolvent can be written as the Neumann series in λ:

$$\hat{R} = 1 + \lambda\hat{K} + \lambda^2\hat{K}^2 + \cdots = \sum_{n=0}^{\infty} \lambda^n \hat{K}^n. \tag{8.6}$$

Here $\hat{K}^n$ is also integral operator with so called *iterated* (repeated) kernel:

$$K_n(x, x') = \int_\Omega dx_1 \int_\Omega dx_2 ... \int_\Omega dx_{n-1} K(x, x_1)K(x_1, x_2)...K(x_{n-1}, x').$$

If $\int_\Omega |K(x, x'))|dx' < M$, then the series $\psi = \sum \lambda^n \hat{K}^n f$ is uniformly convergent within the circle $|\lambda| \leqslant 1/M$. In particular, that means homogeneous equation

$$(1 - \lambda\hat{K})|\psi\rangle = 0 \tag{8.7}$$

has only the trivial solution $|\psi\rangle = 0$. In this case, solution (8.5) is unique.

It is the first case of the theorem, the *Fredholm alternative* for arbitrary linear operator $\hat{A}$ is valid:

1 The inhomogeneous equation

$$\hat{A}|\psi\rangle = |f\rangle \tag{8.8}$$

has the unique solution for each f when and only when the corresponding homogeneous equation

$$\hat{A}|\psi\rangle = 0$$

has only the trivial solution $|\psi\rangle = 0$.

2 If the adjoint problem

$$\langle \psi | \hat{A} = 0$$

has n independent non-zero solutions $\langle \psi_k |$, $k = 1, \ldots, r$, then inhomogeneous equation (8.8) has solutions when and only when the solvability conditions:

$$\langle \psi_k | f \rangle = 0, \quad k = 1, \ldots, r$$

are valid, i.e., the right side is orthogonal to the zero modes of the adjoint problem. At that Eq. (8.8) has the infinite set of solutions.

For Fredholm equation (8.2) with $h = 1$, while $\hat{A} = 1 - \lambda \hat{K}$, the alternative determines that at $\lambda = \lambda_n$, where λ_n is the eigenvalue, for solvability the right side must satisfy r orthogonality conditions. When $\lambda \neq \lambda_n$ the solution is unique.

Method of the *Fredholm determinants* allows one to find the resolvent:

$$R(x, y; \lambda) = \frac{\Delta(x, y; \lambda)}{\Delta(\lambda)}, \quad \Delta(\lambda) \neq 0,$$

where the Fredholm determinants $\Delta(\lambda)$ and Fredholm minor $\Delta(x, y; \lambda)$ can be found as sums of series

$$\Delta(x, y; \lambda) = K(x, y) + \sum_{n=1}^{\infty} \frac{(-1)^n}{n!} \lambda^n B_n(x, y); \quad \Delta(\lambda) = 1 + \sum_{n=1}^{\infty} \frac{(-1)^n}{n!} \lambda^n C_n.$$

Functions $B_n(x, y)$ and coefficients C_n in their turn are defined by relations

$$B_n(x, y) = C_n K(x, y) - n \int_0^1 K(x, t) B_{n-1}(t, y)\, dt,$$

$$C_n = \int_0^1 B_{n-1}(x, x)\, dx, \quad n = 1, 2, \ldots,$$

where $C_0 = 1$, $B_0(x, y) = K(x, y)$.

8.2 Degenerate Kernel

In an important particular case of the *degenerate* (separable) kernel, when

$$K(x, y) = \sum_{n=1}^{N} \psi_n(x) \tilde{\psi}_n(y), \tag{8.9}$$

the solution of Fredholm equation (8.2) is the pure algebraic procedure. Functions $\{\psi_n(x)\}$ and $\{\psi_n(y)\}$ may be considered as independent.

To find the solution to Fredholm equation with degenerate kernel

$$\psi(x) = \lambda \sum_{n=1}^{N} \psi_n(x) \int_{\Omega} \tilde{\psi}_n(y)\psi(y)dy + f(x), \tag{8.10}$$

we present the difference $\varphi(x) = \psi(x) - f(x)$ in terms of functions $\psi_n(x)$,

$$\varphi(x) = \lambda \sum_{n=1}^{N} C_n \psi_n(x),$$

then from (8.10), we get the algebraic set

$$C_n - \lambda \sum_{l=1}^{N} M_{nl} C_l = f_n, \tag{8.11}$$

where $f_l = \langle \tilde{\psi}_l | f \rangle$, $M_{nl} = \langle \tilde{\psi}_n | \psi_l \rangle$. If $\det(1 - \lambda M) \neq 0$, then the solution $C = (1-\lambda M)^{-1} f$ is unique. Here C, f are vectors with components C_n and f_n.

Another particular case that reduces to an algebraic equation is the *difference* kernel $K(x, y) = K(x - y)$. The Fredholm equation of this type can be simplified by the Fourier transform, which turns the convolution into the product. Sometimes the degenerate or difference kernel makes it possible to solve even the non-linear equation, particularly the Urysohn equation:

$$\psi(x) = \int_a^b K(x, s, \psi(x))\, dx$$

or the Hammerstein equation

$$\psi(x) = \int_a^b K(x, s) F(s, \psi(x))\, dx.$$

Here K is the kernel, $\psi(x)$ is the unknown function, $f(x)$ is the given function.

8.3 Symmetric Kernel

For symmetric kernel $K(x, y) = K(y, x)$ the Hilbert–Schmidt theorem is valid. Let function $f(x)$ be given by integral

$$f(x) = \int_{\Omega} K(x, y) g(y) dy,$$

where the symmetric kernel satisfies condition

$$\int_{\Omega}\int_{\Omega} |K(x, y)|^2 \, dxdy < \infty,$$

that means that $K(x, y)$ is L^2 function. Here $g(x)$ is also L^2 function. Then the function can be presented as a series of the orthogonal eigenfunctions $|\psi_n\rangle$ of operator $\hat{K}$:

$$f(x) = \sum_n f_n \psi_n(x) \quad \text{or} \quad |f\rangle = \sum f_n |\psi_n\rangle,$$

where $f_n = \langle\psi_n|f\rangle = g_n/\lambda_n$, $g_n = \langle\psi_n|g\rangle$. The series is convergent absolutely and uniformly.

The representation of kernel as the bilinear form follows from the Rietz theorem:

$$K(x, y) = \sum_{n=1}^{\infty} \lambda_n^{-1} \psi_n(x) \psi_n^*(y) \quad \text{or} \quad \hat{K} = \sum_{n=1}^{\infty} \lambda_n^{-1} |\psi_n\rangle\langle\psi_n|. \tag{8.12}$$

8.4 Inverse Problem for Schrödinger Operator

The *inverse* problem for Schrödinger operator

$$\hat{L} = \frac{\partial^2}{\partial x^2} + U(x)$$

with $U(x) \to 0$ at $|x| \to \infty$ is the problem of potential reconstruction from the scattering data, i.e., from asymptotic states of spectral problem:

$$\hat{L}\Psi = \left(\frac{\partial^2}{\partial x^2} + U(x)\right)\Psi = -k^2\Psi. \tag{8.13}$$

Operator $\hat{L}$, as known, is Hermitian (self-adjoint), and then has the real spectrum: Im $k^2 = 0$. Continuous spectrum of the operator lies at $k^2 > 0$, the discrete spectrum consists of the states with negative energy $E_n = k_n^2 < 0$.

Direct Scattering Problem

For each pair of solutions (8.13) Ψ_1 and Ψ_2 the Wronskian

$$W\{\Psi_1, \Psi_2\} = \begin{vmatrix} \Psi_1 & \Psi_2 \\ \Psi_1' & \Psi_2' \end{vmatrix} = \Psi_1\Psi_2' - \Psi_1'\Psi_2 \tag{8.14}$$

is independent of x. Functions Ψ_1 and Ψ_2 are independent when their Wronskian is non-zero. To determine the scattering data let us consider special solutions, the Jost functions. They are determined by asymptotic at $|x| \to \infty$:

$$\psi(x, k) \to \mathrm{e}^{ikx} \quad \text{at} \quad x \to \infty, \tag{8.15}$$

$$\Phi(x, k) \to \mathrm{e}^{-ikx} \quad \text{at} \quad x \to -\infty. \tag{8.16}$$

Functions ψ and ψ^* are independent. They form the complete basis of solutions at given k^2. Then Φ and Φ^* can be written in terms of basis:

$$\Phi(x, k) = a(k)\psi^*(x, k) + b(k)\psi(x, k), \tag{8.17}$$

$$\Phi^*(x, k) = b^*(k)\psi^*(x, k) + a^*(k)\psi(x, k) \tag{8.18}$$

or in matrix notation:

$$\begin{pmatrix} \Phi \\ \Phi^* \end{pmatrix} = \begin{pmatrix} a & b \\ b^* & a^* \end{pmatrix} \begin{pmatrix} \psi^* \\ \psi \end{pmatrix} \equiv \hat{S} \begin{pmatrix} \psi^* \\ \psi \end{pmatrix}.$$

From (8.17) and (8.18) with the help of (8.14), we obtain

$$a(k) = \frac{1}{2ik} W(\Phi, \psi), \quad b(k) = -\frac{1}{2ik} W(\Phi, \psi^*), \tag{8.19}$$

and show the determinant of scattering matrix $\hat{S}$ to be equal to unity:

$$\det \hat{S} = |a|^2 - |b|^2 = 1. \tag{8.20}$$

The last equality corresponds to the usual conservation law $|r|^2 + |d|^2 = 1$, where $r = b/a$ is the reflection, $d = 1/a$ is the transmission coefficient. From (8.17) and (8.18) at $\operatorname{Im} k = 0$ it follows that

$$a(k) = a^*(-k), \quad b(k) = b^*(-k). \tag{8.21}$$

By virtue of definition (8.15), (8.16), Jost function ψ or Φ is continuable analytically to the upper half-plane k ($\operatorname{Im} k > 0$), while ψ^* and Φ^* are analytic in the lower half-plane (see problem **8.9**). Consequently, function $a(k)$ according to (8.19) is continuable analytically to the upper half-plane. Since at $k \to \infty$ $\psi \to \mathrm{e}^{ikx}$, $\Phi \to \mathrm{e}^{-ikx}$, function $a(k)$ tends to unity at $k \to \infty$. Points in the upper half-plane, where $a(k_n) = 0$, correspond to the discrete spectrum. Wronskian in these points equals zero according to (8.19), i.e., functions Φ and ψ are linearly dependent: $\Phi(k_n) = C_n\psi(k_n)$. On the other hand, there is a solution to Eq. (8.13) damping at $x \to \infty$, as well as at $x \to -\infty$ (by definition (8.15) and (8.16)). Thus the solution describes the bound state, for which zeros $k_n = i\kappa_n$ of function $a(k)$ lie in the imaginary axis. The whole set of values $a(k)$, $b(k)$, κ_n and C_n form the complete set of scattering data for operator $\hat{L}$.

Gelfand–Levitan–Marchenko Equation

For solving the *inverse* problem, i.e., potential $U(x)$ reconstruction from the scattering data, the analytical properties of Jost functions ψ and Φ are of importance. Since ψ is continuable analytically to the upper half-plane $\mathrm{Im}\, k > 0$ (see problem **8.9**), function $\psi(x, k)$ can be presented in the following form:

$$\psi(x, k) = \mathrm{e}^{ikx} + \int_x^{\infty} K(x, y)\mathrm{e}^{iky} dy, \tag{8.22}$$

where $K(x, y)$ is the real kernel. From this equation, called *triangular* representation, immediately follows that ψ is analytic at $\mathrm{Im}\, k > 0$.

Substituting representation (8.22) into Eq. (8.13), we get:

$$U(x)\mathrm{e}^{ikx} - \frac{\partial}{\partial x}[K(x, x)\mathrm{e}^{ikx}] - \frac{\partial}{\partial x}K(x, y)|_{y=x}\mathrm{e}^{ikx} +$$
$$\int_x^{\infty} \frac{\partial^2}{\partial x^2}K(x, y)\mathrm{e}^{iky}\, dy = -(U(x) + k^2)\int_x^{\infty} K(x, y)\mathrm{e}^{iky}\, dy.$$

Changing $-k^2\mathrm{e}^{iky} = \partial^2/\partial y^2 \mathrm{e}^{iky}$ in the former integral after double integration by parts, we find:

$$\int_x^{\infty} \left[\frac{\partial^2}{\partial x^2}K(x, y) - \frac{\partial^2}{\partial y^2}K(x, y) + U(x)K(x, y)\right] \mathrm{e}^{iky}\, dy$$
$$+[U(x) - 2\frac{d}{dx}K(x, x)]\mathrm{e}^{ikx} = 0.$$

This equality should be true for all k. Since functions e^{iky} are independent at different y, the equality is valid if and only if, when

$$\frac{\partial^2}{\partial x^2}K(x, y) - \frac{\partial^2}{\partial y^2}K(x, y) + U(x)K(x, y) = 0, \tag{8.23}$$

$$U(x) = 2\frac{d}{dx}K(x, x). \tag{8.24}$$

Equation (8.23) with boundary condition (8.24) presents so-called the Goursat problem. The Goursat problem is uniquely solvable provided $K(x, x + y) \to 0$ at $x \to \infty$. Moreover, formula (8.24), connecting kernel $K(x, y)$ with potential $U(x)$, also follows from asymptotic expansion of the triangular representation at $k \to \infty$ and from Eq. (8.41) of problem **8.10**.

We turn now to the derivation of the equations for inverse problem, connected kernel $K(x, y)$ with the scattering data. Divide Eq. (8.17) by $a(k)$, subtract the asymptotic at $k \to \infty$ from the left and right sides, then multiply by e^{iky} ($y \geqslant x$) and integrate over k from $-\infty$ to ∞. We obtain

$$\int_{-\infty}^{\infty}\left[\frac{\Phi(x,k)}{a(k)} - e^{-ikx}\right]e^{iky}dk = \int_{-\infty}^{\infty}dk\int_{x}^{\infty}K(x,s)e^{ik(y-s)}ds + \int_{-\infty}^{\infty}r(k)e^{ik(x+y)}dk + \int_{x}^{\infty}K(x,s)\int_{-\infty}^{\infty}r(k)e^{ik(y+s)}dk. \quad (8.25)$$

At $y > x$ the integrand in the left side tends to zero with the growing imaginary part of k, then the integration contour can be closed in the upper half-plane. The integral equals the sum of residues in zeros of function $a(k)$.

When $a(k)$ has no zeros, i.e., bound states are absent, the integral at the left is zero. As a result, we get

$$K(x,y) + F(x+y) + \int_{x}^{\infty}K(x,s)F(s+y)ds = 0, \quad (8.26)$$

where

$$F(x) = \frac{1}{2\pi}\int_{-\infty}^{\infty}r(k)e^{ikx}dk. \quad (8.27)$$

Eq. (8.26) is the required master equation of the inverse scattering problem, often called the Gelfand–Levitan–Marchenko (GLM) equation.

When $a(k)$ has zeros, the left side in Eq. (8.25) equals the sum of residues

$$2\pi i\sum_{n}\frac{\Phi(x,i\kappa_n)}{a'(i\kappa_n)}e^{-\kappa_n y}. \quad (8.28)$$

In points $k = i\kappa_n$ functions Φ and ψ are connected $\Phi(x,i\kappa_n) = C_n\psi(x,i\kappa_n)$. For ψ the triangular representation (8.22) is valid. By grouping all terms in the equation, we finally see that the GLM equation has the same form, but additional sum appears in kernel F:

$$F(x) = \frac{1}{2\pi}\int_{-\infty}^{\infty}r(k)e^{ikx}dk - \sum_{n}\frac{iC_n}{a'(i\kappa_n)}e^{-\kappa_n x}. \quad (8.29)$$

The sum is responsible for the discrete spectrum.

If we continue the first equality (8.21) for complex k $a(k) = a^*(-k^*)$, then we see that function $a(k)$ at imaginary $k = i\kappa$ is real: $a(i\kappa) = a^*(i\kappa)$. Hence the derivative $a'(i\kappa)$ occurs pure imaginary, then coefficient $M_n^2 = -iC_n/a'(i\kappa_n)$ is real. Moreover, the coefficient is positive (see problem **8.31**).

Eq. (8.26) allows one to find the kernel $K(x, y)$, and at the same time reconstruct from the scattering data $a(k)$, $b(k)$, κ_n and C_n, using Eq. (8.24).

We now turn to the simplest solutions to the GLM equation, where the integral term is absent in kernel $F(x)$. For these solutions $r(k) = 0$, that is why the corresponding $U(x)$ is referred to as *non-reflective* potential. The set of $\{\kappa_n\}$, the discrete spectrum of operator $\hat{L}$, and residues M_n^2 define these potentials completely.

Let $F(x)$ be determined by single value κ:

$$F(x) = M^2 \mathrm{e}^{-\kappa x}.$$

Substituting this value to GLM equation, we make sure that the dependence of kernel $K(x, y)$ on variable y can be determined:

$$K(x, y) = \psi_0(x)\mathrm{e}^{-\kappa y}. \tag{8.30}$$

By virtue of (8.23) function $\psi_0(x)$ is the eigenfunction of discrete spectrum:

$$\left(\frac{d^2}{dx^2} + U(x)\right)\psi_0(x) = \kappa^2\psi_0(x).$$

Substituting (8.30) into GLM equation (8.26) and, integrating, we find

$$\psi_0(x) = -\frac{M^2 \mathrm{e}^{-\kappa x}}{1 + \frac{M^2}{2\kappa}\mathrm{e}^{-2\kappa x}} = -\frac{\sqrt{2\kappa}\,M}{2}\cosh^{-1}\kappa(x - q), \quad q = \frac{1}{2\kappa}\ln\frac{M^2}{2\kappa}.$$

Corresponding potential is

$$U(x) = 2\frac{d}{dx}K(x, x) = \frac{2\kappa^2}{\cosh^2\kappa(x - q)}. \tag{8.31}$$

This is the simplest non-reflective potential, having single bound state with energy $E = -\kappa^2$.

Traditional methods solving the integral equations are discussed in textbooks (Kabwal, 1971; Mikhlin, 1960; Pipkin, 1991; Tricomi, 1957). Handbook (Polyanin and Manzhirov, 2008) includes a comprehensive collection of exactly solvable cases. The inverse scattering problem for Schrödinger operator is analyzed in detail in books (Ablowitz and Segur, 1981; Newell, 1985; Novikov *et al.*, 1984).

Worked Examples

8.1. Prove that for integral operators

$$\hat{K}\psi = \int_\Omega K(x, y)\rho(y)\psi(y)dy,$$

where $\rho(x) \geqslant 0$ is real and $K(x, y) = K^*(y, x)$, the eigenfunctions with different eigenvalues are orthogonal with weight $\rho(x)$:

$$\int_\Omega \psi_n^*(x)\psi_m(x)\rho(x)dx = \delta_{nm}.$$

Solution. Eigenfunctions $\psi_n(x)$ and $\psi_m(x)$ satisfy equations:

$$\int_\Omega K(x, y)\rho(y)\psi_n(y)dy = \mu_n\psi_n(x), \quad \int_\Omega K(x, y)\rho(y)\psi_m(y)dy = \mu_m\psi_m(x).$$

Let us show first that all the eigenvalues $\mu_n \equiv \lambda_n^{-1}$ are real. Multiply both sides of the first equation by $\rho(x)\psi_n^*(x)$ and integrate over x. At the left, we get expression:

$$\int_\Omega K(x, y)\rho(y)\rho(x)\psi_n(y)\psi_n^*(x)\, dxdy,$$

being real owing to the symmetry $K(x, y) = K^*(y, x)$. Then the right side and the eigenvalue are real.

Multiply the first equation by $\rho(x)\psi_m^*(x)$, the second by $\rho(x)\psi_n^*(x)$ and integrate both over x. Complex conjugation gives

$$(\mu_n - \mu_m)\int_\Omega \rho(x)\psi_n(x)\psi_m^*(x)dx = 0.$$

Then $\psi_n(x)$ and $\psi_m(x)$ are orthogonal with weight $\rho(x)$.

The Hilbert–Schmidt theorem, especially the bilinear form (8.12), occurs very important in the spectral problem for self-adjoint Sturm–Liouville operator

$$\hat{L}\psi = \frac{d}{dx}\left(p(x)\frac{d\psi}{dx}\right) + q(x)\psi = \lambda\rho(x)\psi \tag{8.32}$$

with the zero boundary conditions: $\psi(a) = \psi(b) = 0$. In this case, Eq. (8.32) can be presented in the form of integral equation

$$\psi(x) = \lambda\int_a^b G(x, y)\rho(y)\psi(y)\, dy, \tag{8.33}$$

where Green's function of a self-adjoint operator is symmetric: $G(x, y) = G(y, x)$. By setting $[\rho(x)]^{1/2}\psi(x) = \Psi(x)$ under the assumption that $\rho(x)$ is non-negative over the interval (a, b), we get from (8.33)

$$\Psi(x) = \lambda\int_a^b \tilde{K}(x, y)\Psi(y)\, dy$$

with symmetric kernel $\tilde{K}(x, y) = [\rho(x)]^{1/2} G(x, y)[\rho(y)]^{1/2}$. Then using Hilbert–Schmidt theorem, we obtain that the eigenfunctions are orthogonal with weight $\rho(x)$. Its eigenfunctions form the complete orthonormal set when $q(x) < 0$, $x \in [a, b]$, i.e., for negative definite operator.

8.2. Find the weight functions for:
(a) Bessel functions $J_n(x)$;
(b) Laguerre polynomials;
(c) Hermite polynomials,
starting from their differential equations.

Solution. (a) Bessel's equation can be written as an eigenvalue problem:

$$\left(x\frac{d}{dx}x\frac{d}{dx} + x^2\right) J_n(x) = n^2 J_n(x).$$

To make the operator self-adjoint, as well as in Eq. (8.32), we must divide the equation by x, then the weight function is $\rho(x) = 1/x$. The relation for Bessel's functions in $[0, \infty)$ is

$$\int_0^\infty \frac{dx}{x} J_n(x) J_m(x) = \frac{2\sin[\pi(m-n)/2]}{\pi(m^2 - n^2)}, \qquad m, n > 0.$$

Then the orthogonality takes place when m, n both are even or odd.

(b) Laguerre equation is

$$e^x x^{-m} \frac{d}{dx} e^{-x} x^{m+1} \frac{d}{dx} L_n^m(x) = -n L_n^m(x).$$

Then the Laguerre polynomials are orthogonal at semiaxis $(0, \infty)$ with weight $\rho(x) = e^{-x} x^m$.

(c) Hermite equation has the form:

$$e^{x^2} \frac{d}{dx} e^{-x^2} \frac{d}{dx} H_n = -2n H_n.$$

Then the Hermite polynomials are orthogonal at real axis $(-\infty, \infty)$ with weight $\rho(x) = e^{-x^2}$.

8.3. Solve equation

$$u(x) = \int_0^x s \ln\left(\frac{s}{x}\right) u(s)\, ds + \frac{x^4}{16}. \tag{8.34}$$

Solution. Calculating two derivatives of Eq. (8.34), we get

$$u'(x) = -\frac{1}{x}\int_0^x su(s)\,ds + \frac{x^3}{4},$$

$$u''(x) = \frac{1}{x^2}\int_0^x su(s)\,ds - u(x) + \frac{3x^2}{4},$$

from which the differential equation follows

$$u'' + \frac{1}{x}u' + u = x^2$$

with boundary condition $u(0) = u'(0) = 0$.

Fundamental set of solutions consists of Bessel and Neumann functions J_0, Y_0 of the zero order. For solving the non-uniform equation let us build up the Green function, that turns to zero at $x = 0$,

$$G(x, y) = \frac{Y_0(y)J_0(x)\Theta(y - x) + J_0(y)Y_0(x)\Theta(x - y)}{W\{J_0(x), Y_0(x)\}}$$
$$= \frac{\pi}{2}y\left[Y_0(y)J_0(x)\Theta(y - x) + J_0(y)Y_0(x)\Theta(x - y)\right],$$

where the Wronskian is equal to

$$W\{J_0(y), Y_0(y)\} = J_0Y_0' - J_0'Y_0 = \frac{2}{\pi y}. \tag{8.35}$$

Now the solution can be presented as

$$u(x) = \int_0^\infty G(x, y)y^2\,dy.$$

Using the recurrence relations after double integration by parts we get the explicit expression:

$$u(x) = x^2 - 4 + 4J_0(x).$$

8.4. Find the resolvent of kernel $K(x, y) = x^2y - xy^2$ by the Fredholm determinants method.

Solution. Sequentially get $C_0 = 1$, $B_0(x, y) = x^2y - xy^2$, $C_1 = 0$,

$$B_1(x, y) = xy\left[-\frac{x + y}{4} + \frac{1}{5} + \frac{xy}{3}\right],$$

$C_2 = 1/120$, $B_2 = 0$, and then all subsequent coefficients C_n and functions B_n turns to zero. The reason for breaking the sequence is the specific form of the kernel $K(x, y) = f(x)g(y) - g(x)f(y)$ (Show that for this class of kernels the following relation is valid

$$\int_0^1\int_0^1 K(x, y)K(y, z)K(z, t)\, dydz = \frac{1}{2}K(x, t)\int_0^1\int_0^1 K(z, y)K(y, z)\, dydz,$$

with the help of which the repeated kernels can be expressed in terms of B_1, B_2.

$$R(x, y; \lambda) = \frac{x^2y - xy^2 - \lambda xy\left[\frac{1}{5} + \frac{1}{3}xy - \frac{1}{4}(x + y)\right]}{1 + \frac{1}{240}\lambda^2}.$$

8.5. Solve the spectral problem

$$\phi(x) = \sqrt{\frac{2}{\pi}}\lambda\int_0^\infty \sin(xy)\phi(y)\, dy.$$

Solution. Denote the sine Fourier transform of $\phi(y)$ as

$$\Phi(x) \equiv \sqrt{\frac{2}{\pi}}\int_0^\infty \sin(xy)\phi(y)\, dy.$$

Applying the transform to both sides, we obtain

$$\Phi(x) = \lambda\phi(x),$$

whence

$$\phi(x) = \lambda\Phi(x) = \lambda^2\phi(x).$$

Consequently, the operator has only two eigenvalues $\lambda = \pm 1$; the eigenfunctions are

$$\phi(x) = g(x) + \lambda G(x),$$

where $g(x)$ is an arbitrary function with existing Fourier transform $G(x)$, for example,

$$\phi(x) = \exp(-ax) \pm \sqrt{\frac{2}{\pi}}\frac{x}{a^2 + x^2}, \quad 0 < x < \infty, \ a > 0.$$

8.6. Find the eigenfunctions decreasing at infinity faster than any power, and the eigenvalues of the integral equation

$$\lambda \int_{-\infty}^{\infty} K(x, y)\phi(y)\, dy = \phi(x)$$

with the Keilson–Storer kernel:

$$K(x, y) = \frac{1}{\sqrt{\pi(1-\alpha^2)}} \exp\left[-\frac{(x-\alpha y)^2}{1-\alpha^2}\right], \quad 0 \leqslant \alpha < 1. \tag{8.36}$$

Solution. Applying the Fourier transform to the initial equation, we get the functional equation for Fourier transform $\Phi(k)$ of function $\phi(x)$

$$\Phi(k) = \lambda\Phi(k\alpha) \exp\left(-(1-\alpha^2)\frac{k^2}{4}\right).$$

Function $g(k) = \Phi(k)\exp(k^2/4)$ stretch by λ times after stretching its argument by α times:

$$g(k) = \lambda g(k\alpha),$$

from which $g(k) = k^\nu$ is a power function, and then

$$\Phi_\nu(k) = k^\nu \exp(-k^2/4), \quad \lambda = \alpha^{-\nu}.$$

At $\nu > -1$ function $\phi(x)$ decreases at infinity. At integer $\nu = n$ the inverse Fourier transform gives the Gauss–Hermite functions

$$\phi_n(x) = H_n(x)\exp(-x^2), \quad \lambda_n = \alpha^{-n}, \quad n = 0, 1, 2, \dots, \tag{8.37}$$

where H_n are the Hermite polynomials.

8.7. Find eigenfunctions of the Fourier transform

$$\lambda \int_{-\infty}^{\infty} \phi(y)\exp(ixy)\frac{dy}{\sqrt{2\pi}} = \phi(x).$$

Solution. Denote the transform of function $\phi(y)$ as

$$\Phi(x) \equiv \frac{1}{\sqrt{2\pi}} \int_{-\infty}^{\infty} \exp(ixy)\phi(y)\, dy.$$

Executing the Fourier transform, we get

$$\Phi(x) = \lambda\phi(-x),$$

then

$$\phi(x) = \lambda\Phi(x) = \lambda^2\phi(-x) = \lambda^3\Phi(-x) = \lambda^4\phi(x).$$

Consequently, the operator has four eigenvalues $\lambda = \pm i, \pm 1$, while the eigenfunctions are

$$\phi(x) = g(x) + \lambda G(x) + \lambda^2 g(-x) + \lambda^3 G(-x),$$

where $g(x)$ is an arbitrary function having the Fourier transform $G(x)$, for example,

$$\phi_n(x) = H_n(x)\exp(-x^2/2).$$

That is the set of eigenfunctions for a quantum mechanical oscillator. Note that the Schrödinger equation for an oscillator:

$$\frac{d^2\psi}{dx^2} - x^2\psi = 2E\psi$$

is transformed into the same equation. Even functions correspond to $\lambda = \pm 1$, and odd modes belong to $\lambda = \pm \mathrm{i}$.

8.8. Find solution $f(y)$ to the Urysohn equation:

(a)

$$2\int_{-\infty}^{+\infty} f(x-y)f(y)y\,dy = x(|x|+1)\mathrm{e}^{-|x|}.$$

Solution. Recall that if $F(\omega)$ is the Fourier transform of function $f(x)$, then the Fourier transform of function $xf(x)$ is $\mathrm{i}F'(\omega)$. After the transform, we get

$$2\mathrm{i}F'(\omega)F(\omega) = 4\mathrm{i}\frac{d}{d\omega}\frac{1}{(1+\omega^2)^2}.$$

The Fourier transform of $F(\omega)$ should tend to zero at $\omega \to \pm\infty$. This condition makes it possible to determine the integration constant.

$$F(\omega) = \pm\frac{2}{1+\omega^2}; \quad f(x) = \pm\mathrm{e}^{-|x|}.$$

(b)

$$\int_{-\infty}^{+\infty} f(x-y)f(y)(1+y^2)\,dy = 1.$$

Solution. Implementing the Fourier transform, we have

$$F^2(\omega) - F''(\omega)f(\omega) = 2\pi\delta(\omega).$$

The solution tending to zero at $\omega \to \pm\infty$ is

$$F(\omega) = \pm\sqrt{\pi}e^{-|\omega|}; \quad f(x) = \pm\frac{1}{\sqrt{\pi}(1+x^2)}.$$

8.9. Show that Jost functions $\psi(x, k)$ and $\Phi(x, k)$ are continuable analytically to the upper half-plane $\mathrm{Im}\, k > 0$.

Solution. The Schrödinger operator determines Green function G that satisfy equation:

$$\left(\frac{d^2}{dx^2} + k^2\right) G = \delta(x - x'), \quad G(x - x') = 0,\ x > x'.$$

We find

$$G(x - x') = \begin{cases} -\dfrac{1}{k} \sin k(x - x') \text{ at } x' > x, \\ 0 \qquad\qquad\qquad\quad \text{at } x > x'. \end{cases}$$

With a help of the Green function the equation for ψ reduces to the Volterra equation

$$\psi(x, k) = e^{ikx} + \frac{1}{k}\int_x^\infty \sin k(x - x')U(x', k)\psi(x', k)\, dx'. \tag{8.38}$$

Consider an auxiliary function $\chi(x, k) = \psi(x, k)\exp(-ikx)$:

$$\chi(x, k) = 1 + \frac{1}{2ik}\int_x^\infty [1 - \exp(2ik(x' - x))]U(x')\chi(x', k)\, dx'. \tag{8.39}$$

At real k the solution to (8.38) exists, i.e., the integral in (8.39) is convergent. In the upper half-plane its convergence improves, then function $\chi(x, k)$ is analytic here. The analytical continuation of function $\Phi(x, k)$ into the upper half-plane can be constructed in the same way.

8.10. Find the relation between the Jost function asymptotic $\psi(x, k)$ at large k and potential $U(x)$.

Solution. From (8.39) it follows that at $k \to \infty$

$$\chi(x, k) = 1 + \frac{1}{2ik} \int_x^\infty U(x')\,dx' + o\left(\frac{1}{k}\right), \tag{8.40}$$

then

$$\frac{d}{dx} \lim_{k\to\infty} 2ik\chi(x, k) = -U(x). \tag{8.41}$$

8.11. Find the general form of the non-reflective potential with discrete levels $k_n = i\kappa_n$ $(n = 1, \ldots, N)$.

Solution. In this case, $a(k)$ has N zeros at $k = i\kappa_n$ and $r(k) \equiv 0$, then F has a form of discrete sum:

$$F(x + y) = \sum_n M_n^2 \exp\left[-\kappa_n(x + y)\right].$$

The y-dependence of kernel $K(x, y)$ has the form:[1]

$$K(x, y) = \sum_n \psi_n(x) e^{-\kappa_n y}.$$

Direct calculation of $\psi_n(x)$ brings the following set:

$$\psi_n(x) + M_n^2 e^{-\kappa_n x} + M_n^2 \sum_{m=1}^N \frac{\psi_m(x) \exp[-(\kappa_n + \kappa_m)x]}{\kappa_n + \kappa_m} = 0. \tag{8.42}$$

or in matrix form

$$\sum_{m=1}^N A_{nm}\psi_m = -M_n^2 e^{-\kappa_n x}, \quad A_{nm} = \delta_{nm} + M_n^2 \frac{\exp[-(\kappa_n + \kappa_m)x]}{\kappa_n + \kappa_m}.$$

The solution to (8.42) can be found by Cramer's rule:

$$\psi_n(x) = \frac{\det B_n(x)}{\det A(x)},$$

where $B_n(x)$ is obtained from matrix $A(x)$ after the replacing of n-th column by column $(-M_n^2 e^{-\kappa_n x})$. Notice that potential $U(x)$ (8.24) is determined by diagonal elements $K(x, x)$, i.e.,

$$\sum_{n=1}^N \psi_n(x) e^{-\kappa_n x} = \frac{\sum_{n=1}^N \det B_n e^{-\kappa_n x}}{\det A(x)}.$$

[1] In this case, Eq. (8.26) has a degenerate kernel.

It is easy to show that $\sum_{n=1}^{N} \det B_n e^{-\kappa_n x} = d(\det A)/dx$. For this purpose, we observe that the derivative of a determinant is a sum of determinants. In n-th determinant n-th column is replaced by its derivative $-M_n^2 \exp[-(\kappa_n + \kappa_m)x]$, then

$$U(x) = 2\frac{d^2}{dx^2} \ln \det A.$$

8.12. Write the expression for $U(x)$ in terms of reflection coefficient $r(k) = b(k)/a(k)$ for small potential, where the discrete spectrum is absent (the first Born approximation).

Solution. In continuous spectrum, we use Eq. (8.27). For small potential, we neglect the integral term in Eq. (8.26). As a result,

$$K(x, x) \approx -F(2x)$$

or for potential

$$U(x) = -2\frac{d}{dx}F(2x) = -\frac{2i}{\pi}\int_{-\infty}^{\infty} r(k)ke^{2ikx}\, dk.$$

8.13. Reconstruct the transmission coefficient $d(k)$ (or $a(k)$) from its absolute value at $\mathrm{Im}\, k = 0$ and the position of discrete spectrum $k_n = i\kappa_n$.

Solution. Owing to the property of $a(k)$: $a(k) = a^*(-k)$ at $\mathrm{Im}\, k = 0$ the function can be presented in the form

$$a(k) = \tilde{a}(k)\prod_{n=1}^{N}\frac{k - i\kappa_n}{k + i\kappa_n},$$

where $\tilde{a}(k)$ is the analytic function in the upper half-plane with no zeros, having limiting value $\tilde{a}(k) \to 1$ at $k \to \infty$.

Consider function $\ln \tilde{a}(k)$ and write out the Kramers–Kronig relation

$$\arg \tilde{a}(k) = -\frac{1}{\pi}\text{v.p.}\int_{-\infty}^{\infty}\frac{\ln|\tilde{a}(k')|}{k' - k}\, dk',$$

where "v.p." means the principal value. Since $\ln|\tilde{a}(k)|$ is even function, its integral is reduced to

$$\text{v.p.}\int_{-\infty}^{\infty}\frac{\ln|\tilde{a}(k')|}{k' - k}\, dk' = -\frac{2}{k}\text{v.p.}\int_{0}^{\infty}\frac{\ln|\tilde{a}(k')|}{1 - k'^2/k^2}dk'.$$

Using equality $\ln|\tilde{a}(k)| = \ln|a(k)|$ at $\operatorname{Im} k = 0$, we get

$$\arg a(k) = \frac{1}{\mathrm{i}}\sum_n \left[\ln\left(1 - \frac{\mathrm{i}\kappa_n}{k}\right) - \ln\left(1 + \frac{\mathrm{i}\kappa_n}{k}\right)\right]$$

$$+\frac{2}{\pi k}\text{v.p.}\int_0^\infty \frac{\ln|a(k')|}{1 - k'^2/k^2}\,dk'.$$

Exercises

8.14. Using Eq. (8.12), find the expression for resolvent.

8.15. The solution to Volterra equation of the second kind

$$\phi(x) - \int_0^x K(x, y)\phi(y)\,dy = f(x) \tag{8.43}$$

can be expressed in terms of the resolvent $R(x, y)$:

$$\phi(x) = f(x) + \int_0^x R(x, y) f(y)\,dy. \tag{8.44}$$

(a) Show that the resolvent satisfies integral equation

$$R(x, y) = K(x, y) + \int_y^x K(x, t)R(t, y)\,dt. \tag{8.45}$$

(b) Prove that one can present the resolvent as a sum:

$$R(x, y) = \sum_{n=1}^\infty K_n(x, y) \tag{8.46}$$

of repeated kernels

$$K_{n+1}(x, y) = \int_y^x K(x, t)K_n(t, y)\,dt, \quad n = 1, 2, \ldots, \qquad K_1(x, y) \equiv K(x, y).$$

8.16. With the help of Eq. (8.46), find the resolvent for following kernels of Volterra equations:

(a) $K(x, y) = 1$;
(b) $K(x, y) = \lambda(x - y),\ \lambda > 0$;
(c) $K(x, y) = \exp(x - y)$;
(d) $K(x, y) = \lambda \exp(x^2 - y^2)$;
(e) $K(x, y) = \lambda \cosh x / \cosh y$.

8.17. With the help of Laplace transform, find the resolvent for following difference kernels of Volterra equations:
(a) $K(x, y) = \sin(x - y)$;
(b) $K(x, y) = \sinh(x - y)$;
(c) $K(x, y) = \cosh(x - y)$.

8.18. Let $u(x)$ be the solution of ordinary second-order equation:

$$\hat{L}u(x) = \frac{d^2u}{dx^2} + p(x)\frac{du}{dx} + q(x)u = f_0(x) \tag{8.47}$$

with initial condition $u(0) = C_0$, $u'(0) = C_1$ and analytic functions $p(x), q(x)$.
(a) Reduce Eq. (8.47) to the Volterra equation for function $\phi(x) = u''(x)$.
(b) Show that, if there is Green's function to Eq. (8.47), satisfying equation

$$\hat{L}G(x, y) = \delta(x - y)$$

with condition $G \equiv 0$ at $x < y$, the resolvent of kernel $K(x, y) = -p(x) - q(x)(x - y)$ at $x > y$ is expressed by formula

$$R(x, y) = \frac{\partial^2 G(x, y)}{\partial x^2}.$$

(c) By reducing to ordinary differential equation find the resolvent of Volterra equation with kernel

$$K(x, y) = \frac{2}{x^2}(x - y).$$

8.19. Show that the resolvent of difference polynomial kernel

$$K(x, s) = a_0(x) + a_1(x)(x - s) + \cdots + \frac{a_{n-1}(x)}{(n-1)!}(x - s)^{n-1}$$

is given by the expression

$$R(x, s) = \frac{\partial^n g}{\partial x^n},$$

where $g(x, s)$ satisfies equation

$$g^{(n)} - \left[a_0(x)g^{(n-1)} + a_1(x)g^{(n-2)} + \cdots + a_{n-1}(x)g\right] = \delta(x - s)$$

with zero boundary conditions at infinity.

8.20. Solve an equation with singular kernel (the Abel equation):

$$\int_0^x \frac{\phi(s)}{\sqrt{x - s}}\, ds = x.$$

8.21. Solution to the Fredholm equation of the second kind:

$$\phi(x) - \lambda \int_0^1 K(x, y)\phi(y)\, dy = f(x) \tag{8.48}$$

can be expressed in terms of resolvent $R(x, y; \lambda)$

$$\phi(x) = f(x) + \lambda \int_0^1 R(x, y; \lambda) f(y)\, dy.$$

(a) Show that the resolvent obeys the integral equation

$$R(x, y; \lambda) = K(x, y) + \lambda \int_0^1 K(x, t)R(t, y; \lambda)\, dt.$$

(b) Show that the resolvent can be written as the Neumann series

$$R(x, y; \lambda) = \sum_{n=1}^{\infty} \lambda^{n-1} K_n(x, y), \tag{8.49}$$

where repeated kernels are determined from recurrence relations:

$$K_1(x, y) \equiv K(x, y), \quad K_{n+1}(x, y) = \int_0^1 K(x, t)K_n(t, y)\, dt, \; n = 1, 2, \ldots$$

(c) Show that any repeated kernel obeys formula

$$K_n(x, y) = \int_0^1 K_m(x, t)K_{n-m}(t, y)\, dt, \; n = 2, 3, \ldots$$

(d) Show that for Fredholm kernel satisfying condition

$$\|K\|^2 = \int_0^1 \int_0^1 |K(x, y)|^2 \, dx dy < \infty,$$

the convergence radius of Neumann series (8.49) obeys inequality $|\lambda| < 1/\|K\|$.

8.22. By direct Neumann series summation find the resolvent of following Fredholm kernels:

(a) $K(x, y) = 1$;
(b) $K(x, y) = \exp(x - y)$;
(c) $K(x, y) = xy$.

8.23. Find the solution to the Fredholm equation with degenerate kernels:

(a)

$$\int_{-\infty}^{\infty} \exp(-x^2 - s^2)\phi(s) \, ds = \phi(x) + x^2;$$

(b)

$$\int_{-1}^{1} x^2 s^2 \phi(s) \, ds = \phi(x) + x^2;$$

(c)

$$\phi(x) - \lambda \int_0^1 (1 + x^2 y^2)\phi(y) \, dy = x^3;$$

(d)

$$\phi(x) - \lambda \int_0^1 \exp(x - y)(1 + xy)\phi(y) \, dy = \exp(x).$$

8.24. Show that if the Green function $G(x, y)$ is known for boundary problem

$$\hat{L}u = p_0(x)u^{(n)} + p_1(x)u^{(n-1)} + \cdots + p_n(x)u = f_0(x),$$
$$V_k(u) = \alpha_k^{(0)} u(0) + \alpha_k^{(1)} u'(0) + \cdots + \alpha_k^{(n-1)} u^{(n-1)}(0)$$
$$+\beta_k^{(0)} u(1) + \beta_k^{(1)} u'(1) + \cdots + \beta_k^{(n-1)} u^{(n-1)}(1) = 0,$$

where $\alpha_k^{(j)}$, $\beta_k^{(j)}$ are given constants, $k = 1, 2, \ldots, n$, then the Fredholm equation

$$u(x) = \lambda \int_0^1 G(x, y)u(y)\, dy + f(x).$$

reduces to the eigenvalue problem:

$$\hat{L}u = \lambda u + f_0(x), \quad V_k(u) = 0,$$

when $V_k(f) = 0$. How functions $f(x)$ and $f_0(x)$ are related?

8.25. By reducing to differential equations solve the Fredholm equation:

(a)

$$\phi(x) + \lambda \int_0^1 K(x, x')\phi(x')\, dx' = x^2 - x, \quad K(x, x') = x_<(x_> - 1), \quad \lambda > 0;$$

(b)

$$\phi(x) + \int_{-1}^1 |x - s|\phi(s)\, ds = \frac{x^3}{3} - x.$$

8.26. Find the solution to equation:

$$\phi(x) = \mathrm{e}^{-x} + \lambda \int_0^\infty \cos(2xs)\phi(s)\, ds.$$

8.27. Solve the spectral problem

$$\phi(x) = \lambda \int_{-\infty}^\infty \exp(-|x - y|)\, \phi(y)\, dy.$$

8.28. Solve the Cauchy problem with initial condition $u(x, 0) = 2\exp(-x^2) \times x^2$ for integro-differential equation

$$\frac{\partial u}{\partial t} = -\nu u(x, t) + \nu \int_{-\infty}^\infty K(x, s)u(s, t)\, ds$$

with Keilson–Storer kernel (8.36).

8.29. Find the solution $\phi(s)$ to the Urysohn equation:

(a)

$$\int_{-\infty}^{\infty} \phi(s)\phi(x-s)\,ds = e^{-x^2};$$

(b)

$$\int_{0}^{x} \phi(s)\phi(x-s)\,ds = \sin x.$$

8.30. Find all the solutions to the Hammerstein equation:

(a)

$$\phi(x) = \int_{0}^{1} xs\phi^2(s)\,ds + \frac{3}{4}x;$$

(b)

$$\phi(x) = \int_{0}^{x} \frac{\cos^2 s}{\cos x}\phi^2(s)\,ds + \frac{1}{\cos x}, \qquad x \in (0, 1).$$

8.31. Show that in kernel $F(x+y)$, defined by Eq. (8.29), the coefficient

$$M_n^2 = \frac{-\mathrm{i}C_n}{a'(\mathrm{i}\kappa_n)}$$

is positive.

8.32. For potential $U(x) = 2\kappa^2/\cosh^2 \kappa x$, find the eigenfunctions of continuous spectrum with the help of triangular representation (8.22).

8.33. Find the transmission coefficient $d(k) = 1/a(k)$ for potential $U(x) = 2\kappa^2/\cosh^2 \kappa x$.

8.34. Prove that the kernel $K(x, y)$ of triangular representation (8.22) for a non-reflective potential is expressed in form:

$$K(x, y) = \sum \psi_n(x)e^{-\kappa_n y},$$

where $\psi_n(x)$ is the eigenfunction of n-th discrete level.

Answers

8.14. $R(x, y) = \sum_{n=1}^{\infty} \frac{\lambda_n}{\lambda_n - \lambda} \psi_n(x)\psi_n^*(y)$.

8.15(a). **Hint.** Substitute solution (8.44) into the Volterra equation (8.43) and, changing the integration order, get the resolvent (8.45).

8.16(a). $R(x, y) = \exp(x - y)$.

8.16(b). $R(x, y) = \sqrt{\lambda} \sinh \sqrt{\lambda}(x - y)$.

8.16(c). $R(x, y) = \exp 2(x - y)$.

8.16(d). $R(x, y) = \lambda \exp[\lambda(x - y) + x^2 - y^2]$.

8.16(e). $R(x, y) = \lambda \exp[\lambda(x - y)] \cosh x / \cosh y$.

8.17(a). $R(x, y) = x - y$.

8.17(b). $R(x, y) = \frac{1}{\sqrt{2}} \sinh \sqrt{2}(x - y)$.

8.17(c).

$$R(x, y) = \exp\left(\frac{x - y}{2}\right)\left(\cosh \tau + \frac{1}{\sqrt{5}} \sinh \tau\right), \quad \tau = \frac{\sqrt{5}(x - y)}{2}.$$

8.18(a). Differential equation reduces to the Volterra equation (8.43) with kernel $K(x, y) = -p(x) - q(x)(x - y)$ and right side

$$f(x) = f_0(x) - p(x)C_1 - q(x)(C_1 x + C_0).$$

8.18(c).

$$R(x, y) = \frac{2}{3y}\left(1 - \frac{y^3}{x^3}\right).$$

8.20. **Hint.** The equations can be solved by the Abel transformation. Redesignate $x \rightarrow t$, multiply by $1/\sqrt{x - t}$, integrate over dt from 0 to x, change the order of integration and use the known integral $\int_s^x dt\,(x - t)^{-1/2}(t - s)^{-1/2} = \pi$.

$$\phi(x) = 2\sqrt{x}/\pi.$$

8.22(a). $R(x, y; \lambda) = 1/(1 - \lambda)$.

8.22(b). $R(x, y; \lambda) = \exp(x - y)/(1 - \lambda)$.

8.22(c). $R(x, y; \lambda) = xy/(1 - \lambda/3)$.

8.23(a). **Hint.** The Fredholm equation with degenerate kernel reduces to the set of algebraic equations.

$$\phi(x) = -x^2 + \frac{\sqrt{\pi}/2}{\sqrt{\pi}/2 - 1}\exp(-x^2).$$

8.23(b). $\phi(x) = -5x^2/3$.

8.23(c).

$$\phi(x) = x^3 + \frac{\lambda\left(1 - \frac{1}{2}\lambda\right)}{6\lambda}x^2 + \frac{\lambda\left(1 + \frac{1}{45}\lambda\right)}{4\Delta},$$
$$\Delta = \frac{1}{45}\left(4\lambda^2 - 54\lambda + 45\right) \neq 0.$$

8.23(d).

$$\phi(x) = e^x\left[1 + \frac{\lambda}{\Delta}\left(1 - \frac{\lambda}{12} + \frac{x}{2}\right)\right], \quad \Delta = 1 - \frac{4}{3}\lambda + \frac{\lambda^2}{12} \neq 0.$$

8.24.

$$f(x) = \int_0^1 G(x, y) f_0(y)\, dy.$$

8.25(a). Kernel $K(x, x')$ is the Green function of boundary problem:

$$u'' = f_0(x), \quad u(0) = u(1) = 0,$$

and then function $f(x)$ satisfies the same boundary condition. The Fredholm equation is equivalent to boundary problem

$$\phi'' + \lambda\phi = f''(x) = 2, \quad \phi(0) = \phi(1) = 0.$$

The Green function is built up in problem **7.26**(a):

$$\phi(x) = \frac{2}{k^2}\left[1 - \cos kx - \frac{1 - \cos k}{\sin k}\sin kx\right], \quad k = \sqrt{\lambda} \neq \pi n, \; n = 1, 2, \ldots$$

8.25(b). **Hint.** Equation reduces to differential one $\phi'' + 2\phi = 2x$ with boundary conditions $\phi'(-1) + \phi'(1) = 0$, $\phi(0) + \phi(1) = \phi'(-1) - \phi'(1)$. The solution is

$$\phi(x) = x - \frac{\sin\sqrt{2}x}{\sqrt{2}\cos\sqrt{2}}.$$

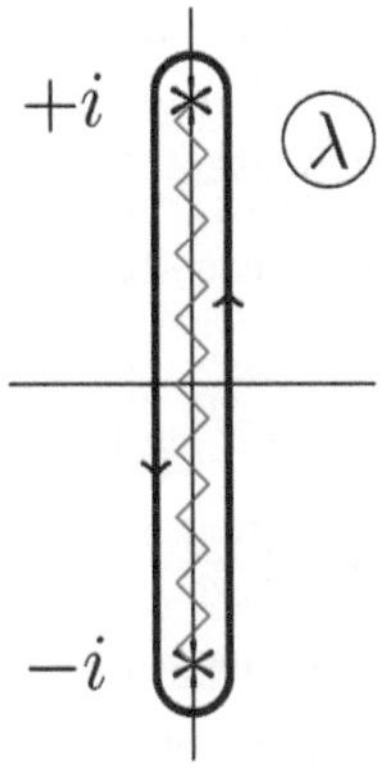

Figure 8.1 Integration path C. The cut, connecting the branching points of the integrand, is shown by zigzag. The direction of integration is shown by an arrow.

8.26.

$$\phi(x) = \frac{\exp(-x) + \dfrac{\lambda}{1+4x^2}}{1 - \dfrac{\pi\lambda^2}{4}}, \quad \lambda \neq \pm\frac{2}{\sqrt{\pi}}.$$

8.27. Operator has no eigenfunctions in class $L^2(-\infty, \infty)$. There are solutions in class of non-normable limited functions:

$$\lambda = \frac{1+k^2}{2}, \quad \phi(x) = \exp(ikx).$$

Here $\operatorname{Im} k = 0$, else the solution is not limited in the real axis. Each real $\lambda \geqslant 1/2$ is an eigenvalue.

8.28. Hint. Separate variables x and t and use the Fourier method: decompose the solution over eigenfunctions (8.37).

$$u(x,t) = \left[1 + \left(-1 + 2x^2\right)\exp\left(-\nu t(1-\alpha^2)\right)\right]\exp(-x^2).$$

8.29(a). Hint. The Fredholm–Urysohn equation of the convolution type can be solved by the Fourier transform.

$$\phi(x) = \pm\frac{2^{1/2}}{\pi^{1/4}}\exp(-2x^2).$$

8.29(b). Hint. Implement the Laplace transformation and solve the equation. Inverse transform is given by contour integral

$$\phi(x) = \pm\frac{1}{2\pi\mathrm{i}}\int_C \frac{\exp(\lambda x)}{\sqrt{1+\lambda^2}}\,d\lambda, \tag{8.50}$$

where the path can be closed. Deform it into the contour C going along the cut edges (Fig. 8.1) between $\lambda = \pm\mathrm{i}$. After replacement $\lambda = \mathrm{i}\sin\varphi$, Eq. (8.50) reduces to the Bessel integral representation, then $\phi(x) = \pm J_0(x)$.

8.30(a). **Hint.** Using the same procedure as in **8.23**(a), we reduce the problem to a quadratic equation.

$$\phi_1(x) = x, \quad \phi_2(x) = 3x.$$

8.30(b). $\phi(x) = 1/(1 - x)\cos x$.

8.31. Hint. Use the formula of perturbation theory for the energy shift $\delta E = -\langle\psi|\delta U|\psi\rangle$ of the Schrödinger operator.

8.32.

$$\psi(k, x) = \mathrm{e}^{\mathrm{i}kx}\frac{\mathrm{i}\kappa\tanh\kappa x + k}{\mathrm{i}\kappa + k}.$$

8.33. $a(k) = (k - \mathrm{i}\kappa)/(k + \mathrm{i}\kappa)$.

Chapter 9

Groups and Representations

9.1 Groups

The *group* is a set of elements $G = \{g_1, g_2, \dots\}$ with the binary operation. The operation mapping pair $g_1, g_2 \in G$ to the element $g_3 \in G$ is called "product" and denoted $g_3 = g_1 g_2$. The operation must satisfy the following conditions:

(a) $(ab)c = a(bc)$ (associativity);
(b) the identity element 1 exists, such as for any $a \in G$, $1a = a1 = a$ (existence of unity);
(c) for $a \in G$ there exists element $a^{-1} \in G$, such as $a^{-1}a = aa^{-1} = 1$ (existence of inverses).

A group consisting of a finite number n of elements is a *finite* group, otherwise, it is infinite. The cardinality of set G is the *order* of a group and denoted $|G|$. Subset H of the group G is a *subgroup*, if it is a group itself. It is denoted $H < G$. If for all the elements $g_1, g_2 \in G$ equality $g_1 g_2 = g_2 g_1$ takes place, then the group is *Abelian* (commutative). *Order* (period) of element x is the smallest natural number n, for which $x^n = 1$.

Mapping $G \to H$ of group G to group H is the *homomorphism* of G into H, if it saves the operation, namely from $g_1 \to h_1$, $g_2 \to h_2$, it follows that $g_1 g_2 \to h_1 h_2$. Homomorphism $G \to H$ is *isomorphism* when the mapping is one-to-one. Let $H < G$ be the subgroup. A set of elements of the form hg,

Mathematical Methods of Physics: Problems with Solutions
Igor V. Kolokolov, Evgeny A. Kuznetsov, Alexander I. Milstein, Evgeny V. Podivilov, Alexander I. Chernykh, David A. Shapiro, and Elena G. Shapiro

ISBN 978-981-5129-21-2 (Hardcover), 978-1-003-51360-5 (eBook)
www.jennystanford.com

where $h \in H$ is any element of subgroup and $g \in G$ is a fixed element of the group, is called *right coset* and denoted Hg. Similarly, gh is called *left coset.* The number r of different right cosets is called *index* of H in G and denoted $r = |G : H|$. Let $H, Hx_2, \ldots, Hx_r$ be the set of right cosets that exhausts the group. It is denoted $G = H + Hx_2 + \cdots + Hx_r$, where the plus sign means the union of subsets here.

We say that element $x, y \in G$ are *conjugate* (equivalent), if exists $z \in G$ such that $y = zxz^{-1}$. The set of elements equivalent to x is called the *conjugacy class* including x. If $g^{-1}Hg = H$ for all $g \in G$, then the subgroup H is called *invariant* (normal) and denoted $H \lhd G$.

The cosets give rise to a new group. Let $T \lhd G$ and $G = T + Tx_2 + \cdots + Tx_r$. Choose the cosets Tx_i as the elements of a new group $\overline{G}$. Define the product in $\overline{G}$ by the formula

$$(Tx_i)(Tx_j) = T(x_ix_j). \tag{9.1}$$

Group $\overline{G}$ is called *factor group* G over T and denoted $\overline{G} = G/T$.

9.2 Representations

Call the *representation* of group G any homomorphism α of G into some group W. *Matrix* representation of degree (dimension) n over the field of complex numbers $\mathbb{C}$ is homomorphism $\alpha : g \to T(g)$ of the group G into subgroup $W < GL(n, \mathbb{C})$, where $GL(n, \mathbb{C})$ is the group of complex matrices of dimension n with nonzero determinant. In other words, a representation is a mathematical group action on a vector space.

The representation is *faithful,* if the mapping α is the isomorphism. A matrix representation is faithful if from $T(g) = E$, it follows that $g = 1$, where E is the identity matrix.

The *trivial* (unity) representation comes out when all the group elements are mapping into 1. Here 1 is considered as a matrix 1×1. Let $G = \{g_1, \ldots, g_n\}$ be a finite group of the order n, V is the vector space with basis $v_1, \ldots, v_n$. For each i, $1 \leqslant i \leqslant n$ there exists the unique number j, $1 \leqslant j \leqslant n$ for which $gg_i = g_j$. Assume $T(g)v_i = v_j$, then $T(g)$ is the matrix of transformation in space V that converts vector v_i number i to vector v_j number j. Element in (j, i)-th cell of this matrix is 1, other elements of i-th column and j-th raw in $T(g)$ are equal to zero. Mapping $g \to T(g)$ is called *regular* representation of the group G and occurs faithfully.

We say that two representations T and T' are *equivalent* if matrix $S \in GL(n, \mathbb{C})$ exists, such as $T'(g) = ST(g)S^{-1}$ for each $g \in G$. Matrix representation $U : G \to GL(n, \mathbb{C})$ is *reducible* if being equivalent to representation:

$$T(g) = \begin{pmatrix} T_1(g) & V(g) \\ 0 & T_2(g) \end{pmatrix}, \tag{9.2}$$

where $T_j(g)$, $j = 1, 2$ is matrix $r_j \times r_j$, $r_1 + r_2 = n$. Otherwise the representation is called *irreducible*. When $V(g) = 0$ the representation is *completely* (fully) reducible. Any reducible representation of a finite group occurs completely reducible. *Character* $\chi(g)$ of representation Q of the group G is a trace: $\chi(g) = \mathrm{tr}\, Q(g)$. The character of irreducible representation is called *irreducible character*.

The number of non-equivalent irreducible representations is equal to the number of conjugacy classes. Let $\sigma_1, \ldots, \sigma_s$ be the classes, $D^{(1)}(g), \ldots, D^{(s)}(g)$ are matrices of all the irreducible representations. Then the *orthogonality* relation is valid:

$$\sum_{g \in G} D_{ij}^{(\alpha)*}(g) D_{kl}^{(\beta)}(g) = \frac{|G|}{n_\alpha} \delta_{\alpha\beta} \delta_{ik} \delta_{jl},$$
$$\alpha, \beta = 1, \ldots, s; \; i, j = 1, \ldots, n_\alpha; \; k, l = 1, \ldots, n_\beta, \tag{9.3}$$

where $n_\alpha = \dim D^{(\alpha)}(g)$, s is the number of different representations.

Consider some representation. Character values are not changed within a conjugacy class, then the table columns are marked by classes σ_i:

	σ_1	$\ldots$	σ_s
$\chi_1(g)$	$\chi_1(\sigma_1)$	$\ldots$	$\chi_1(\sigma_s)$
$\chi_s(g)$	$\chi_s(\sigma_1)$	$\ldots$	$\chi_s(\sigma_s)$

It is called the irreducible character table or simply "table of characters." Here $\chi_1(g), \ldots, \chi_s(g)$ are all the irreducible non-equivalent characters. From (9.3) the orthogonality of rows follows:

$$\sum_{i=1}^{s} h_i \chi_\alpha^*(\sigma_i) \chi_\beta(\sigma_i) = |G| \delta_{\alpha\beta}. \tag{9.4}$$

Here $h_i = |\sigma_i|$ is the number of elements in class σ_i. The character table considered as a square matrix has also the orthogonal columns:

$$\sum_{\alpha=1}^{s} \chi_\alpha^*(\sigma_i)\chi_\alpha(\sigma_j) = \frac{|G|}{h_i}\delta_{ij}. \tag{9.5}$$

The unity element solely forms a separate conjugacy class. A character of this class is equal to the degree of representation $\chi_\alpha(1) = n_\alpha$, then for n_α the relation follows

$$\sum_{\alpha=1}^{s} n_\alpha^2 = |G|, \tag{9.6}$$

that connects the degrees of irreducible representations and the order of the group.

Arbitrary representation $T(g)$ of a finite group G can be decomposed into the direct sum of irreducible representations

$$T(g) = c_1 T^{(1)}(g) \oplus \cdots \oplus c_s T^{(s)}(g). \tag{9.7}$$

Here $\oplus$ sign denotes the direct addition of two matrices T, D into block-diagonal one:

$$T \oplus D = \begin{pmatrix} T & 0 \\ 0 & D \end{pmatrix}. \tag{9.8}$$

Integers c_α show how many times irreducible representation $T^{(\alpha)}$ enters to $T(g)$. Block $T^{(1)}(g)$ occurs c_1 times, block $T^{(2)}(g)$ occurs c_2 times etc. Coefficients of the decomposition can be found from the orthogonality relation (9.4):

$$c_\alpha = \frac{1}{|G|}\sum_i h_i \chi_\alpha^*(\sigma_i)\chi(\sigma_i) \equiv \langle \chi_\alpha(g)|\chi\rangle_G. \tag{9.9}$$

Here $\chi(g)$ is the character of representation $T(g)$, $\langle\ldots\rangle_G$ means the average over the group, where the bra-vector in triangular brackets is implied complex conjugated.

Some comprehensive publications on the group theory consider the irreducible representations, see (Hall, 1999). Books in mathematical physics (Arfken and Weber, 1999; Hamermesh, 1989; Heine, 1977; Mathews and Walker, 1970; Petrashen, 1969; Wigner, 1959) consider it in more details.

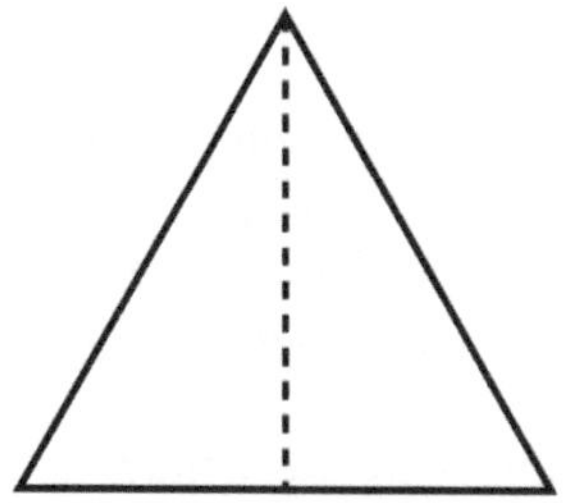

Figure 9.1 Enumeration scheme of the triangle vertices. A second-order axis is shown by dashed line.

Worked Examples

9.1. Build up the isomorphism of symmetry group $\mathbf{D}_3$ of the regular triangle and permutation group S_3:

$$S_3 = \left\{ \begin{pmatrix} 1 & 2 & 3 \\ i_1 & i_2 & i_3 \end{pmatrix} \right\}.$$

Solution. Let us enumerate the vertices in the counterclockwise direction, as shown in Fig. 9.1. Let r be 120° rotation, p be 180° rotation inverted vertices 1 and 3 and leaving 2 in place. Then

$$r \to \begin{pmatrix} 1\,2\,3 \\ 2\,3\,1 \end{pmatrix} \; r^2 \to \begin{pmatrix} 1\,2\,3 \\ 3\,1\,2 \end{pmatrix} \; p \to \begin{pmatrix} 1\,2\,3 \\ 3\,2\,1 \end{pmatrix}$$

$$rp \to \begin{pmatrix} 1\,2\,3 \\ 2\,1\,3 \end{pmatrix} \; r^2 p \to \begin{pmatrix} 1\,2\,3 \\ 1\,3\,2 \end{pmatrix} \; 1 \to \begin{pmatrix} 1\,2\,3 \\ 1\,2\,3 \end{pmatrix}.$$

It is easy to recognize that we have built the isomorphism.

9.2. Prove that two left (or right) cosets of a group G do not intersect or coincide (Lagrange's theorem).

Solution. While Hx and Hy have no mutual element, then there is nothing to prove. Then let the mutual element z exists, i.e., $z \in Hx$ and $z \in Hy$. Then $z = h_1 x = h_2 y$, hence $x = h_1^{-1} h_2 y$ and $hx = hh_1^{-1} h_2 y = h' y$. Then $Hx \subseteq Hy$. Analogously, $Hy \subseteq Hx$. Consequently, $Hx = Hy$.

9.3. Let G finite, $H < G$ be a subgroup. Prove that $|G| = |H| \cdot |G : H|$.

Solution. $G = H + Hx_2 + \cdots + Hx_r$, cosets Hx_i and Hx_j involve an equal number of elements $|H|$, a number r of different cosets is $r = |G : H|$.

9.4. (a) Let $C_G(x)$ be the set of elements from group G that commute with any $x \in G$. Prove that $C_G(x)$ is a subgroup of G.

(b) Let σ_i be a conjugacy class, $x \in \sigma_i$. Then number $|\sigma_i|$ of elements in class σ_i equals index $|G : C_G(x)|$ of subgroup $C_G(x)$.

Solution. Let $C = C_G(x)$ and elements conjugate to x with help of $g_1, g_2 \in G$ are equal, i.e., $g_1^{-1}xg_1 = g_2^{-1}xg_2$. Denote $c = g_1g_2^{-1}$, then

$$c^{-1}xc = g_2g_1^{-1}xg_1g_2^{-1} = x \Rightarrow xc = cx.$$

Consequently, $c \in C \Rightarrow Cg_1 = Cg_2$, i.e., g_1, g_2 belong to the same right coset.

Let g_1, g_2 lie in the same right coset with respect to subgroup C, i.e.,

$$Cg_1 = Cg_2 \Rightarrow c = g_1g_2^{-1} \in C.$$

Then

$$x = c^{-1}xc = (g_1g_2^{-1})^{-1}xg_1g_2^{-1} \Rightarrow g_1^{-1}xg_1 = g_2^{-1}xg_2,$$

i.e., elements, conjugate to x with the help of g_1, g_2, are equal. Thus the number of different elements conjugate to x is equal to the number of right cosets of group G with respect to subgroup C.

9.5. Find the order of the cube symmetry group.

Solution. Enumerate the apexes of a cube, as shown in Fig. 9.2(a). The elements generating the cube symmetry are rotations with respect to axes of the third and fourth order, shown in Fig. 9.2(b). Together with the mirror reflection in the horizontal midplane, they can be written as apex permutations:

$$a = \begin{pmatrix} 1\,2\,3\,4\,5\,6\,7\,8 \\ 2\,3\,4\,1\,6\,7\,8\,5 \end{pmatrix}, \quad b = \begin{pmatrix} 1\,2\,3\,4\,5\,6\,7\,8 \\ 1\,4\,8\,5\,2\,3\,7\,6 \end{pmatrix},$$

$$c = \begin{pmatrix} 1\,2\,3\,4\,5\,6\,7\,8 \\ 5\,6\,7\,8\,1\,2\,3\,4 \end{pmatrix}.$$

Elements a and b generate group G_1, mapping apex 1 into any other, as evident from the diagram

$$1 \xrightarrow{a} 2 \xrightarrow{a} 3 \xrightarrow{a} 4 \xrightarrow{b} 5 \xrightarrow{a} 6 \xrightarrow{a} 7 \xrightarrow{a} 8.$$

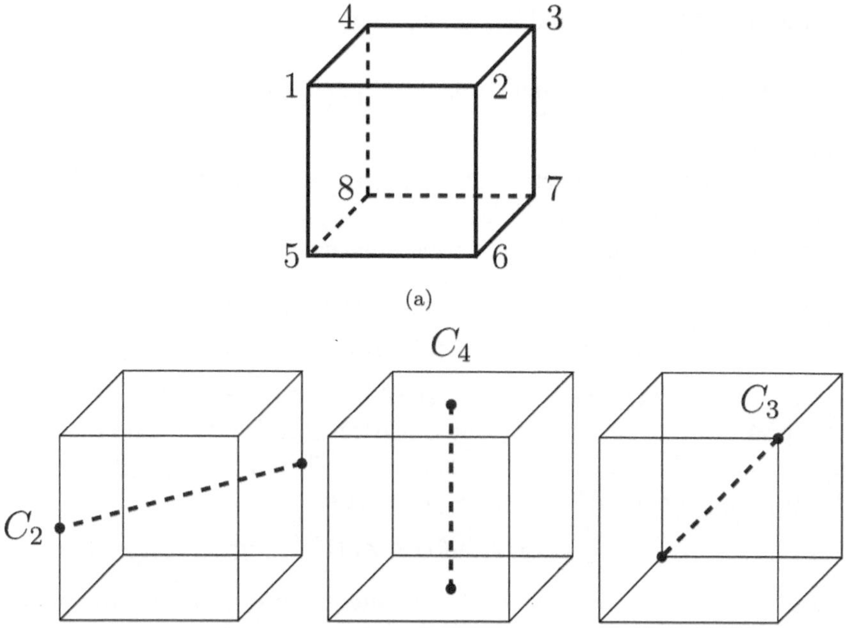

Figure 9.2 Enumeration of cube apexes (a) and elements of cube symmetry: axes of the second, fourth and third order (b).

Here $i \xrightarrow{x} j$ means that element x transforms i to j. Elements leaving 1 in place form subgroup $H_1 < G_1$. The right cosets are

$$G_1 = H_1 + H_1x_2 + H_1x_3 + H_1x_4 + H_1x_5 + H_1x_6 + H_1x_7 + H_1x_8.$$

Since all the elements from coset H_1x_i transform 1 to i and the number of apexes is 8, we write down all the right cosets with respect to H_1, then $|G_1 : H_1| = 8$. The subgroup consists of three elements: $H_1 = \{1, b, b^2\}$. Then $|G_1| = |H_1| \cdot |G : H_1| = 3 \cdot 8 = 24$. Reflection c is not contained in G_1. Since $c^2 = 1, ca = ac, cb = a^2ba^2c$, we see that $G = G_1 + G_1c$, then $|G| = 48$.

9.6. Show that the subgroup of index 2 is invariant.

Solution. If $G = H + Hx$, then $G = H + xH$. Consequently $Hx = xH$.

9.7. Check the product in factor group (9.1) to be uniquely defined.

Solution. Let $t_1x_i \in Tx_i$ and $t_2x_j \in Tx_j$. Then $t_1x_it_2x_j = t_1(x_it_2x_i^{-1})(x_ix_j) = t_3x_ix_j$, $t_3 \in T$, since $T \lhd G$. If $x_ix_j = x_k \in Tx_k$, then $t_3x_ix_j \in Tx_k$. Thus all the products of elements from Tx_i and Tx_j fall into coset Tx_k.

9.8. Show that:

(a) character is a function of conjugacy class, i.e., the characters of conjugate elements are equal;

(b) equivalent representations have equal characters.

Solution. If A is $n \times n$ matrix then its characteristic polynomial is

$$f(\lambda) = |A - \lambda E| = (-1)^n[\lambda^n - a_1\lambda^{n-1} + \cdots + (-1)^n a_n],$$

where $a_1 = \text{tr}\, A$. Let T be a matrix, then $|T^{-1}AT - \lambda E| = |T^{-1}(A - \lambda E)T| = |T^{-1}| \cdot |A - \lambda E| \cdot |T| = |A - \lambda E|$. Thus, matrices A and $T^{-1}AT$ have the same polynomial and all the more the same trace. Then matrices $T(g_1^{-1}gg_1) = T^{-1}(g_1)T(g)T(g_1)$ and $T(g)$ have equal traces:

$$\chi(g_1^{-1}gg_1) = \chi(g).$$

9.9. Show that any representation of a finite group is equivalent to unitary (matrix representation is unitary if it includes solely unitary matrices).

Solution. Let $T(g)$, $g \in G$ be a matrix representation of degree n and let $g_1 = 1, g_2, \ldots, g_t$ are elements of finite group G. Then

$$M = E + T^\dagger(g_2)T(g_2) + \cdots + T^\dagger(g_t)T(g_t)$$

is a positively defined Hermitian matrix, since each term is Hermitian and positive. Remind that the sign $\dagger$ denotes the Hermitian conjugation: $T^\dagger = (T^*)^T$.

Notice that for each $g \in G$

$$T^\dagger(g)MT(g) = \sum_{i=1}^{t} T^\dagger(g)T^\dagger(g_i)T(g_i)T(g)$$

$$= \sum_{i=1}^{t} T^\dagger(g_ig)T(g_ig) = \sum_{l=1}^{t} T^\dagger(g_l)T(g_l) = M. \qquad (9.10)$$

In linear algebra, it is proven that for any positively defined Hermitian matrix M there exists a non-degenerate matrix C, for which $C^\dagger MC = E$. Hence we get $M = \left(C^{-1}\right)^\dagger C^{-1}$. Substituting this expression for M into equality (9.10) we get

$$T^\dagger(g)\left(C^{-1}\right)^\dagger C^{-1}T(g) = \left(C^{-1}\right)^\dagger C^{-1},$$

hence

$$\left(C^{-1}T(g)C\right)^{\dagger}\left(C^{-1}T(g)C\right) = E.$$

That is why representation $T'(g) = C^{-1}T(g)C$ is unitary.

9.10. Construct the character table for group of triangle $\mathbf{D}_3$.

Solution. Group $\mathbf{D}_3 = \{1, r, r^2, p, pr, pr^2\}$, where $r^3 = p^2 = (pr)^2 = 1$, consists of three conjugacy classes:

$$\sigma_1 = \{1\}, \quad \sigma_2 = \{r, r^2\}, \quad \sigma_3 = \{r, pr, pr^2\}.$$

Then there are three irreducible representations of degree n_1, n_2 and n_3. Using Eq. (9.6) get

$$n_1^2 + n_2^2 + n_3^2 = |\mathbf{D}_3| = 6.$$

This equation has a unique solution $n_1 = 1,\ n_2 = 1,\ n_3 = 2$. Group D_3 has two irreducible representations of the first degree and one of the second degree:

(a) unity representation $\chi_1(g) = 1$ is the homomorphism of $\mathbf{D}_3$ into the unity element;
(b) odd representation $\chi_2(g) = \pm 1$ is the homomorphism of D_3 into group $C_2 = \{1, -1\}$. Its kernel is a subgroup generated by $2\pi/3$ rotation, i.e., $C_3 = \{1, r, r^2\} \to 1; \{p, pr, pr^2\} \to -1$.
(c) second-degree representation is generated by two 2×2 matrices, e.g.,

$$r = \begin{pmatrix} -\frac{1}{2} & -\frac{\sqrt{3}}{2} \\ \frac{\sqrt{3}}{2} & -\frac{1}{2} \end{pmatrix}, \quad p = \begin{pmatrix} -1 & 0 \\ 0 & 1 \end{pmatrix}.$$

Character $\chi_3(g)$ consists of their traces. As a result, we have a character table.

	σ_1	σ_2	σ_3
χ_1	1	1	1
χ_2	1	1	-1
χ_3	2	-1	0

Check the orthogonality of characters.

9.11. Find the character table for group $\mathbf{T}$ of regular tetrahedron rotations.

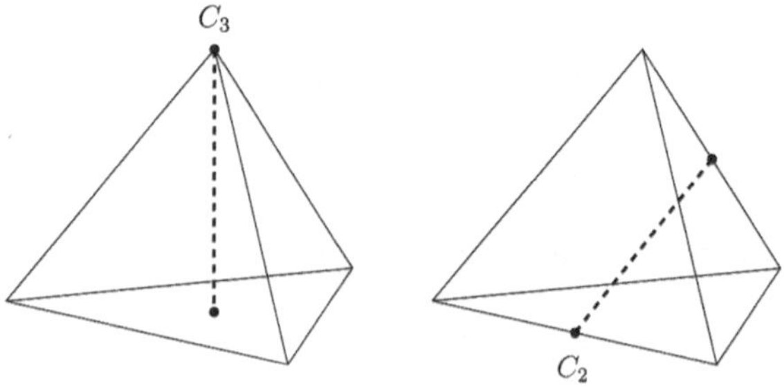

Figure 9.3 Axes of the third and the second order in the regular tetrahedron.

Solution. The group is of the order $|\mathbf{T}| = 12$ and generated by elements of the second- and third-order, shown in Fig. 9.3. Denote these elements p, r, respectively. Then the group can be written as

$$\mathbf{T} = \{1, r, r^2, p, pr, rp, pr^2, r^2p, rpr, r^2pr^2, rpr^2, r^2pr\},$$

where $p^2 = r^3 = (pr)^3 = 1$. Two auxiliary commutation relations follow from these conditions:

$$rpr = pr^2p, \quad r^2pr^2 = prp.$$

On the basis of geometric criteria from problems **9.20** and **9.21** we find four conjugacy classes:

$\sigma_1 = \{1\}$, $\sigma_2 = \{p, rpr^2, r^2pr\}$, $\sigma_3 = \{r, pr, rp, r^2pr^2\}$, $\sigma_4 = \{r^2, pr^2, r^2p, rpr\}$,

namely, unity element, 3 rotations by angle π, 4 rotations by $2\pi/3$ and 4 by $-2\pi/3$. From Eq. (9.6) we find the only decomposition into a sum of 4 squares:

$$12 = 1^2 + 1^2 + 1^2 + 3^2.$$

The unity element together with three second-order rotations form the invariant subgroup:

$$H = \sigma_1 + \sigma_2 = \{1, p, rpr^2, r^2pr\} \lhd \mathbf{T}.$$

The factor group consisting of cosets is isomorphic to the cyclic group $C_3 = \{1, r, r^2\}$, $r^3 = 1$. Its table of characters consists of cubic roots from 1, namely, 1, ε, ε^2, where $\varepsilon = e^{2\pi i/3}$:

	1	r	r^2
χ_1	1	1	1
χ_2	1	ε	ε^2
χ_3	1	ε^2	ε

Three first-degree representations of **T** are given by the factor group C_3. The third-degree representation is generated by 3 × 3 matrices:

$$1 = \begin{pmatrix} 1 & 0 & 0 \\ 0 & 1 & 0 \\ 0 & 0 & 1 \end{pmatrix}, \quad r = \begin{pmatrix} -\frac{1}{2} & -\frac{\sqrt{3}}{2} & 0 \\ \frac{\sqrt{3}}{2} & -\frac{1}{2} & 0 \\ 0 & 0 & 1 \end{pmatrix}, \quad p = \begin{pmatrix} -1 & 0 & 0 \\ 0 & -1 & 0 \\ 0 & 0 & 1 \end{pmatrix}.$$

Hence the character table is

	σ_1	σ_2	σ_3	σ_4
χ_1	1	1	1	1
χ_2	1	1	ε	ε^2
χ_3	1	1	ε^2	ε
χ_4	3	-1	0	0

Enumerate the tetrahedron apexes and write the permutations corresponding to representatives of each conjugacy class.

Exercises

9.12. Show that in any group $(ab)^{-1} = b^{-1}a^{-1}$ and in general

$$(a_1 a_2 \dots a_{n-1} a_n)^{-1} = a_n^{-1} a_{n-1}^{-1} \dots a_2^{-1} a_1^{-1}.$$

9.13. Show that the orders of a and a^{-1} coincide.

9.14. Show that if x and y are conjugate, then they are of the same order.

9.15. Show that ab and ba are of equal order.

9.16. Show that the symmetries of the regular triangle form a group. Build up its table of multiplication and find subgroups.

9.17. Let G be a group and $H < G$ be its subgroup. Show the numbers of the left and right cosets to be equal.

9.18. Show that:

(a) element $x \in G$ is conjugate to itself;
(b) if x is conjugate to y, then y is conjugate to x;
(c) if x is conjugate to y, and y is conjugate to z, then x is conjugate to z;
(d) group G is the union of non-overlapping conjugacy classes:

$$G = C_1 + \dots + C_s.$$

9.19. Show that the only finite group with two conjugacy classes is the group of order 2.

9.20. Show that in a transformation group, two rotations by angle φ regarding different axes are conjugate if the transformation exists in the group transforming one axis into another.

9.21. Show that rotations by angle φ and $-\varphi$ regarding axis OO' are conjugate if one of the following symmetry exists:

(a) a second-order axis perpendicular to OO';

(b) a mirror plane passing through axis OO'.

9.22. Find conjugacy classes for the following symmetry groups:

(a) regular triangle;

(b) square;

(c) regular tetrahedron.

9.23. Show that $H < G$ is invariant if and only if, when the right coset Hx is also the left coset xH.

9.24. Let H be the homomorphic image of group G. Consider its kernel, the set T of elements $t \in G$ that transfer into $1 \in G$. Show that $T \lhd G$.

9.25. Show that the cube rotation group **O** is homomorphic to group S_3, the permutation group of 3 symbols.

9.26. Show that if the order of a group is not a prime number, then the group has non-trivial[1] subgroups.

9.27. Set of elements $Z(G)$, that commute with all the elements of group G, is called *center* of G. Show that if $G/Z(G)$ is a cyclic group, then G is Abelian.

9.28. Show that all the irreducible representations of an Abelian group are of the first degree.

9.29. Let $G = D_4$ be the square symmetry group.

(a) Show that it is generated by two elements a and b, that satisfy relations $a^4 = 1,\ b^2 = 1,\ (ab)^2 = 1$.

(b) Let matrices $T(a)$ and $T(b)$ obey the relations:

$$[T(a)]^4 = E, \quad [T(b)]^2 = E, \quad [T(a)T(b)]^2 = E. \qquad (9.11)$$

[1]Trivial subgroups are the unity and the group itself.

Then, evidently, mappings $a^i b^j \to [T(a)]^i [T(b)]^j$ form the representation of G. Construct examples of matrices satisfying conditions (9.11).

9.30. Let $T(G)$ be the representation of a finite group by matrices over $\mathbb{C}$. Show that:

(a) character $\chi(g)$ is equal to the sum of roots of unity;

(b) $\chi(g^{-1}) = \chi(g)^*$.

9.31. Show that each representation of factor group G/H $(H \lhd G)$ is also the representation of group G.

9.32. Matrices of a representation of group S_3 are obtained by corresponding permutations of rows in the unity matrix 3×3:

$$1 \to \begin{pmatrix} 1 & 0 & 0 \\ 0 & 1 & 0 \\ 0 & 0 & 1 \end{pmatrix}, \quad p \to \begin{pmatrix} 0 & 1 & 0 \\ 1 & 0 & 0 \\ 0 & 0 & 1 \end{pmatrix}, \quad r \to \begin{pmatrix} 0 & 1 & 0 \\ 0 & 0 & 1 \\ 1 & 0 & 0 \end{pmatrix}.$$

Decompose the representation over irreducible components.

9.33. Build up the regular representation of S_3.

9.34. Decompose the regular representation of S_3 over irreducible components.

Answers

9.15. Hint. They are conjugate.

9.16. Use the notation from problem **9.1**. Then the multiplication table is

	1	r	r^2	p	rp	r^2p
1	1	r	r^2	p	rp	r^2p
r	r	r^2	1	rp	r^2p	p
r^2	r^2	1	r	r^2p	p	rp
p	p	r^2p	rp	1	r^2	r
rp	rp	p	r^2p	r	1	r^2
r^2p	r^2p	rp	p	r^2	r	1

There are four non-trivial subgroups: $H = \{1, r, r^2\}$, $F_1 = \{1, p\}$, $F_2 = \{1, rp\}$, $F_3 = \{1, r^2p\}$.

9.17. Hint. Map left coset gH to right coset $(gH)^{-1} \to Hg^{-1}$.

9.19. Hint. Let $\sigma_1 = 1$ and σ_2 be the conjugacy classes of group G. Then $G = \sigma_1 + \sigma_2$. If $|G| = n$ and $|\sigma_2| = m$, then $n = 1 + m$, at that m divides n (problem **9.4**(b)).

9.25. Hint. Let x, y, z be three straight lines connecting the centers of opposite cube faces. Then, the cube's symmetries induce a permutation group S_3 of system (x, y, z).

9.26. Hint. Let $g \in G$ and $g \neq 1$. Consider the subgroup involving all the powers of g: $1, g, g^2, \ldots, g^{n-1}, g^n = 1$.

9.28. Hint. Consider vector space V, where the Abelian group G is realized as a group of linear transformations. Let $g \in G$ and $g \neq 1$, then the space of eigenvectors of g, belonging to a fixed eigenvalue λ is G-invariant space, Then it coincides with V. Transformation g corresponds to a matrix proportional to the unity matrix (scalar matrix).

9.29(b). An example is:

$$T(a) = \begin{pmatrix} 0 & -1 \\ 1 & 0 \end{pmatrix}, \quad T(b) = \begin{pmatrix} 0 & 1 \\ 1 & 0 \end{pmatrix}, \quad T(ab) = \begin{pmatrix} -1 & 0 \\ 0 & 1 \end{pmatrix}.$$

9.32. $T(g) = T^{(1)}(g) \oplus T^{(3)}(g)$.

9.33.

$$T\begin{pmatrix} 1 & 2 & 3 \\ 3 & 1 & 2 \end{pmatrix} = \begin{pmatrix} 0&0&1&0&0&0 \\ 1&0&0&0&0&0 \\ 0&1&0&0&0&0 \\ 0&0&0&0&0&1 \\ 0&0&0&1&0&0 \\ 0&0&0&0&1&0 \end{pmatrix}, \quad T\begin{pmatrix} 1 & 2 & 3 \\ 1 & 3 & 2 \end{pmatrix} = \begin{pmatrix} 0&0&0&1&0&0 \\ 0&0&0&0&0&1 \\ 0&0&0&0&1&0 \\ 1&0&0&0&0&0 \\ 0&0&1&0&0&0 \\ 0&1&0&0&0&0 \end{pmatrix}.$$

9.34.

$$T(g) = T^{(1)}(g) \oplus T^{(2)}(g) \oplus 2T^{(3)}(g).$$

Chapter 10

Continuous Groups

10.1 Lie Groups and Algebras

Consider a group G of linear transformations. Let matrix elements be analytic functions of real parameters $\alpha = (\alpha_1, \ldots, \alpha_r)$. There exists a one-to-one correspondence between the coordinates' origin together with its neighborhood in the r-dimensional space and neighborhood of the identity element of the group. The zero of parameters $\alpha = 0$ corresponds to the identity element. Denote $g(\alpha_1, \alpha_2, \ldots, \alpha_r)$ an element of group G. If

$$g(\alpha_1, \alpha_2, \ldots, \alpha_r)g(\alpha'_1, \alpha'_2, \ldots, \alpha'_r) = g(\alpha''_1, \alpha''_2, \ldots, \alpha''_r),$$

then the coordinates of element α'' are functions of α, α':

$$\alpha''_k = \phi_k(\alpha_1, \alpha_2, \ldots, \alpha_r; \alpha'_1, \alpha'_2, \ldots, \alpha'_r), \quad k = 1, \ldots, r.$$

When functions ϕ_k are infinitely differentiable by all the arguments then G is called the *Lie group*.[1] Number r of independent parameters is its *dimension*, $r = \dim G$. The choice of parameters (parametrization) is ambiguous, while the dimension r is single. In this chapter, we consider the matrix Lie groups.

Intensional examples are classical matrix groups. Some of the classical Lie groups are: the complete linear group $\mathbf{GL}(n, \mathbb{C})$ or $\mathbf{GL}(n, \mathbb{R})$ (all the reversible

[1] In other words, the Lie group is a smooth manifold with group laws given by smooth (infinitely differentiable) maps.

Mathematical Methods of Physics: Problems with Solutions
Igor V. Kolokolov, Evgeny A. Kuznetsov, Alexander I. Milstein, Evgeny V. Podivilov, Alexander I. Chernykh, David A. Shapiro, and Elena G. Shapiro

ISBN 978-981-5129-21-2 (Hardcover), 978-1-003-51360-5 (eBook)
www.jennystanford.com

complex or real $n \times n$ matrices), the special group $\mathbf{SL}(n, \mathbb{C})$ or $\mathbf{SL}(n, \mathbb{R})$ (all the unimodular matrices, i.e., with unity determinant), the unitary group $\mathbf{U}(n, \mathbb{C})$ (complex unitary matrices), the orthogonal group $\mathbf{O}(n, \mathbb{R})$ (real orthogonal matrices), the special unitary group $\mathbf{SU}(n, \mathbb{C})$ (unimodular complex unitary matrices) the special orthogonal group $\mathbf{SO}(n, \mathbb{R})$ (unimodular real orthogonal matrices), the special pseudo-orthogonal group $\mathbf{SO}(p, q; \mathbb{R})$, $p + q = n$ (unimodular real pseudo-orthogonal matrices, that conserves form $\sum_{i=1}^{p} x_i^2 - \sum_{i=p+1}^{n} x_i^2$). The second argument, the field of matrix elements, is usually omitted for the sake of shortening the notation

Lie algebra is the linear space L, supplied by binary operation $[x, y]$ (Lie braces), that satisfies conditions for all $x, y, z \in L,\ a \in \mathbb{C}$:

(a) $[ax + y, z] = a[x, z] + [y, z]$ (linearity);
(b) $[x, y] = -[y, x]$ (antisymmetry);
(c) $[x, [y, z]] + [z, [x, y]] + [y, [z, x]] = 0$ (Jacobi identity).

Examples of Lie braces are the commutator of square matrices, the Poisson bracket between functions of coordinates and momenta in classical mechanics, and the vector product in $\mathbb{R}^3$ (see problem **10.10**).

Consider the Lie group G of linear transformations with matrix elements $g_{ik} = g_{ik}(\alpha_1, \ldots, \alpha_r)$. A partial derivative with respect to parameter α_l in the coordinate origin gives a matrix with elements

$$(I_l)_{ik} = \left(\frac{\partial g_{ik}}{\partial \alpha_l}\right)_{\alpha=0}.$$

Matrix I_l is called the *generator* (infinitesimal operator) of group G. Commutator $[I_k, I_s] = I_k I_s - I_s I_k$ reduces to a linear combination of basic matrices

$$[I_k, I_s] = \sum_l c_{ks}^l I_l.$$

Since the commutator satisfies properties a,b,c, the linear shell of generators forms a Lie algebra called the *Lie algebra of Lie group G*. Coefficients c_{ks}^l are a set of constants known as the *structure constants* the Lie algebra.

Its Lie algebra can help to reconstruct the Lie group. The reconstruction procedure is unambiguous when the group is connected and simply connected.[2] For instance, one can reconstruct the one-parametric group by

[2] M is a connected manifold if one can join any two points by a smooth curve, belonging to M. Manifold M is simply connected if we can pull each closed curve to a point.

exponential mapping:

$$\exp\theta I = 1 + \theta I + \frac{1}{2!}\theta^2 I^2 + \dots,$$

where θ is the parameter (see problem **10.1**).

10.2 Representations of the Rotation Group

A *matrix representation* of the Lie group G is a homomorphism of G to the group of linear transformations $T(g)$ of linear space V. The representation in space V is *reducible* if V contains an invariant subspace with respect to G (different from V and zero subspace).

Linear space of representation may be also infinite dimensional, e.g., the Hilbert space. Denote $\mathcal{F}$ the space of infinitely differentiable functions $f(x)$, $x \in \mathbb{R}^3$, defined at a sphere. For each element of $g \in \mathbf{SO}(3)$ we assign linear transformation

$$T(g) : f(x) \to f'(x) = f(g^{-1}x), \tag{10.1}$$

that maps $\mathcal{F}$ to itself. Equation (10.1) determines *action of element g on function f*. If a new transform $T(g_2)$ follows $T(g_1)$, namely

$$T(g_2) : f'(x) \to f''(x) = f'(g_2^{-1}x),$$

then the result is their composition

$$T(g_2)T(g_1) : f(x) \to f''(x) = f'(g_2^{-1}x) = f(g_1^{-1}(g_2^{-1}x)) = f((g_2g_1)^{-1}x),$$

hence $T(g_2)T(g_1) = T(g_2g_1)$. Then accordance $g \to T(g)$ is a representation of the group of rotations $\mathbf{SO}(3)$ in the infinite-dimensional function space.

In order to find out the irreducible representations of group $\mathbf{SO}(3)$, we consider its Lie algebra satisfying commutation relations:

$$[I_\alpha, I_\beta] = e_{\alpha\beta\gamma}I_\gamma,$$

where infinitesimal operator I_α is the generator of rotation with respect to axis $\alpha = x, y, z$. In quantum mechanics, the operator $\mathbf{J} = i\mathbf{I}$ is the angular momentum. We find the finite-dimensional irreducible representations of the rotation group. First, we consider the corresponding representations of its Lie algebra. This problem reduces to seeking the spectrum of J^2 and its eigenvectors. As a result each irreducible representation $D^{(j)}$ is specified by the eigenvalue $j(j+1)$ of operator J^2. Its dimension at given integer j is equal to the number of possible projections of the angular momentum to the quantization axis z, namely dim $D^{(j)} = 2j + 1$.

Let V and U be vector spaces above field $\mathbb{C}$ of a finite dimension m and n, respectively. Let $\{v_i\}$ and $\{u_i\}$ be the bases of spaces V and U. For arbitrary $v \in V, u \in U$ consider the set of formal sums of products $v \otimes u$, with the following properties:

$$v \otimes (u' + u'') = v \otimes u' + v \otimes u'',$$
$$(v' + v'') \otimes u = v' \otimes u + v'' \otimes u,$$
$$\lambda(v \otimes u) = (\lambda v) \otimes u = v \otimes (\lambda u), \quad \lambda \in \mathbb{C}.$$

The set constructed in that way is a vector space over vectors $v_i \otimes u_j = w_{ij}$, where $i = 1, \ldots, m$, $j = 1, \ldots, n$. It is called *tensor product* of spaces V and U and denoted $W = V \otimes U$. When V and U are the spaces with a scalar product, then the scalar product in $V \otimes U$ can be defined by the formula

$$(v_1 \otimes u_1, v_2 \otimes u_2) = (v_1, v_2)(u_1, u_2).$$

We introduce the concept of *direct* (tensor) product of square matrices. Let there be two square matrices $T = \|t_{ij}\|$ and $R = \|r_{ij}\|$ of the order n and m, respectively. Direct product of matrices T and R is the following square matrix of order $n \times m$:

$$T \otimes R = \begin{pmatrix} t_{11}R & t_{12}R & \ldots & t_{1n}R \\ t_{21}R & t_{22}R & \ldots & t_{2n}R \\ \ldots & \ldots & \ldots & \ldots \\ t_{n1}R & t_{n2}R & \ldots & t_{nn}R \end{pmatrix}.$$

Its trace is $\mathrm{tr}\,(T \otimes R) = \mathrm{tr}\,T \cdot \mathrm{tr}\,R$.

Let two representations $T(g)$ and $T'(g)$ be specified for group G, i.e., linear operators $T(g)$ in space V and $T'(g)$ in space U correspond to each $g \in G$, so that $T(g_1g_2) = T(g_1)T(g_2)$, $T'(g_1g_2) = T'(g_1)T'(g_2)$ for all $g_1, g_2 \in G$. Define the *tensor* (direct) product of the group representations. Choose basis of spaces V and U: $V = \langle v_1, v_2, \ldots, v_{\bar{n}}\rangle$, $U = \langle u_1, u_2, \ldots, u_m\rangle$. Let in this bases matrix $D(g)$ corresponds to operator $T(g)$ and matrix $D'(g)$ corresponds to operator $T'(g)$, i.e.,

$$T(g)v_i = \sum_{j=1}^{n} D_{ji}(g)v_j \quad T'(g)u_k = \sum_{l=1}^{m} D_{lk}(g)u_l.$$

Define linear operators $T''(g)$ in space $V \otimes U$ as

$$T''(g)(v_i \otimes u_k) = T(g)v_i \otimes T'(g)u_k = \sum_j D_{ji}(g)v_j \otimes \sum_l D'_{lk}(g)u_l$$

$$= \sum_{j,l} D_{ji}(g)D'_{lk}(g)\left(v_j \otimes u_l\right). \quad (10.2)$$

It is easy to check that the matrix corresponding to operator $T''(g)$, is the direct product $D(g) \otimes D'(g)$ of their matrices. It is the *direct* (tensor) product of group representations.

The direct product can be decomposed over irreducible representations. For group **SO**(3) this problem corresponds to angular momenta coupling in quantum mechanics. Let $D^{(j_1)}$ and $D^{(j_2)}$ be two irreducible representations with angular momenta j_1 and j_2, respectively. Then the direct product $D^{(j_1)} \otimes D^{(j_2)}$ can be written as a direct sum

$$D^{(j_1)} \otimes D^{(j_2)} = D^{(j_1+j_2)} \oplus D^{(j_1+j_2-1)} \oplus \cdots \oplus D^{(|j_1-j_2|)} \quad (10.3)$$

(see problem **10.29**). It is the *Clebsch–Gordan decomposition* for group **SO**(3).

From decomposition (10.3) it follows that the bases of the representations $D^{(J)}$: $|J\,M\rangle(-J \leq M \leq J)$ can be written in terms of the basis of direct product $|j_1 m_1\rangle \otimes |j_2 m_2\rangle$:

$$|J\,M,\, j_1 j_2\rangle = \sum_{m_1+m_2=M} C^{J\,M}_{j_1 m_1\; j_2 m_2} |j_1 m_1\rangle \otimes |j_2 m_2\rangle.$$

The coefficients of decomposition $C^{J\,M}_{j_1 m_1\; j_2 m_2}$ are called the Clebsch–Gordan coefficients. Books (Barut and Raczka, 1977; Dubrovin *et al.*, 1992; Richtmyer, 1978; Zhelobenko, 2006) describe the Lie groups and their irreducible representations.

Worked Examples

10.1. Consider one-parameter group $G = g(\theta)$. Choose the parameter so that

$$g(\theta_1)g(\theta_2) = g(\theta_1 + \theta_2); \quad (10.4)$$

$$g(0) = E. \quad (10.5)$$

Let $I_\theta = (dg/d\theta)_{\theta=0}$ be the infinitesimal matrix corresponding to parameter θ. Show that

$$g(\theta) = \exp I_\theta \theta. \quad (10.6)$$

Solution. Differentiating both sides of Eq. (10.4) by θ_1 and setting $\theta_1 = 0$, $\theta_2 = \theta$, we get $dg(\theta)/d\theta = I_\theta g(\theta)$. These equations have the unique solution satisfying initial condition (10.5).

10.2. Find the generator of group **SO**(2) in the representation by functions $f(x, y)$ of two real variables. Write down all the irreducible representations and check up the orthogonality between their characters.

Solution. Let $g(\alpha)$ be a rotation, α is the angle. Transformation of function f is given by Eq. (10.1):

$$g(\alpha)f(x, y) = f(x\cos\alpha + y\sin\alpha, -x\sin\alpha + y\cos\alpha).$$

Differentiating both sides at $\alpha = 0$ gives the generator I:

$$If = y\frac{\partial f}{\partial x} - x\frac{\partial f}{\partial y}.$$

In polar coordinates, $I = -\partial/\partial\varphi$, where φ is the azimuthal angle.

Its eigenfunctions are $\psi_m(\varphi) = e^{im\varphi}$, where m occurs integer, as follows from 2π-periodicity. Each function ψ_m is the basis of an irreducible representation. The group is Abelian, then all the representations are one-dimensional: $D_m(\alpha) = e^{im\alpha}/\sqrt{2\pi}$. We consider D_m, $m = 0, \pm 1, \ldots$ as an 1×1 matrix, then the representations coincide with characters:

$$\chi_m(\alpha) = \frac{1}{\sqrt{2\pi}}e^{im\alpha}.$$

Testing the orthogonality, we get

$$\langle\chi_m|\chi_n\rangle_G = \overline{\chi_m^*(\alpha)\chi_n(\alpha)} = \frac{1}{2\pi}\int_{-\pi}^{\pi} e^{-im\alpha+in\alpha}\,d\alpha = \delta_{mn},$$

where a bar denotes the averaging over a period.

10.3. Let $g(\alpha_1, \alpha_2, \alpha_3)$ be the matrix corresponding to rotation with respect to axis $\boldsymbol{\alpha} = (\alpha_1, \alpha_2, \alpha_3)$ by angle $\theta = (\alpha_1^2 + \alpha_2^2 + \alpha_3^2)^{1/2}$. Show that

$$g(\alpha_1, \alpha_2, \alpha_3) = \exp\left(I_1\alpha_1 + I_2\alpha_2 + I_3\alpha_3\right),$$

where I_1, I_2, I_3 are infinitesimal matrices of group **SO**(3).

Solution. Consider the one-parametric group of rotations with respect to axis $\boldsymbol{\alpha}$. Using Eq. (10.6), we get

$$g(\theta) = \exp I_\theta\theta = \exp\theta\sum_{j=1}^{3} I_j \left.\frac{d\alpha_j}{d\theta}\right|_{\theta=0}.$$

Here

$$\alpha_1 = \theta \cos(\widehat{Ox, \boldsymbol{\alpha}}), \quad \alpha_2 = \theta \cos(\widehat{Oy, \boldsymbol{\alpha}}), \quad \alpha_3 = \theta \cos(\widehat{Oz, \boldsymbol{\alpha}}).$$

Consequently

$$g(\alpha_1, \alpha_2, \alpha_3) = \exp\left(I_1\alpha_1 + I_2\alpha_2 + I_3\alpha_3\right).$$

10.4. Find the infinitesimal matrices of rotation group **SO**(3), with parametrization from problem **10.3**.

Solution. Choose components $\alpha_1, \alpha_2, \alpha_3$ of vector $\boldsymbol{\alpha}$ as the parameters of group **SO**(3):

$$g(\alpha_1, 0, 0) = \begin{pmatrix} 1 & 0 & 0 \\ 0 & \cos\alpha_1 & -\sin\alpha_1 \\ 0 & \sin\alpha_1 & \cos\alpha_1 \end{pmatrix}.$$

Hence

$$I_1 = \left.\frac{\partial g(\alpha_1, 0, 0)}{\partial \alpha_1}\right|_{\alpha_1=0} = \begin{pmatrix} 0 & 0 & 0 \\ 0 & 0 & -1 \\ 0 & 1 & 0 \end{pmatrix}.$$

Analogously

$$I_2 = \begin{pmatrix} 0 & 0 & 1 \\ 0 & 0 & 0 \\ -1 & 0 & 0 \end{pmatrix}, \quad I_3 = \begin{pmatrix} 0 & -1 & 0 \\ 1 & 0 & 0 \\ 0 & 0 & 0 \end{pmatrix}.$$

Calculating commutators between the matrices we obtain the structure constants of Lie algebra:

$$[I_i, I_j] = e_{ijk} I_k. \qquad (10.7)$$

10.5. A possible parametrization of 2×2 unitary matrix $g \in$ **SU**(2) is $g = e^{i\mathbf{a}\cdot\boldsymbol{\sigma}}$, where

$$\mathbf{a}\cdot\boldsymbol{\sigma} = \begin{pmatrix} a_3 & a_1 - ia_2 \\ a_1 + ia_2 & -a_3, \end{pmatrix}$$

$\mathbf{a} = (a_1, a_2, a_3)$ are the parameters. Find generators of the group and the Lie algebra.

Solution. The generators are partial derivatives

$$L_k = \left.\frac{\partial g}{\partial a_k}\right|_{\mathbf{a}=0}.$$

Hence $L_k = i\sigma_k$, $[L_i, L_j] = -2e_{ijk}L_k$. To obtain relation (10.7) all the generators have to be scaled by -2 times: $I_k = -L_k/2 = -i\sigma_k/2$.

10.6. Find the explicit form of matrix $g \in \mathbf{SU}(2)$ in the standard parametrization $\boldsymbol{\varphi} = \varphi\mathbf{n}$, where $\mathbf{n}$ is a unit vector, φ is the angle of rotation around $\mathbf{n}$. Calculate its character.

Solution. The matrix is given by formula

$$g(\boldsymbol{\varphi}) = \exp\left[-\frac{i\boldsymbol{\sigma}\cdot\boldsymbol{\varphi}}{2}\right]$$
$$= \begin{pmatrix} \cos\frac{\varphi}{2} - in_3\sin\frac{\varphi}{2} & (-in_1 + n_2)\sin\frac{\varphi}{2} \\ (-in_1 - n_2)\sin\frac{\varphi}{2} & \cos\frac{\varphi}{2} + in_3\sin\frac{\varphi}{2} \end{pmatrix}.$$

The character is $\chi(\varphi) = 2\cos(\varphi/2)$.

Introducing notations $x_0 = \cos(\varphi/2)$, $x_i = n_i\sin(\varphi/2)$, we get the invariant

$$x_0^2 + x_1^2 + x_2^2 + x_3^3 = 1,$$

the equation of the unit sphere in four-dimensional space. The sphere denoted S^3 in topology, is connected and simply connected manifold of the group $\mathbf{SU}(2)$.

10.7. Find generators of the group $\mathbf{SU}(2)$ and the Casimir operator[3] in a representation on functions of two complex variables $f(u, v)$. Test the commutative relations.

Solution. Consider the infinitesimal rotations by angle α around axes n_i, $i = 1, 2, 3$:

$$I_1 f = \left.\frac{dg_{n1}(\alpha)}{d\alpha}\right|_{\alpha=0} f(u, v)$$
$$= \frac{d}{d\alpha} f\left(u\cos\frac{\alpha}{2} + iv\sin\frac{\alpha}{2}, v\cos\frac{\alpha}{2} + iu\sin\frac{\alpha}{2}\right)\Bigg|_{\alpha=0}$$
$$= \frac{i}{2}\left(u\frac{\partial}{\partial v} + v\frac{\partial}{\partial u}\right) f(u, v),$$
$$I_2 f = \left.\frac{dg_{n2}(\alpha)}{d\alpha}\right|_{\alpha=0} f(u, v)$$
$$= \frac{d}{d\alpha} f\left(u\cos\frac{\alpha}{2} + v\sin\frac{\alpha}{2}, v\cos\frac{\alpha}{2} - u\sin\frac{\alpha}{2}\right)\Bigg|_{\alpha=0}$$

[3] A quadratic combination of generators which commutes with all the generators.

$$= \frac{i}{2}\left(v\frac{\partial}{\partial u} - u\frac{\partial}{\partial v}\right) f(u, v),$$

$$I_3 f = \left.\frac{dg_{n3}(\alpha)}{d\alpha}\right|_{\alpha=0} f(u, v) = \left.\frac{d}{d\alpha} f\left(ue^{i\alpha/2}, ve^{-i\alpha/2}\right)\right|_{\alpha=0}$$

$$= \frac{i}{2}\left(u\frac{\partial}{\partial u} - v\frac{\partial}{\partial v}\right) f(u, v).$$

The Casimir operator $C = I_1^2 + I_2^2 + I_3^2$ can be simplified and written in the form:

$$C = -L^2 - L, \quad L = \frac{1}{2}\left(u\frac{\partial}{\partial u} + v\frac{\partial}{\partial v}\right).$$

The commutation relations can be derived using formula for the commutator of a product $[A, BC] = A[B, C] + [A, B]C$ and ultimately coincide with (10.7).

10.8. Find the irreducible representations of the group **SU**(2) with a finite dimension.

Solution. Instead of the Casimir operator we may look for eigenfunctions of operator L. Its eigenfunctions turn to be the uniform polynomials $\Pi_{nk} = u^k v^{n-k}$. For generator I_3 the eigenvalues are $I_3\Pi_{nk} = i(k - n/2)\Pi_{nk}$. For the Casimir operator they are $C\Pi_{nk} = -n(n+2)\Pi_{nk}/4$. The dimension of the corresponding subspace is a number of uniform polynomials of degree n, namely $\dim \Pi_n = n + 1$.

For the finite dimension, n must be an integer. Test this property building up the raising and lowering operators

$$I_\pm = I_1 \pm I_2.$$

Denote $l = n/2$, then integer l correspond to irreducible representations of **SO**(3), while half-integer l correspond to the representations of group **SU**(2) only. Quantum mechanics explains the reason for the double number of irreducible representations of the group **SU**(2) compared to **SO**(3). The wave functions ψ_i, ψ_f for a particle with spin 1/2 are transformed by a representation of group **SU**(2), i.e., changes its sign after rotation by angle 2π. Whereas the observable matrix elements of Hermitian operator $\langle\psi_f|\hat{V}|\psi_i\rangle$ must be unchanged after the rotation. Then **SO**(3) serves as the transformation group of observable values.

10.9. Build up a homomorphism of group **SU**(2) to group **SO**(3).

Solution. Matrices from **SU**(2) have the form

$$\begin{pmatrix} \alpha & \beta \\ -\beta^* & \alpha^* \end{pmatrix},$$

where α and β are arbitrary complex numbers, satisfying only the condition $\alpha\alpha^* + \beta\beta^* = 1$. The group **SU**(2) of all matrices of this form is the set of linear transformations

$$u = \alpha u' + \beta v', \quad v = -\beta^* u' + \alpha^* v', \tag{10.8}$$

conserving the form $uu^* + vv^*$. With the help of complex parameters u, v one can define three real parameters

$$x_1 = u^* v + v^* u, \quad x_2 = -\mathrm{i}(u^* v - v^* u), \quad x_3 = uu^* - vv^*. \tag{10.9}$$

They satisfy a condition

$$x_1^2 + x_2^2 + x_3^2 = (u^* u + v^* v)^2. \tag{10.10}$$

A linear transformation (10.8), applied to Eq. (10.9), induces a linear transformation of real variables x_1, x_2, x_3. According to equality (10.10) the transformation belongs to the orthogonal group **SO**(3).

Write down this linear transformation:

$$\begin{aligned} x_1 = {} & \frac{\alpha^2 + \alpha^{2*} - \beta^2 - \beta^{2*}}{2} x_1' \\ & + \frac{\alpha^2 - \alpha^{2*} + \beta^2 - \beta^{2*}}{2\mathrm{i}} x_2' - (\alpha\beta + \alpha^*\beta^*) x_3', \\ x_2 = {} & \frac{\alpha^{2*} - \alpha^2 + \beta^2 - \beta^{2*}}{2\mathrm{i}} x_1' \\ & + \frac{\alpha^2 + \alpha^{2*} + \beta^2 + \beta^{2*}}{2} x_2' - \mathrm{i}(\alpha\beta - \alpha^*\beta^*) x_3', \\ x_3 = {} & (\alpha\beta^* + \alpha^*\beta) x_1' + \mathrm{i}(\alpha^*\beta - \alpha\beta^*) x_2' + (\alpha\alpha^* - \beta\beta^*) x_3'. \end{aligned}$$

We have mapped an element of group **SU**(2) into group **SO**(3). The built mapping is not one-to-one, since

$$\begin{pmatrix} 1 & 0 \\ 0 & 1 \end{pmatrix} \quad \text{and} \quad \begin{pmatrix} -1 & 0 \\ 0 & -1 \end{pmatrix}$$

are represented by the same identity matrix from a group **SO**(3).

10.10. Prove that three-dimensional vector space $\mathbb{R}^3$ supplied by the vector product is the Lie algebra. What the Lie group does this algebra correspond?

Solution. $\mathbb{R}^3$ is the vector space with vector product: $[\mathbf{a}\times\mathbf{b}] \in \mathbb{R}^3$. The vector product obeys all the algebra Lie properties. Then $\mathbb{R}^3$ is the Lie algebra with dimension 3. In order to reconstruct the corresponding Lie group we may use the one-to-one correspondence between vector $\mathbf{a} = (a_1, a_2, a_3)$ and 3×3 antisymmetric matrix

$$\mathbf{a} \to \begin{pmatrix} 0 & -a_1 & a_3 \\ a_1 & 0 & -a_2 \\ -a_3 & a_2 & 0 \end{pmatrix}.$$

Thus, $\mathbb{R}^3$ is isomorphic to the algebra of 3×3 antisymmetric matrices. In this space, the vector product reduces to the commutator of matrices. We could get the unity vectors as a basis that correspond to generators I_1, I_2, I_3. Their commutation relations are given by Eq. (10.7). Hence, the group consists of 3×3 orthogonal matrices; this is the fundamental[4] representation $D^{(1)}$ of the rotation group **SO**(3) with angular momentum $j = 1$.

10.11. Let $f(\theta, \varphi)$ be a function given on the sphere of unit radius, g is the rotation by angle α with respect to axis Oz corresponding to matrix $T(g)$. Find $I_3 f(\theta, \varphi)$, where I_3 is the generator of rotation with respect to axis z.

Solution. Since $T(g)f(\theta, \varphi) = f(\theta, \varphi - \alpha)$, at small α

$$T(g)f(\theta, \varphi) \approx f(\theta, \varphi) - \alpha\frac{\partial f(\theta, \varphi)}{\partial \varphi}, \text{ then } I_3 f(\theta, \varphi) = -\frac{\partial f(\theta, \varphi)}{\partial \varphi}.$$

10.12. Let A_1, A_2, A_3 be the anti-Hermitian matrices obeying relations

$$A_1A_2 - A_2A_1 = A_3, \quad A_2A_3 - A_3A_2 = A_1, \quad A_3A_1 - A_1A_3 = A_2. \quad (10.11)$$

Consider their linear combinations

$$H_+ = \mathrm{i}A_1 - A_2, \quad H_- = \mathrm{i}A_1 + A_2, \quad H_3 = \mathrm{i}A_3.$$

(a) Show that

$$[H_+, H_-] = 2H_3, \quad [H_3, H_\pm] = \pm H_\pm.$$

Solution. The commutation relations are obtained by the direct calculation.

---

[4]The fundamental is the exact irreducible representation of a minimal dimension. For group **SO**(3) it is vector representation $j = 1$.

(b) Let v_λ be the eigenvector of matrix H_3 corresponding to its eigenvalue λ. Show that H_+v_λ is the eigenvector of H_3 corresponding to eigenvalue $\lambda+1$ and H_-v_λ is the eigenvector of H_3 corresponding to eigenvalue $\lambda-1$. Find the norm of vectors $H_\pm v_\lambda$, when $\|v_\lambda\| = 1$.

Let v_λ be the eigenvector with eigenvalue λ, i.e., $H_3 v_\lambda = \lambda v_\lambda$. Then, following commutation relation between $H_\pm$ and H_3,

$$H_3 H_\pm v_\lambda = H_\pm (H_3 \pm 1) v_\lambda = (\lambda \pm 1) H_\pm v_\lambda.$$

Therefore, H_+, H_- are *raising and lowering* operators, respectively. Thus

$$H_+ v_\lambda = \beta_\lambda v_{\lambda+1}, \qquad H_- v_\lambda = \alpha_\lambda v_{\lambda-1},$$

where coefficients α_λ and β_λ are determined by normalization condition $|v_\lambda| = 1$ for all λ. Since matrices H_+ and H_- are Hermitian conjugate,

$$\begin{aligned}(H_+ v_\lambda, v_{\lambda+1}) &= \beta_\lambda (v_{\lambda+1}, v_{\lambda+1}) = \beta_\lambda, \\ (v_\lambda, H_- v_{\lambda+1}) &= \alpha_{\lambda+1} (v_\lambda, v_\lambda) = \alpha_{\lambda+1},\end{aligned}$$

then

$$\beta_\lambda = \alpha_{\lambda+1}. \tag{10.12}$$

Generators A_1, A_2, A_3 form closed algebra. They allow us to construct the Casimir operator $C = -\left(A_1^2 + A_2^2 + A_3^2\right)$. Express the operator in terms of matrices H:

$$\begin{aligned}C = &\ \tfrac{1}{2}(H_+H_- + H_-H_+) + H_3^2 = \\ = &\ H_3^2 + H_-H_+ + H_3 = H_3^2 + H_+H_- - H_3.\end{aligned} \tag{10.13}$$

Let J be the maximal eigenvalue of operator H_3, i.e., $H_+v_J = 0$. Then from (10.13) we obtain

$$(v_J, Cv_J) = J(J+1).$$

Among the eigenvectors of H_3, there are eigenvectors of C with eigenvalues $J(J+1)$ them from vector v_J after a series of H_- actions.

Let λ be an eigenvalue that is less than J. Then

$$H_+ v_{\lambda-1} = \frac{1}{\alpha_\lambda} H_+ H_- v_\lambda = \frac{1}{\alpha_\lambda}(H_-H_+ + 2H_3) v_\lambda =$$

$$\frac{1}{\alpha_\lambda}(\alpha_{\lambda+1}\beta_\lambda + 2\lambda) v_\lambda = \beta_{\lambda-1} v_\lambda.$$

Hence, taking into account Eq. (10.12), we get

$$\beta_\lambda^2 + 2\lambda = \beta_{\lambda-1}^2. \tag{10.14}$$

If $\lambda = J$, then $H_+ v_\lambda = 0$ and, consequently, $\beta_{J-1}^2 = 2J$. Exploiting Eq. (10.14), it is possible to derive by induction that $\beta_\lambda^2 = J(J+1) - \lambda(\lambda+1)$. Then from Eq. (10.12) we have $\alpha_\lambda^2 = J(J+1) - \lambda(\lambda-1)$. Thus

$$\begin{aligned} H_3 v_\lambda &= \lambda v_\lambda, \\ H_- v_\lambda &= \sqrt{J(J+1) - \lambda(\lambda-1)}\, v_{\lambda-1}, \\ H_+ v_\lambda &= \sqrt{J(J+1) - \lambda(\lambda+1)}\, v_{\lambda+1}. \end{aligned} \tag{10.15}$$

10.13. Prove that all eigenvalues of the matrix H_3 are integer or half-integer numbers: $-j, -j+1, \ldots, j-1, j$.

Solution. Let us act successively by the powers of the lowering operator H_- on eigenvector of highest weight v_J. We obtain a sequence of eigenvectors $v_J, v_{J-1}, \ldots$ The last vector in this set is v_{-J}, since $H_- v_{-J} = 0$. The number of all eigenvectors in the sequence with the given J is equal to $2J+1$. So J is an integer or half-integer number.

10.14. Find the Clebsch–Gordan coefficients

$$C^{jm}_{j_1 m_1\, j_2 m_2} = \langle j\, m | j_1\, m_1\, j_2\, m_2 \rangle,$$
$$|j_1\, m_1\, j_2\, m_2\rangle \equiv |j_1 m_1\rangle \otimes |j_2 m_2\rangle, \quad |jm\rangle \equiv |jm,\, j_1 j_2\rangle$$

for $j_1 = 1$, $j_2 = 2$, $j = 1$, $m = 0$.

Solution. We begin with the eigenvector of the highest weight $m = 1$. Decompose it by all the eigenvectors with angular momenta j_1 and j_2, using relation $m = m_1 + m_2$,

$$|1\,1\rangle = \alpha|1\,0\rangle|2\,1\rangle + \beta|1\,1\rangle|2\,0\rangle + \gamma|1\,-1\rangle|2\,2\rangle, \tag{10.16}$$

where we have to find coefficients α, β, γ. Denote V the linear space with basis

$$|1\,-1\rangle, |1\,0\rangle, |1\,1\rangle,$$

and U the space with basis

$$|2\,-2\rangle, |2\,-1\rangle, |2\,0\rangle, |2\,1\rangle, |2\,2\rangle.$$

Then the raising operator J_+ acting in space $V \otimes U$ can be presented as

$$J_+ = J_+^1 \otimes E^2 + E^1 \otimes J_+^2,$$

where E^1, E^2 are identity operators, J_+^1, J_+^2 are raising operators in spaces V, U.

Acting on both sides of Eq. (10.16) by raising operator J_+ and exploiting formula (10.15) for its matrix elements

$$J_\pm|jm\rangle = \sqrt{j(j+1) - m(m \pm 1)}|j\, m \pm 1\rangle,$$

we obtain

$$0 = \sqrt{2}\alpha|1\,1\rangle|2\,1\rangle + 2\alpha|1\,0\rangle|2\,2\rangle + \sqrt{6}\beta|1\,1\rangle|2\,1\rangle + \sqrt{2}\gamma|1\,0\rangle|2\,2\rangle.$$

From this equality, we get

$$\alpha\sqrt{2} + \beta\sqrt{6} = 0, \quad 2\alpha + \gamma\sqrt{2} = 0.$$

The third relation is given by the normalization condition $\alpha^2 + \beta^2 + \gamma^2 = 1$, whence $\alpha = \sqrt{3/10}$, $\beta = -1/\sqrt{10}$, $\gamma = -\sqrt{3/5}$.

Now we act on the eigenvector of highest weight (10.16) by lowering operator and get the decomposition:

$$\sqrt{2}|1\,0\rangle = (\alpha\sqrt{2}+2\gamma)|1\,-1\rangle|2\,1\rangle + (\alpha\sqrt{6}+\beta\sqrt{2})|1\,0\rangle|2\,0\rangle + \beta\sqrt{6}|1\,1\rangle|2\,-1\rangle.$$

As a result, we find all three desired Clebsch–Gordan coefficients:

$$C^{10}_{1-1\,21} = C^{10}_{11\,2-1} = -\sqrt{\frac{3}{10}}, \quad C^{10}_{10\,20} = \frac{2}{\sqrt{10}}.$$

10.15. What transformations of four-dimensional linear space conserve the quadratic form

$$s^2 = x_0^2 - x_1^2 - x_2^2 - x_3^2,$$

the Lorentz invariant in Minkowski spacetime?

Solution. Let us introduce the column vector **X** and matrix $\hat{F}$:

$$\mathbf{X} = \begin{pmatrix} x_0 \\ x_1 \\ x_2 \\ x_3 \end{pmatrix}, \quad \hat{F} = \begin{pmatrix} 1 & 0 & 0 & 0 \\ 0 & -1 & 0 & 0 \\ 0 & 0 & -1 & 0 \\ 0 & 0 & 0 & -1 \end{pmatrix}.$$

Write the interval conservation in matrix form:

$$s^2 = \mathbf{X}^T \hat{F} \mathbf{X} = \mathbf{Y}^T \hat{F} \mathbf{Y}, \quad \mathbf{Y} = \hat{\Lambda}\mathbf{X}, \tag{10.17}$$

where $\hat{F}$ is the Minkowski metric tensor. Substituting vector **Y** into the quadric, we get

$$\mathbf{Y}^T \hat{F} \mathbf{Y} = (\hat{\Lambda}\mathbf{X})^T \hat{F} \hat{\Lambda}\mathbf{X} = \mathbf{X}^T \hat{\Lambda}^T \hat{F} \hat{\Lambda}\mathbf{X},$$

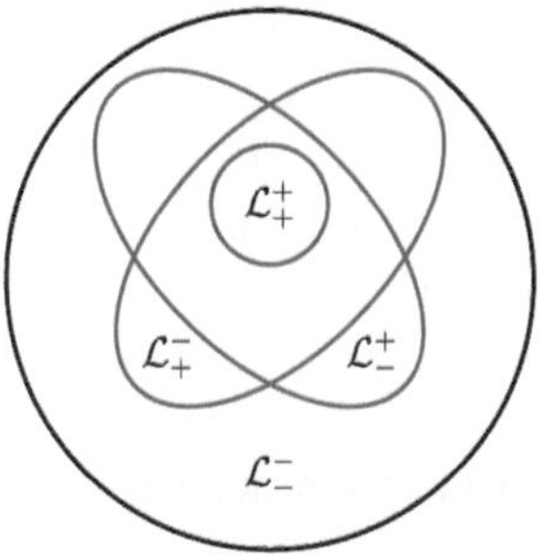

Figure 10.1 Subsets $\mathcal{L}_\pm^\pm$ of the general Lorentz group: each ellipse marks a group.

hence

$$\hat{F} = \hat{\Lambda}^T \hat{F} \hat{\Lambda}. \tag{10.18}$$

A more general set satisfies condition (10.18) than **SO**(1, 3) (the restricted Lorentz group). The set is called the general Lorentz group. Calculate the determinants of Eq. (10.18) both sides and get $(\det \hat{\Lambda})^2 = 1$. Consequently, the general Lorentz group splits up by sets $\mathcal{L} = \mathcal{L}_+ \cup \mathcal{L}_-$ with determinants ± 1. The set $\mathcal{L}_+$ with positive determinant includes the identity matrix; then, it is a group. The set $\mathcal{L}_-$ is not a group.

Now write the element at the intersection of the first column and first row in (10.18):

$$\Lambda_{00}^2 - \Lambda_{10}^2 - \Lambda_{20}^2 - \Lambda_{30}^2 = 1.$$

Then $\Lambda_{00}^2 \geqslant 1$, whence $\Lambda_{00} \geqslant 1$ or $\Lambda_{00} \leqslant -1$. We denote the sign of Λ_{00} by the superscript. In these notations, the general group splits up into four subsets:

$\mathcal{L}_+^+$	$\mathcal{L}_+^-$	$\mathcal{L}_-^+$	$\mathcal{L}_-^-$
$\Lambda_{00} \geqslant 1$ $\det \hat{\Lambda} = 1$	$\Lambda_{00} \leqslant -1$ $\det \hat{\Lambda} = 1$	$\Lambda_{00} \geqslant 1$ $\det \hat{\Lambda} = -1$	$\Lambda_{00} \leqslant -1$ $\det \hat{\Lambda} = -1$

Notice that subsets $\mathcal{L}_+^-$ and $\mathcal{L}_-^+$ are not subgroups since do not include 1. However, they form subgroups after union with the restricted Lorentz group $\mathcal{L}_+^+ = \mathbf{SO}(1, 3)$, as shown in Fig. 10.1.

10.16. Establish the correspondence of the restricted Lorentz group **SO**(1, 3) and the rotation group **SO**(4) of 4-dimension Euclidian space $\mathbb{R}^4$.

Solution. Introduce the parametrization of Lorentz group choosing the angles of rotations in planes (x_1x_2), (x_1x_3), (x_2x_3) and parameters of hyperbolic rotations in planes (x_0x_1), (x_0x_2), (x_0x_3),. The rotation group in **SO**(4) in four-dimensional space has also 6 real parameters. The invariant of rotation is positively defined quadric

$$x_0^2 + x_1^2 + x_2^2 + x_3^2. \tag{10.19}$$

Consider a rotation in plane (x_0x_1) by angle φ

$$x_0' = x_0 \cos\varphi - x_1 \sin\varphi, \quad x_1' = x_0 \sin\varphi + x_1 \cos\varphi. \tag{10.20}$$

Then consider Eq. (10.20) at imaginary x_0, φ, i.e., change $x_0 \to ix_0$, $x_0' \to ix_0'$, $\varphi \to i\varphi$. Rotation (10.20) transforms into hyperbolic transformation

$$x_0' = x_0 \cosh\varphi + x_1 \sinh\varphi, \quad x_1' = x_0 \sinh\varphi + x_1 \cosh\varphi. \tag{10.21}$$

Analogously, rotations from **SO**(4) in planes (x_0x_2), (x_0x_3) transform into boosts from $\mathcal{L}$. Rotations in other planes (x_1x_2), (x_1x_3), (x_2x_3) are the same in both the groups. Thus we have built up the mapping from the rotations to the Lorentz group. It is one-to-one mapping conserving the product of matrices. At the same time their manifolds are different: the rotation group **SO**(4) is compact, whereas the restricted Lorentz group **SO**(1, 3) is not compact.

10.17. Find the generators of the group **SO**(1, 3) and their commutation relations.

Solution. A matrix from the Lorentz group can be written as the product of 6 matrices:

$$\hat{R}_1 = \begin{pmatrix} 1 & 0 & 0 & 0 \\ 0 & 1 & 0 & 0 \\ 0 & 0 & \cos\varphi_1 & -\sin\varphi_1 \\ 0 & 0 & \sin\varphi_1 & \cos\varphi_1 \end{pmatrix}, \quad \hat{H}_1 = \begin{pmatrix} \cosh b_1 & \sinh b_1 & 0 & 0 \\ \sinh b_1 & \cosh b_1 & 0 & 0 \\ 0 & 0 & 1 & 0 \\ 0 & 0 & 0 & 1 \end{pmatrix},$$

$$\hat{R}_2 = \begin{pmatrix} 1 & 0 & 0 & 0 \\ 0 & \cos\varphi_2 & 0 & \sin\varphi_2 \\ 0 & 0 & 1 & 0 \\ 0 & -\sin\varphi_2 & 0 & \cos\varphi_2 \end{pmatrix}, \quad \hat{H}_2 = \begin{pmatrix} \cosh b_2 & 0 & \sinh b_2 & 0 \\ 0 & 1 & 0 & 0 \\ \sinh b_2 & 0 & \cosh b_2 & 0 \\ 0 & 0 & 0 & 1 \end{pmatrix},$$

$$\hat{R}_3 = \begin{pmatrix} 1 & 0 & 0 & 0 \\ 0 & \cos\varphi_3 & -\sin\varphi_3 & 0 \\ 0 & \sin\varphi_3 & \cos\varphi_3 & 0 \\ 0 & 0 & 0 & 1 \end{pmatrix}, \quad \hat{H}_3 = \begin{pmatrix} \cosh b_3 & 0 & 0 & \sinh b_3 \\ 0 & 1 & 0 & 0 \\ 0 & 0 & 1 & 0 \\ \sinh b_3 & 0 & 0 & \cosh b_3 \end{pmatrix}. \tag{10.22}$$

The generators are partial derivatives of the matrices:

$$\hat{L}_i = \left.\frac{\partial \hat{R}_i}{\partial \varphi_i}\right|_{\varphi_i=0}, \qquad \hat{M}_i = \left.\frac{\partial \hat{H}_i}{\partial b_i}\right|_{b_i=0}.$$

The generators of rotations in **SO**(1, 3) and **SO**(4) in planes (x_2x_3), (x_1x_3), (x_1x_2) are equal. Introduce the vector of rotation: $\hat{\mathbf{L}} = (\hat{L}_1, \hat{L}_2, \hat{L}_3)$.

$$\hat{L}_1 = \begin{pmatrix} 0&0&0&0\\ 0&0&0&0\\ 0&0&0&-1\\ 0&0&1&0 \end{pmatrix}, \quad \hat{L}_2 = \begin{pmatrix} 0&0&0&0\\ 0&0&0&1\\ 0&0&0&0\\ 0&-1&0&0 \end{pmatrix}, \quad \hat{L}_3 = \begin{pmatrix} 0&0&0&0\\ 0&0&-1&0\\ 0&1&0&0\\ 0&0&0&0 \end{pmatrix}.$$

The generators of boosts,

$$\hat{M}_1 = \begin{pmatrix} 0&1&0&0\\ 1&0&0&0\\ 0&0&0&0\\ 0&0&0&0 \end{pmatrix}, \quad \hat{M}_2 = \begin{pmatrix} 0&0&1&0\\ 0&0&0&0\\ 1&0&0&0\\ 0&0&0&0 \end{pmatrix}, \quad \hat{M}_3 = \begin{pmatrix} 0&0&0&1\\ 0&0&0&0\\ 0&0&0&0\\ 1&0&0&0 \end{pmatrix}$$

are also arranged into single vector $\hat{\mathbf{M}} = (\hat{M}_1, \hat{M}_2, \hat{M}_3)$. As a result, we get 18 commutation relations:

$$[\hat{L}_i, \hat{L}_j] = e_{ijk}\hat{L}_k, \quad [\hat{M}_i, \hat{M}_j] = -e_{ijk}\hat{L}_k, \quad [\hat{M}_i, \hat{L}_j] = e_{ijk}\hat{M}_k. \qquad (10.23)$$

To obtain a unitary representation, we have to use anti-Hermitian matrices. Generators $\hat{\mathbf{L}}$ changes its sign after the transposition, while $\hat{\mathbf{M}}$ does not change. Then introduce new combined anti-Hermitian generators:

$$\hat{\mathbf{A}} = \frac{\hat{\mathbf{L}} + \mathrm{i}\hat{\mathbf{M}}}{2}, \quad \hat{\mathbf{B}} = \frac{\hat{\mathbf{L}} - \mathrm{i}\hat{\mathbf{M}}}{2}.$$

The representation becomes unitary. We obtain the commutation relation between new generators from Eq. (10.23):

$$[\hat{A}_i, \hat{A}_j] = e_{ijk}\hat{A}_k, \quad [\hat{B}_i, \hat{B}_j] = e_{ijk}\hat{B}_k, \quad [\hat{A}_i, \hat{B}_j] = 0. \qquad (10.24)$$

10.18. Derive the Clebsh–Gordan formula for the Lorentz group.

Solution. An elegant form of relations (10.24) has a simple interpretation. In fact, we treated group **SO(4)** $\approx$ **SO(3)** $\times$ **SO(3)**, being isomorphic to the direct product of two three-dimensional rotation groups. Then we could use the known irreducible representations of the rotation group with angular momenta j, j', if choose the squares of commutating "moments" $\hat{\mathbf{A}}^2$, $\hat{\mathbf{B}}^2$ as the Casimir operators. Denote this representation as $D^{(jj')} = D^{(j)} \otimes D^{(j')}$, its dimension is

$$\dim D^{(jj')} = (2j+1)(2j'+1).$$

The basis consists of vectors $|jj'mm'\rangle$.

The decomposition of a direct product into the direct sum of irreducible representations is given by known Clebsh–Gordan formulas for the 3D rotation group

$$D^{(j_1 j_1')} \otimes D^{(j_2 j_2')} = \bigoplus_{j=|j_1-j_2|}^{j_1+j_2} \bigoplus_{j'=|j_1'-j_2'|}^{j_1'+j_2'} D^{(jj')}. \tag{10.25}$$

Thus the decomposition reduces to a quantum mechanical problem of addition of two independent angular momenta.

10.19. Show that the permutation of indices commutes with the action of the group G in the space of rank r tensors.

Solution. One proves this fact by direct verification. Let the matrix D_{ij} correspond to the element $g \in G$, then

$$\begin{aligned}(\sigma T')_{\alpha_1\alpha_2\ldots\alpha_r} &= T'_{\alpha_{\sigma(1)}\alpha_{\sigma(2)}\ldots\alpha_{\sigma(r)}} = D_{\alpha_{\sigma(1)}k_{\sigma(1)}} \ldots D_{\alpha_{\sigma(r)}k_{\sigma(r)}} T_{k_{\sigma(1)}\ldots k_{\sigma(r)}} = \\ &= D_{\alpha_1 k_1} \ldots D_{\alpha_r k_r} (\sigma T)_{k_1\ldots k_r} = (\sigma T)'_{\alpha_1\ldots\alpha_r}.\end{aligned}$$

10.20. The representation D^* of a finite group G can be obtained from given representation D by the complex conjugation.[5] Consider direct product $D = D^{(\alpha)*} \otimes D^{(\beta)}$ of two irreducible representations of group G. Show that the trivial representation enters only if $D^{(\alpha)} = D^{(\beta)}$.

Solution. Decompose the direct product into the direct sum: $D = \bigoplus_\lambda c_\lambda D^{(\lambda)}$. Find how many times the trivial representation enters D. For this purpose multiply character $\chi(g) = \chi^{(\alpha)*}(g)\chi^{(\beta)}(g)$ of representation D by the character of trivial one $\chi^{(1)*}(g) = 1$ and sum up over the group:

$$c_1 = \frac{1}{|G|} \sum_g \chi^{(1)*}(g)\chi^{(\alpha)*}(g)\chi^{(\beta)}(g) = \delta_{\alpha\beta}.$$

The last equation follows from the orthogonality relation (9.4).

[5] If D^* and D coincide, then D is called real representation.

10.21. Prove the Campbell–Hausdorff formula:

$$F(\mu) = e^{\mu B} A e^{-\mu B} = A + \mu[B, A] + \frac{\mu^2}{2!}[B, [B, A]] + \ldots, \qquad (10.26)$$

where μ is a real parameter, A, B are the linear operators.

Solution. Differentiating function $F(\mu)$, we get repeated commutators:

$$\frac{dF}{d\mu} = [B, F], \quad \frac{d^2F}{d\mu^2} = [B, [B, F]], \ldots$$

Using the obtained derivatives at $\mu = 0$ and identity $F(0) = A$, we come to the Taylor series (10.26).

10.22. The rotation of vector $\mathbf{r}$ by angle φ is defined by formula:

$$(\mathbf{r}'\boldsymbol{\sigma}) = e^{\mathrm{i}\mathbf{n}\boldsymbol{\sigma}\varphi/2}(\mathbf{r}\boldsymbol{\sigma})e^{-\mathrm{i}\mathbf{n}\boldsymbol{\sigma}\varphi/2},$$

where $\mathbf{n}$ is the unity vector, $\boldsymbol{\sigma} = (\sigma_1, \sigma_2, \sigma_3)$ is the set of Pauli matrices. Since this is a linear transformation, it can be presented as $r'_i = R_{ij}r_j$. With the help of Campbell–Hausdorff formula, we find the matrix of rotation R_{ij}.

Solution. Using Eq. (10.26) and commutator $[\sigma_i, \sigma_j] = 2i e_{ijk}\sigma_k$, we get the series

$$(\mathbf{r}'\boldsymbol{\sigma}) = (\mathbf{r}\boldsymbol{\sigma}) + \mathrm{i}\frac{\varphi}{2}[\mathbf{n}\boldsymbol{\sigma}, \mathbf{r}\boldsymbol{\sigma}] + \frac{1}{2!}\left(\frac{\mathrm{i}\varphi}{2}\right)^2 [\mathbf{n}\boldsymbol{\sigma}, [\mathbf{n}\boldsymbol{\sigma}, \mathbf{r}\boldsymbol{\sigma}]] + \ldots$$

With the help of identity $e_{ikm}e_{jlm} = \delta_{ij}\delta_{kl} - \delta_{il}\delta_{kj}$ obtain the expressions for repeated commutators:

$$[\mathbf{n}\boldsymbol{\sigma}, \mathbf{r}\boldsymbol{\sigma}] = 2i(\mathbf{n} \times \mathbf{r})\boldsymbol{\sigma},$$

$$[\mathbf{n}\boldsymbol{\sigma}, [\mathbf{n}\boldsymbol{\sigma}, \mathbf{r}\boldsymbol{\sigma}]] = (2i)^2 \{(\mathbf{n}\boldsymbol{\sigma})(\mathbf{r}\mathbf{n}) - (\mathbf{r}\boldsymbol{\sigma})\}.$$

The subsequent repeated commutators have the same structure: the odd terms are proportional to the Levi–Civita symbol $e_{ijk}n_k$, the even reduce to transverse projector $n_i n_j - \delta_{ij}$, then

$$\mathbf{r}' = \mathbf{r}\left(1 - \frac{\varphi^2}{2!} + \ldots\right) + (\mathbf{n} \times \mathbf{r})\left(\varphi - \frac{\varphi^3}{3!} + \ldots\right)$$
$$+\mathbf{n}(\mathbf{n}\mathbf{r})\left(\frac{\varphi^2}{2!} - \frac{\varphi^4}{4!} + \ldots\right)$$
$$= \mathbf{r}\cos\varphi + (\mathbf{n} \times \mathbf{r})\sin\varphi + \mathbf{n}(\mathbf{n}\mathbf{r})(1 - \cos\varphi),$$

whence the matrix of finite rotation is

$$R_{ij} = \delta_{ij}\cos\varphi + e_{ijk}n_k\sin\varphi + n_i n_j(1 - \cos\varphi).$$

Since $g = e^{\mathrm{i}\mathbf{n}\boldsymbol{\sigma}\varphi/2} \in \mathbf{SU}(2)$, this formula presents one more homomorphism $\mathbf{SU}(2) \to \mathbf{SO}(3)$.

Exercises

10.23. Do the transformations:

(a) rotations in a plane;

(b) stretching and shear in a plane;

(c) the Lorentz transformations;

(d) linear fractional transformations of complex plane $z' = (az+b)/(cz+d)$ form a continuous group?

10.24. Find the dimension of the following matrix Lie group:

(a) the complete linear group $\mathbf{GL}(n, \mathbb{C})$;

(b) the unimodular group $\mathbf{SL}(n, \mathbb{C})$, involving all the matrices with unity determinant;

(c) the unitary group $\mathbf{U}(n, \mathbb{C})$;

(d) the orthogonal group $\mathbf{O}(n, \mathbb{R})$;

(e) the rotation group $\mathbf{SO}(3, \mathbb{R})$.

10.25. Check that for $\mathbf{SO}(3)$ generators from problem **10.3**, the following commutation relations are valid:

$$I_1 I_2 - I_2 I_1 = I_3, \quad I_2 I_3 - I_3 I_2 = I_1, \quad I_3 I_1 - I_1 I_3 = I_2. \tag{10.27}$$

10.26. Let $f(\theta, \varphi)$ be a function defined on the unit sphere, g is the rotation by angle α with respect to axis Oz, which corresponds to matrix $T(g)$.

(a) Let $f = Y_{lm}(\theta, \varphi)$. How does operator $\mathrm{i}I_3$ act on f?

(b) Consider the linear space over vectors

$$Y_{l,-l}, \ Y_{l,-l+1}, \ldots, Y_{ll}.$$

Find the matrix in this basis corresponding to the rotation by angle α with respect to axis Oz.

10.27. Demonstrate that all the rotations by the same angle around different axes are equivalent in group $\mathbf{SO}(3)$.

10.28. Find the character of irreducible $(2l+1)$-dimensional representation of group $\mathbf{SO}(3)$ (from problem **10.26**(b)).

10.29. Let $D^{(j)}$ and $D^{(j')}$ be irreducible representations of group $\mathbf{SO}(3)$. Show that

$$D^{(j)} \otimes D^{(j')} = D^{(j+j')} \oplus D^{(j+j'-1)} \oplus \cdots \oplus D^{(|j-j'|)}$$

(Clebsch–Gordan decomposition).

10.30. Find the Clebsch–Gordan coefficients for decomposition of the product $D^{(1/2)} \otimes D^{(1/2)}$ over irreducible representations.

10.31. Decompose the second-rank tensor over the irreducible components in group $G = \mathbf{SO}(3)$.

10.32. Let $\mathcal{F}$ be the space of all the infinitely differentiable functions $f(x)$ in $\mathbb{R}^3$. For each matrix $T(g)$, $g \in \mathbf{SO}(3)$ assume that $(gf)(x) = f\left(T^{-1}(g)x\right)$. Find generators of this representation.

10.33. Consider the set of uniform polynomials of degree n of the form:

$$P_n(x, y, z) = \sum_{k+l+m=n} a_{klm} x^k y^l z^m, \quad k, l, m \geq 0.$$

All the polynomials form a linear space.

(a) Find the dimension of the space.

(b) Find the dimension of the subspace of harmonic polynomials (i.e., polynomials obeying the Laplace equation $\triangle P_n = 0$).

(c) Show that the representation of group $\mathbf{SO}(3)$ in the subspace of harmonic polynomials is irreducible.

10.34. Let $g = g_{\xi,\eta,\theta} \in E_2$ be a mapping of the plane to itself. We denote E_2 the group of transformations of the Euclidian two-dimensional space $\mathbb{R}^2$, including rotations $T \in \mathbf{SO}(2)$ and translations:

$$g : \begin{pmatrix} x \\ y \end{pmatrix} \to \begin{pmatrix} x' \\ y' \end{pmatrix} = \begin{pmatrix} x\cos\theta - y\sin\theta + \xi \\ x\sin\theta + y\cos\theta + \eta \end{pmatrix}.$$

There is a correspondence of E_2 elements and 3×3 matrices:

$$g \to \begin{pmatrix} \cos\theta & -\sin\theta & \xi \\ \sin\theta & \cos\theta & \eta \\ 0 & 0 & 1 \end{pmatrix}. \qquad (10.28)$$

Check that (10.28) is a representation of group E_2.

10.35. Let $\mathcal{F}$ be the space of infinitely differentiable functions defined for all x, y. The representation from problem **10.34** on functions from $\mathcal{F}$ can be obtained by transformation

$$(gf)(x, y) = f\left((x-\xi)\cos\theta + (y-\eta)\sin\theta, -(x-\xi)\sin\theta + (y-\eta)\cos\theta\right).$$

(a) Show that for the generators $L_{1,2,3}$ the following equalities are valid

$$L_1 = -\frac{\partial}{\partial x}, \quad L_2 = -\frac{\partial}{\partial y}, \quad L_3 = y\frac{\partial}{\partial x} - x\frac{\partial}{\partial y}.$$

(b) Find commutators $[L_1, L_2]$, $[L_2, L_3]$, $[L_3, L_1]$.

(c) Let r and φ be the polar coordinates in the plane x, y, $L^{\pm} = L_1 \pm \mathrm{i}L_2$. Make sure that

$$L^{\pm} = -\mathrm{e}^{\pm \mathrm{i}\varphi}\left(\frac{\partial}{\partial r} \pm \frac{\mathrm{i}}{r}\frac{\partial}{\partial \varphi}\right),$$

$$L_3 = -\frac{\partial}{\partial \varphi},$$

$$L^+L^- = L^-L^+ = \triangle.$$

(d) Test that the Laplacian commutes with all the generators L_1, L_2, L_3, i.e., $\triangle$ is the Casimir operator.

10.36. Consider the set of functions

$$\Psi_m(x, y) = \mathrm{e}^{\mathrm{i}m\varphi} J_m(\alpha r),$$

where $J_m(\alpha r)$ is the Bessel function of the order m, $\alpha \neq 0$. Show that the irreducible representation of group E_2 is realized in space $\mathcal{F}$ with basis Ψ_m with integer m.

Answers

10.23(d). Yes, if $\begin{vmatrix} a & b \\ c & d \end{vmatrix} \neq 0$.

10.24(a). $2n^2$ real variables.

10.24(b). $2n^2 - 2$.

10.24(c). n^2.

10.24(d). $n(n-1)/2$.

10.24(e). 3.

10.26(a). $\mathrm{i}I_3 f = mf$.

10.26(b).

$$D(g) = \exp(I_3\alpha) =$$

$$= \exp \begin{pmatrix} il\alpha & 0 & 0 & 0 & \dots & 0 \\ 0 & i(l-1)\alpha & 0 & 0 & \dots & 0 \\ 0 & 0 & i(l-2)\alpha & 0 & \dots & 0 \\ 0 & 0 & 0 & i(l-3)\alpha & \dots & 0 \\ \dots & & & & & \\ 0 & 0 & 0 & 0 & \dots & -il\alpha \end{pmatrix} =$$

$$= \begin{pmatrix} e^{il\alpha} & 0 & 0 & 0 & \dots & 0 \\ 0 & e^{i(l-1)\alpha} & 0 & 0 & \dots & 0 \\ 0 & 0 & e^{i(l-2)\alpha} & 0 & \dots & 0 \\ 0 & 0 & 0 & e^{i(l-3)}\alpha & \dots & 0 \\ \dots & & & & & \\ 0 & 0 & 0 & 0 & \dots & e^{-il\alpha} \end{pmatrix}.$$

10.28.

$$\chi^{(l)}(\varphi) = \frac{\sin\left(l + \frac{1}{2}\right)\varphi}{\sin\frac{\varphi}{2}}.$$

10.29. Hint. Exploit the property valid for each $g \in SO(3)$

$$\mathrm{Tr}(D^{(j)} \otimes D^{(j')}) = \mathrm{Tr}D^{(j)}\mathrm{Tr}D^{(j')}.$$

10.30. The decomposition of representation $D^{(1/2)} \otimes D^{(1/2)}$ over irreducible representations is equivalent to the problem about adding two spins $s_1 = s_2 = 1/2$. There are only two multiplets in the result: the triplet $S = 1$ with parallel spins and the singlet $S = 0$ with antiparallel spins:

$$D^{(1/2)} \otimes D^{(1/2)} = D^{(1)} \oplus D^{(0)}.$$

The Clebsch–Gordan coefficients are the coefficients in the wave function decomposition. For $S = 1$

$$|1\,1\rangle = \left|\frac{1}{2}\,\frac{1}{2}\right\rangle_1 \left|\frac{1}{2}\,\frac{1}{2}\right\rangle_2,$$

$$|1\,0\rangle = \frac{1}{\sqrt{2}}\left(\left|\frac{1}{2}\,\frac{1}{2}\right\rangle_1 \left|\frac{1}{2} - \frac{1}{2}\right\rangle_2 + \left|\frac{1}{2} - \frac{1}{2}\right\rangle_1 \left|\frac{1}{2}\,\frac{1}{2}\right\rangle_2\right),$$

$$|1 - 1\rangle = \left|\frac{1}{2} - \frac{1}{2}\right\rangle_1 \left|\frac{1}{2} - \frac{1}{2}\right\rangle_2.$$

For $S = 0$

$$|0\,0\rangle = \frac{1}{\sqrt{2}}\left(\left|\frac{1}{2}\,\frac{1}{2}\right\rangle_1 \left|\frac{1}{2} - \frac{1}{2}\right\rangle_2 - \left|\frac{1}{2} - \frac{1}{2}\right\rangle_1 \left|\frac{1}{2}\,\frac{1}{2}\right\rangle_2\right).$$

10.31. $D^{(1)} \otimes D^{(1)} = D^{(0)} \oplus D^{(1)} \oplus D^{(2)}$, representation $D^{(2)}$ is realized in the subspace of symmetric traceless rank 2 tensors. $D^{(0)}$ is realized in the subspace of scalar tensors $c\delta_{ij}$, where c is a number.

10.32.

$$L_1 = z\frac{\partial}{\partial y} - y\frac{\partial}{\partial z}, \quad L_2 = x\frac{\partial}{\partial z} - z\frac{\partial}{\partial x}, \quad L_3 = y\frac{\partial}{\partial x} - x\frac{\partial}{\partial y}.$$

10.33(a). $(n+1)(n+2)/2$.

10.33(b). $2n+1$.

10.33(c). **Hint.** Using the Rodriguez formula, we get

$$r^n Y_{nm}(\theta, \varphi) \propto r^n e^{im\varphi}(1 - \cos^2\theta)^{m/2} \left(\frac{d}{d\cos\theta}\right)^{n+m} (\cos^2\theta - 1)^n,$$

$$m = -n, -n+1, \ldots, n.$$

Coming to the Cartesian coordinates

$$r\cos\theta = z, \quad r\sin\theta\, e^{i\varphi} = x + iy, \quad r^2 = x^2 + y^2 + z^2,$$

we found that $r^n Y_{nm}$ is a uniform polynomial of power n. Each limited solution to equation

$$\triangle\left[r^n f(\theta, \varphi)\right] = 0 = r^{n-2}\left(n(n+1) + \Delta_\Omega\right) f,$$

for function f, where Δ_Ω is the angular part of the Laplacian operator, can be written as a linear combination of spherical harmonics $Y_{lm}(\theta, \varphi)$. Then the space over polynomials $r^n Y_{nm}$, $m = -n, -n+1, \ldots, n$ coincides with the whole space of the harmonic polynomials.

10.35(b). $[L_1, L_2] = 0$, $[L_1, L_3] = -L_2$, $[L_2, L_3] = L_1$.

10.36. Hint. Space $\mathcal{F}$ has to transform to itself at the action of each operator L_1, L_2, L_3. Note that

$$L_3\Psi_m = -im\Psi_m,$$
$$L^+\Psi_m = -\alpha\Psi_{m+1},$$
$$L^-\Psi_m = \alpha\Psi_{m-1}.$$

It follows from recurrence relations $J_{m+1}(z) = -J_m'(z) + mJ_m(z)/z$, $J_{m-1}(z) = J_m'(z) + mJ_m(z)/z$. Since L_1, L_2, L_3 is a linear combination of operators L^+, L^-, L_3, we obtain space $\mathcal{F}$ is transformed to itself under the action of group E_2. Acting by operator L^+ on function Ψ_m for arbitrary m, we get a non-breaking series, then the representation has the infinite dimension. To prove the representation being irreducible we should exploit the facts that the set $\{c\Psi_m\}$, $m = 0, \pm 1, \pm 2, \ldots, c \in \mathbb{C}, c \neq 0$, exhausts all eigenvectors of L_3 in space $\mathcal{F}$, and that in any E_2-invariant subspace contains the eigenvector of L_3. Further, acting operators L^+ and L^-, we get the whole set of functions Ψ_m, $m = 0, \pm 1, \pm 2, \ldots$

Chapter 11

Group Theory in Physics

The theory of groups and their irreducible representations have numerous applications in different fields of physics, including atomic and molecular spectroscopy, solid-state physics, nuclear physics, optics, quantum mechanics, and physics of elementary particles. The results of group theory are especially impressive in problems with no small parameter. While the equations are non-solvable, the symmetry consideration remains valid.

We consider several applications here: calculation of degeneration multiplicity of molecular vibrations, splitting of energy levels due to decreasing the symmetry, selection rules in quantum transitions between the energy levels, and number of independent components of an invariant tensor. These and other applications are discussed in books (Hamermesh, 1989; Heine, 1977; Landau and Lifshitz, 1977; Mathews and Walker, 1970; Messiah, 1962; Petrashen, 1969; Wigner, 1959).

11.1 Molecular Oscillations

In classical mechanics, the Hamiltonian of a molecule consisting of n atoms with masses $M_l, l = 1, \ldots, n$, in harmonic approximation has the form

$$H = \sum_{\alpha=1}^{3}\sum_{l=1}^{n} \frac{\left(p_l^\alpha\right)^2}{2M_l} + \frac{1}{2}\sum_{\alpha,\beta=1}^{3}\sum_{l,m=1}^{n} V_{lm}^{\alpha,\beta} q_l^\alpha q_m^\beta. \tag{11.1}$$

Mathematical Methods of Physics: Problems with Solutions
Igor V. Kolokolov, Evgeny A. Kuznetsov, Alexander I. Milstein, Evgeny V. Podivilov, Alexander I. Chernykh, David A. Shapiro, and Elena G. Shapiro

ISBN 978-981-5129-21-2 (Hardcover), 978-1-003-51360-5 (eBook)
www.jennystanford.com

Here q_l^α is the $\alpha = x, y, z$-component of l-th atom small displacement from its equilibrium position, p_l^α is the corresponding momentum, $V_{lm}^{\alpha,\beta}$ are the constants. Further for simplifying we combine the index of the Cartesian components α and the number of an atom l in one index $k = 1, \ldots, 3n$. The Newton equations of motion are

$$M_k \ddot{q}_k = -\sum_{j=1}^{3n} V_{kj} q_j. \tag{11.2}$$

Let us seek the solution to Eq. (11.2) in the form $q_k = a_k \cos(\omega t + \delta)$ (*normal oscillations*). We get the system of equations:

$$\omega^2 M_k a_k = \sum_{j=1}^{3n} V_{kj} a_j$$

that defines the set of normal frequencies ω_s^2 and amplitudes $a_k^{(s)}$. In terms of normalized amplitudes $b_k = \sqrt{M_k}\, a_k$ the problem reduces to the diagonalization of the Hermitian matrix B_{kj}:

$$\omega^2 b_k = \sum_{j=1}^{3n} B_{kj} b_j, \quad B_{kj} = \frac{V_{kj}}{\sqrt{M_k M_j}}. \tag{11.3}$$

The symmetry of our system is a permutation of atoms with the same mass, rotations, and reflections that conserve the Hamiltonian (11.1). In other words, the symmetry group is a subgroup of direct product:

$$G < S_n \times O(3).$$

Recall that the permutation in linear space is performed by linear transformation:

$$B \to T_{(kl)}^{-1} B T_{(kl)}, \quad T_{(kl)}^2 = E, \tag{11.4}$$

where the matrix $T_{(kl)}$ is obtained from the identity matrix E by transposition k-th and l-th columns. When acting on a given vector b_j the matrix $T_{(kl)}$ transposes its k-th and l-th components. Thus, in the space of components of $3n$-dimensional amplitude vectors $b = (b_1, \ldots, b_{3n})$, there is a linear representation D of the symmetry group of a molecule, and permutations of atoms correspond to the transformation T, along with rotations and reflections. This representation is called *initial*.

The symmetry of our system with respect to a permutation T, rotation R or their composition Q means that

$$B = Q^{-1}BQ \quad \text{or} \quad [B, Q] = 0, \tag{11.5}$$

where $Q \in D$ is any matrix from the initial representation from the symmetry group. It consists of matrices corresponding to permutations, rotations, and reflections of the $3n$-dimensional amplitude vectors. This representation is generally reducible. On the other hand, equality (11.5) means that if b is an eigenvector of the matrix B, then Qb is also the eigenvector corresponding to the same eigenvalue. Consequently, any subspace V_s of eigenvectors, corresponding to eigenvalue ω_s^2:

$$\omega_s^2 b_i^\alpha = \sum_{j=1}^{3n} B_{ij} b_j^\alpha, \quad \alpha = 1, \ldots, r, \quad b^\alpha \in V_s,$$

is an invariant subspace with respect to action of operator Q:

$$Qb^\alpha = \sum_{\beta=1}^{r} q_{\alpha\beta} b^\beta, \quad b^\alpha,\ b^\beta \in V_s.$$

Then operators Q implement some r-dimensional representations of a symmetry group in V_s, and $q_{\alpha\beta}$ are matrices of this representation. So, breaking the original representation of the symmetry group into irreducible ones, we can define the dimensions of eigenvalues subspaces, i.e., the multiplicity of normal molecular vibration.

If one uses normal vibration modes as a basis in the $3n$-dimensional space, the Hamiltonian takes the form diagonal quadratic form. It becomes the sum of the Hamiltonians, where each depends on the amplitude of only one normal mode. Turning to the quantum-mechanical consideration, we note that it is possible to produce quantization immediately within the basis of normal oscillations. Then the level structure and the type of wave functions will become apparent: the system is a collection of independent linear oscillators. The eigenfunctions of Hamiltonians are stationary wave functions of each oscillator. If r oscillators have a coinciding frequency, then we see the r-fold degeneration of oscillations in the classical system. So it is possible to find the multiplicity of molecule oscillations using the group theory. To do this, we need to decompose the initial representation in the $3n$-dimensional space, using the characters' method.

The initial representation arises when we consider the action of the elements g of this subgroup on the amplitude vectors of atomic displacements. The character $\chi(g)$ of the initial representation, where transformation g being a rotation or a mirror reflection, is the sum of three-dimensional characters of the vector representations in subspaces of vectors of small displacements of individual atoms that do not move at transforming g. The matrix of rotation R by angle θ around the z axis is the same for a vector (V) or pseudovector (PV):

$$R_V(\theta) = R_{PV}(\theta) = R(\theta) = \begin{pmatrix} \cos\theta & -\sin\theta & 0 \\ \sin\theta & \cos\theta & 0 \\ 0 & 0 & 1 \end{pmatrix}.$$

The reflection matrix with respect to the plane xy is different for these objects. Denote $R_V(\sigma)$ the reflection matrix for vector, $R_{PV}(\sigma)$ for pseudovector, then

$$R_V(\sigma) = \begin{pmatrix} 1 & 0 & 0 \\ 0 & 1 & 0 \\ 0 & 0 & -1 \end{pmatrix}, \quad R_{PV}(\sigma) = \begin{pmatrix} -1 & 0 & 0 \\ 0 & -1 & 0 \\ 0 & 0 & 1 \end{pmatrix}.$$

The character of rotation by angle θ around the arbitrary axis for one atom is $2\cos\theta + 1$. Only atoms at the axis contribute to the character:

$$\chi(\theta) = N_A \chi(R(\theta)) = N_A(2\cos\theta + 1),$$

where N_A is the number of atoms at the axis. The character of reflection for an atom is 1, hence

$$\chi(\sigma) = N_P \chi(R_V(\sigma))$$

where N_P is the number of atoms in the mirror plane. The character of rotation that is accompanied by the reflection in the plane perpendicular to the axis (*mirror rotation*) s for an atom is $2\cos\theta - 1$, hence

$$\chi(s) = N_S(2\cos\theta - 1),$$

where N_S is the number of atoms at the point of the axis and plane intersection. Its possible values are $N_S = 0, 1$.

For a free molecule consisting of n atoms in three-dimensional space, there are degrees of freedom corresponding to movements of the molecule as a whole: three degrees of freedom correspond to translation, and three to the rotation of the molecules as a whole. They correspond to 6 normal modes with zero frequency (zero modes).[1] It is convenient to exclude these degrees of freedom from the beginning, leaving only oscillatory degrees. Initial $3n$-dimensional representation of $D(g)$ can be decomposed into the direct sum of the six-dimensional representation of $D_0(g)$ in the subspace zero modes and $(3n-6)$-dimensional vibrational representation of $D_v(g)$ in the subspace orthogonal to zero modes $D = D_0 \oplus D_v$. To find a character $\chi_v(g)$ in such a *vibrational* subspace, we need to subtract the character $\chi_0(g)$ from the character $\chi(g)$ of the initial representation.

A three-dimensional vector defines the translation, and the rotation is a three-dimensional pseudovector and both are invariant concerning the permutation T of the numbers of atoms. So the action of the element RT of the symmetry group of the molecule on the zero modes coincides with the effect of a single linear transformation R. Therefore, the character $\chi_0(g)$ is equal to the sum of the three-dimensional vector V and pseudovector PV.

The character of the rotation by angle θ around arbitrary axis equals

$$\chi_0(\theta) = \chi(R_V(\theta)) + \chi(R_{PV}(\theta)) = 2(1 + 2\cos\theta).$$

The character of the reflection in an arbitrary plane is

$$\chi_0(\sigma) = \chi(R_V(\sigma)) + \chi(R_{PV}(\sigma)) = 0.$$

The character of mirror reflection is

$$\chi_0(\sigma\theta) = \chi(R_V(\sigma)R_V(\theta)) + \chi(R_{PV}(\sigma)R_{PV}(\theta)) = 0.$$

Thus, the formulas for the character in the vibrational representation (denoted by subscript v) are

$$\begin{aligned} \chi_v(1) &= 3n - 6, & \chi_v(\theta) &= (N_A - 2)(2\cos\theta + 1), \\ \chi_v(\sigma) &= N_P, & \chi_v(s) &= N_S(2\cos\theta - 1). \end{aligned} \tag{11.6}$$

For reflection and mirror rotation the formulas remain the unchanged, compared to the initial representation.

Ultimately, we get the classical degeneration multiplicity that depends on the molecular symmetry. There is a classification of symmetries in molecular spectroscopy: low, medium, and high. The medium group has up to two-dimensional irreducible representations. The high symmetry means group G to have three-dimensional irreducible representations or higher. Then

[1] For a linear molecule (rotator), there are only two rotational degrees of freedom, which means 5 zero modes.

the molecular oscillations have at least 3-fold degeneration multiplicity. Note that in quantum mechanics, the degeneration multiplicity of the state depends on its vibrational quantum number.

11.2 Level Splitting

In quantum mechanics, there is a group of unitary transformations in linear space of states. Let us consider Hamiltonian of the form

$$\hat{H} = \hat{H}_0 + \hat{V},$$

where $\hat{H}_0$ is the main (unperturbed) Hamiltonian, and $\hat{V}$ is a perturbation. Let the Hamiltonian $\hat{H}_0$ have a symmetry group G, that is, $\hat{H}_0$ commutes with each of the representation operators of this group. Then d-dimensional irreducible representation of the group G acts in any d-dimensional invariant subspace.

Consider one such subspace with $d > 1$. Then we say that the corresponding eigenvalue E of the operator $\hat{H}_0$ is d-fold degenerate due to the symmetry. If symmetry $\hat{V}$ is lower than the symmetry of main Hamiltonian $\hat{H}_0$, then the perturbation partially removes the degeneracy. Indeed, in this case, the symmetry group F of full Hamiltonian $\hat{H}$ is a subgroup of G. As a result, some irreducible representations of the group G become reducible representations of the subgroup F. That leads to splitting of the corresponding degenerate levels E to sublevels E_j. Each of E_j is an eigenvalue of the operator $\hat{H}$ in its subspace in which the irreducible representation of F acts.

Let $\|\hat{V}\| \to 0$; then, we can take into account $\hat{V}$ in frames of the perturbation theory. To do this in the beginning, we find the eigenfunctions of the $\hat{H}_0$ within d-dimensional invariant subspace, and choose some basis $\{\psi_n\}$ of this subspace:

$$\hat{H}_0\psi_n = E\psi_n, \quad n = 1, 2, \ldots, d. \tag{11.7}$$

Let us choose the linear combinations of ψ_n to provide

$$\left(\Psi_i, \hat{V}\Psi_j\right) = \Delta E_j \delta_{ij}, \quad \Psi_j = \sum_{n=1}^{d} \alpha_{jn}\psi_n.$$

Values ΔE_j are defined from the *secular* equation:

$$|V_{mn} - \Delta E\,\delta_{mn}| = 0,$$

where $V_{mn} = (\psi_m, \hat{V}\psi_n)$ are matrix elements of the perturbation in the unperturbed basis $\{\psi_n\}$. Substituting Ψ_j to the Schrödinger equation, we get the eigenvalues of $\hat{H}$ up to linear terms in $\hat{V}$ inclusively:

$$\left(\hat{H}_0 + \hat{V}\right)\Psi_j = \left(E + \Delta E_j\right)\Psi_j. \tag{11.8}$$

Corrections to energy ΔE_j are the eigenvalues of matrix V_{mn}. At $\|\hat{V}\| \to 0$ the eigenfunctions of $\hat{H}$ tend to Ψ_j.

A partial degeneration may, nevertheless, remain. Since the operator $\hat{V}$ commutes with the operators of F, then, in the subspace over functions ψ_j; $j = 1, 2, \ldots, d$, matrix V_{mn} commutes with the operators of the initial d-dimensional representations of the group F. Formally, the task becomes equivalent to the problem of linear vibrations of a molecule with d degrees of freedom discussed in the previous section. Thus, the multiplicity degeneration of sublevels E_j is equal to the dimension of the corresponding irreducible representation.

11.3 Selection Rules

It is often required to find the matrix element $O_{fi} = \langle f|\hat{O}|i\rangle$ of operator $\hat{O}$. We do not need to calculate a lot of matrix elements because they equal zero due to the system's symmetry. The rules, by which we can determine in advance which of the matrix elements go to zero, called *selection rules*. In quantum mechanics, the square of a matrix element of the operator (11.8) is proportional to the probability of transition. If $V_{fi} = 0$, we say that the transition from the initial state $|i\rangle$ to the final state $|f\rangle$ is *forbidden*. Else the transition is called *allowed*.

Suppose in the Hilbert space $\mathcal{L}$ of quantum states, the representation of the group G acts: an element $g \in G$ corresponds to the unitary transformation $U(g)$. It is said that the set $\{\hat{O}_i\}$ of operators are transformed by the representation Φ of the group G if

$$\tilde{\hat{O}}_j = U(g)\hat{O}_j U^{-1}(g) = \Phi_{jk}(g)\hat{O}_k. \tag{11.9}$$

The set $\{\hat{O}_i\}$ is referred to as *tensor* operators.

Let $U(R)$ be the unitary operator in the Hilbert space corresponding to rotation R in three-dimensional space ($\mathbf{r}' = R\mathbf{r}$ or $r'_\alpha = R_{\alpha\beta}r_\beta$). According to the definition the transformed wave functions are

$$\tilde{\psi}(\mathbf{r}') = U(R)\psi(\mathbf{r}') = \psi(\mathbf{r}) = \psi(R^{-1}\mathbf{r}'). \tag{11.10}$$

Omitting the primes we write these equalities in the form

$$U(R)\psi(\mathbf{r}) = \psi(R^{-1}\mathbf{r}), \quad U^{-1}(R)\psi(\mathbf{r}) = \psi(R\mathbf{r}). \tag{11.11}$$

For arbitrary states $\tilde{\psi}$ and $\tilde{f}$, that are obtained from ψ under the action of operator $U(R)$, we have

$$\int d\mathbf{r}\tilde{f}^*(\mathbf{r})r_\alpha\tilde{\psi}(\mathbf{r}) = \int d\mathbf{r}f^*(R^{-1}\mathbf{r})r_\alpha\psi(R^{-1}\mathbf{r}) =$$
$$= \int d\mathbf{r}f^*(\mathbf{r})(R\mathbf{r})_\alpha\psi(\mathbf{r}).$$

Here we used variable substitution $R^{-1}\mathbf{r} \to \mathbf{r}$ and the invariance of integration measure. So we prove that the transformational properties of the operators r_α coincide with transformational properties of the corresponding matrix element:

$$\langle\tilde{f}|r_\alpha|\tilde{\psi}\rangle = R_{\alpha\beta}\langle f|r_\beta|\psi\rangle. \tag{11.12}$$

In the same way, this property can be proven for other operators.

Example. Let $G = \mathbf{SO}(3)$ be the rotation group of three-dimensional space. Irreducible representations of this group have dimension $2l + 1$ $(l = 0, 1, \ldots)$, where l is called orbital momentum in the physical literature. Then

(a) the Laplacian operator $\triangle$ is invariant with respect to rotations, i.e., transformed according to the scalar representation ($l = 0$);
(b) three operators r_α ($\alpha = 1, 2, 3$), the components of radius-vector $\mathbf{r}$, are transformed by the vector representation with orbital momentum $l = 1$;
(c) three operators, the projections of angular momentum $l_\alpha = -\mathrm{i}[\mathbf{r} \times \nabla]_\alpha$, are also transformed by the representation with $l = 1$;
(d) five operators, the components of quadruple moment tensor $Q_{\alpha\beta} = r_\alpha r_\beta - \frac{1}{3}\delta_{\alpha\beta}r^2$, is transformed by the representation with $l = 2$.

Consider now two sets of wave functions $\{\psi_n\}$ and $\{\phi_n\}$, that being transformed by irreducible representations V and W, respectively:

$$U(g)\psi_n = V_{nm}\psi_m, \quad U(g)\phi_n = W_{nm}\phi_m, \tag{11.13}$$

and the set of operators $\{\hat{O}_j\}$ transformed by representation (11.9). Let us derive the sufficient condition of the allowed transition. For evidence consider identity

$$M^j_{nm} = \langle\psi_n|U^{-1}(g)U(g)\hat{O}_jU^{-1}(g)U(g)|\phi_m\rangle$$
$$= \langle U(g)\psi_n|U(g)\hat{O}_jU^{-1}(g)|U(g)\phi_m\rangle.$$

In its derivation we use the unitarity of operator $U(g)$. Taking into account the transformation properties ψ_n, ϕ_m (11.13) and $\hat{O}_j$ (11.9), we obtain

$$M^j_{nm} = V^*_{nn'}(g)\Phi_{jj'}(g)W_{mm'}(g)M^{j'}_{n'm'} \equiv T^{jj'}_{nn',mm'}(g)M^{j'}_{n'm'} \,. \tag{11.14}$$

This equality holds for an arbitrary element $g \in G$, and the matrix element M^j_{nm} in this equality is independent of g. The matrices $T^{jj'}_{nn',mm'}(g)$ form a representation of the group G, which is a direct product of irreducible representations $V^* \otimes \Phi \otimes W$. We decompose this representation into a direct sum of irreducibles according to (9.7)

$$T(g) = \sum_\lambda c_\lambda D^{(\lambda)}(g) \tag{11.15}$$

and sum up both sides of Eq. (11.14) over all $g \in G$. The sum of matrices $D^{(\lambda)}(g)$, corresponding to a representation different from the trivial $D^{(0)}$ gives zero. It is the consequence of orthogonality relation (9.3). It follows that the matrix element M^j_{nm} can be nonzero only if the decomposition of the direct product $T = V^* \otimes \Phi \otimes W$ over the irreducibles includes the trivial representation, i.e., $c_0 \neq 0$ in Eq. (11.15).

According to the result of the problem **10.20**, one can simplify the selection rule as follows: so that the matrix element $\langle\psi_n|\hat{O}_j|\phi_m\rangle$ be nonzero, it is necessary (but not sufficient!) in the decomposition of $\Phi \otimes W$ into irreducibles V meet at least once. It is only a consequence of the transformational properties of the wave functions, operators, and the measure invariance. If the operator is reducible, then it should be decomposed into irreducible components. Then one finds the set of allowed transitions for each element separately.

11.4 Invariant Tensors

Let there be a matrix representation of group G with dimension n, i.e., matrix $D_{ik}(g)$ corresponds to each element $g \in G$. The set of numbers $T = T_{i_1 i_2 \ldots i_r} \in \mathbb{C}$ (or $\mathbb{R}$), $i_k = 1, \ldots, n$, $k = 1, \ldots, r$ is called *tensor* of rank r with respect to group G, if it is transformed according to the law:

$$T'_{i_1 i_2 \ldots i_r} = D_{i_1 k_1}(g)D_{i_2 k_2}(g)\ldots D_{i_r k_r}(g)T_{k_1 k_2 \ldots k_r} \tag{11.16}$$

We imply the summation by repeated indices. Numbers $T_{i_1 i_2 \ldots i_r}$ are *components* (coordinates) of the tensor. Tensors of rank r form the vector space of dimension n^r. Tensor T is called *invariant* in a group G if its components do not change under all the transformations from G. The number of independent components of the invariant tensor is equal to the number of trivial (unity) representations.

It is an important property of tensors that the permutation of its indices commutes with action of group G. Let the permutation group S_r of r symbols consist of elements

$$\sigma = \begin{pmatrix} 1 & 2 \ldots & r \\ i_1 & i_2 \ldots & i_r \end{pmatrix}. \tag{11.17}$$

Each permutation determines the linear transformation of tensor space according to the formula $\sigma T_{\alpha_1 \alpha_2 \ldots \alpha_r} = T_{\alpha_{i_1} \alpha_{i_2} \ldots \alpha_{i_r}} = T_{\alpha_{\sigma(1)} \ldots \alpha_{\sigma(r)}}$. These linear transformations commute with $q \in G$. A tensor is called *symmetric* if $\sigma T = T$ for all $\sigma \in S_r$. It is *antisymmetric* (skew-symmetric) when $\sigma T = T$ for even σ and $\sigma T = -T$ for odd σ. Then a symmetric tensor is the eigenvector of every $\sigma \in S_r$ with eigenvalue 1. Thus the symmetric tensors form an invariant subspace with respect to group G. The set of antisymmetric tensors also form an invariant space.

If $G = \mathbf{SO}(3)$ is the rotation group, $T_{i_1 i_2 \ldots i_r}$ are the tensors of rank r in three-dimensional Euclidian space (precisely speaking they are elements of linear subspace $\mathbb{R}^3 \otimes \cdots \otimes \mathbb{R}^3$), then $D(g)$ in Eq. (11.16) is the standard representation of the rotation group with $j = 1$ (vector representation). In this case $D^{(1)}_{i_1 k_1} D^{(1)}_{i_2 k_2} \ldots D^{(1)}_{i_r k_r}$ in Eq. (11.16) is the tensor product of representations with $j = 1$:

$$D_T(g) = \underbrace{D^{(1)}(g) \otimes \cdots \otimes D^{(1)}(g)}_{r \text{ times}}.$$

We can decompose this product into irreducibles. Each irreducible representation of a given r-rank tensor distinguishes its invariant part.

For example, $D^{(1)} \otimes D^{(1)} = D^{(0)} \oplus D^{(1)} \oplus D^{(2)}$. Then the second-rank tensor a_{ik} can be written as

$$a_{ik} = \frac{1}{2}\left(a_{ik} + a_{ki} - \frac{2}{3} a_{ll} \delta_{ik}\right) + \frac{1}{2}(a_{ik} - a_{ki}) + \frac{1}{3} a_{ll} \delta_{ik}. \tag{11.18}$$

Each term in the sum is an invariant tensor with respect to group $\mathbf{SO}(3)$. The first term in Eq. (11.18) has the zero trace and respectively 5 independent components; it corresponds to $j = 2$. The second term corresponds to a vector $j = 1$ with 3 components. The latter is scalar $j = 0$ with 1 component.

Consider tensors of rank 2 in three-dimensional Euclidean space $\mathbb{R}^3$. In the case of Euclidean space, we may write all the indices below. Denote a vector of symmetrized basis as

$$|\Psi^s_{ik}\rangle = \frac{1}{2}\left(|i\rangle \otimes |k\rangle + |k\rangle \otimes |i\rangle\right). \tag{11.19}$$

A symmetrized representation is the representation in one invariant subspace of the tensor product of p copies $\mathbb{R}^3 \otimes \cdots \otimes \mathbb{R}^3$. From (11.19) it follows that the second-rank tensor in the symmetrized basis has components:

$$[D(g) \otimes D(g)]\,|\Psi^s_{ik}\rangle = \frac{1}{2}\sum_{j,l=1}^{N} |j\rangle \otimes |l\rangle \left[D_{ji}(g)D_{lk}(g) + D_{li}(g)D_{jk}(g)\right]. \tag{11.20}$$

To find the symmetric character we equate indices $i = j$, $k = l$ or formally multiply the square bracket on the right in Eq. (11.20) by $\delta_{ij}\delta_{kl}$ and take a sum over repeated indices:

$$\chi_s(g) = \frac{1}{2}\left(\chi^2(g) + \chi(g^2)\right). \tag{11.21}$$

The subspace of symmetric rank l tensors is isomorphous to the space of uniform polynomials of degree l: $T_{i_1,i_2,\ldots,i_l} \sim x_{i_1}x_{i_2}\ldots x_{i_l}$, where $\sim$ denotes "is transformed as." Then for symmetric tensors, one can use the results for uniform polynomials. Instantly, in **SO**(3) the decomposition

$$D^{(l)}_s = D^{(l)} \oplus D^{(l-2)} \oplus D^{(l-4)} + \ldots \tag{11.22}$$

takes place.

For other symmetries, it is necessary to use formulas for the character of a symmetrized representation. In addition to the fully symmetric and antisymmetric parts, tensors of high rank $p \geqslant 3$ have subspaces with partial symmetry. Eq. (11.21), (11.34), (11.35), (11.36) for the symmetric characters can be obtained as the projectors into the invariant subspace of permutation group S_p. Then a character table on page 333 allows one to write the projectors for rank p tensor. These formulas are helpful when we have to calculate the number of independent components of an invariant tensor with a definite symmetry for its indices permutation.

In order to find the character of tensor representation $\chi_T(\varphi)$ in **SO**(3) group, we start from the vector representation $\chi(\varphi) = 1 + 2\cos\varphi$. In group **SO**(3) the number of trivial (scalar) representations in the decomposition is equal to

$$c_0 = \langle\chi_T\rangle_G = \frac{1}{2\pi}\int_{-\pi}^{\pi} \chi_T(\varphi)p(\varphi)\,d\varphi, \quad p(\varphi) = 2\sin^2\frac{\varphi}{2}. \tag{11.23}$$

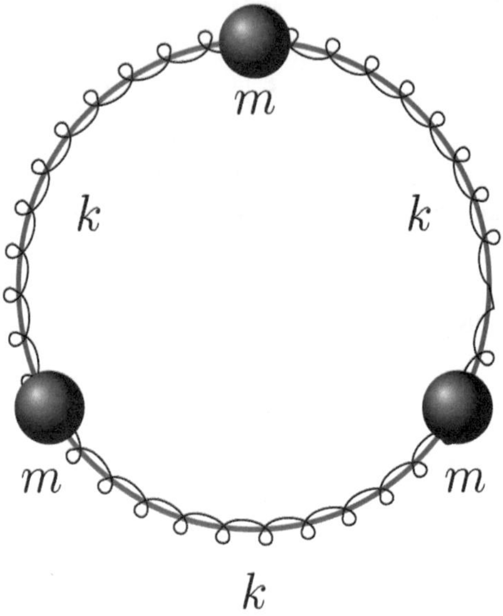

Figure 11.1 Masses at the ring, the model of one-dimensional "molecule."

Then the average over **SO**(3) has to be calculated with weight function $p(\varphi) = 1 - \cos\varphi$. In group **SO**(2) the integration is carried out with the unity weigh $p(\varphi) = 1$ (see problem **10.2**), and then reduces to the averaging over angle φ: $c_0 = \overline{\chi_T}$. In a point group it is the sum:

$$\langle \chi_T \rangle_G = \frac{1}{|G|} \sum_{g \in G} \chi_T(g).$$

Worked Examples

11.1. Find the degeneracy for small oscillations of a "molecule" consisting of three identical "atoms," which can only move along a circle (Fig. 11.1) with the Hamiltonian:

$$H = \frac{1}{2m} \sum_{j=1}^{3} p_j^2 + \frac{k}{2} \sum_{j=1}^{3} (x_j - x_{j-1})^2, \qquad x_3 \equiv x_0, \tag{11.24}$$

where x_j is the displacement of j-th atom along the ring with respect to its equilibrium position, p_j is the momentum.

Solution. The symmetry group is S_3 (or isomorphic $\mathbf{D}_3$). It consists of 3 conjugacy classes: e, the identity element, σ_2, the cyclic permutations of 3 atoms (2 elements), σ_3, the transpositions of two atoms (3 elements). There are three nonequivalent irreducible representations of this group. Since each atom's motion is one-dimensional, the original representation of group $\mathbf{D}_3$ acts in three-dimensional space of amplitude vectors. To find the degeneracy, decompose the initial representation. For this, we find the character of this representation. It is equal to the number of atomic amplitude vectors, unaffected by the action of this element of groups:

$$\chi(e) = 3, \quad \chi(\sigma_2) = 0, \quad \chi(\sigma_3) = 1.$$

Decompose the character into irreducible ones:

$$\chi(g) = \sum_l c_l \chi_l(g),$$

where $\chi_l(g)$ is the character of element g in l-th irreducible representation, c_l is the natural number of equal irreducible representations in the decomposition. Using the orthogonality of irreducible characters, we find the coefficients: $c_1 = 1$, $c_2 = 0$, $c_3 = 1$. Then the initial representation is decomposed by one trivial and one two-dimensional irreducible representation. That means the "molecule" has two normal frequencies, one of which is doubly degenerate.

Let us find the frequencies. For the trivial representation, the eigenvector does not change at any group transformation, i.e., permutations of atoms. It is obviously $b^t = (1, 1, 1)$. From equations of motion, the algebraic equations follow:

$$\begin{aligned} -\frac{m}{k}\omega^2 b_1 &= -2b_1 + b_2 + b_3, \\ -\frac{m}{k}\omega^2 b_2 &= b_1 - 2b_2 + b_3, \\ -\frac{m}{k}\omega^2 b_3 &= b_1 + b_2 - 2b_3. \end{aligned} \tag{11.25}$$

We find that the frequency of mode b^t is equal to zero (the uniform translation along the circle does not require the compression of spring). The orthogonal addition to b^t is a two-dimensional subspace. Choose some vector within this subspace, for instance $b^d = (0, 1, -1)$, and substitute it to Eq. (11.25). As a result, we get the second frequency: $\omega = \sqrt{3k/m}$. It is not difficult to verify this expression. At this motion, an atom and the middle of its opposite spring are motionless. Then the total rigidity of spring is $3k$ while the mass of atom is m.

11.2. The same about four equal masses with equilibrium positions at the vertices of a square. We get the Hamiltonian from Eq. (11.24) by replacing the upper limit of the sum over j by 4 and the periodic boundary condition $x_4 \equiv x_0$.

Solution. The symmetry group of the system is $\mathbf{D}_4$, the group of square. The latter consists of 8 elements and contains 5 classes of conjugate elements: e is the identity element, σ_2 is the rotation by 180° around axis C_4, σ_3 is the rotation by 180° around a middle line, σ_5 is the rotation by 90° or −90° around axis C_4. Dimensions of irreducible representations can be found from the condition (9.6). Only one sum of five squares gives the number 8, namely $1^2 + 1^2 + 1^2 + 1^2 + 2^2 = 8$. So there are 4 one-dimensional representations E, A_2, A_3, A_4 and one two-dimensional representation B. The corresponding character table is

	e	σ_2	σ_3	σ_4	σ_5
E	1	1	1	1	1
A_2	1	1	-1	-1	1
A_3	1	1	1	-1	-1
A_4	1	1	-1	1	-1
B	2	-2	0	0	0

The character of initial four-dimensional representation $T(g)$ in the space of amplitude vectors is

$$\chi(e) = 4, \quad \chi(\sigma_2) = 0, \quad \chi(\sigma_3) = 2, \quad \chi(\sigma_4) = 0, \quad \chi(\sigma_5) = 0.$$

Using the orthogonality of irreducible characters, we obtain coefficients:

$$T = E \oplus A_3 \oplus B.$$

It means that there are two simple frequencies and one double. As well as in the previous problem, the trivial representation corresponds to zero frequency: $\omega^t = 0$, $b^t = (1, 1, 1, 1)$. To identify frequencies of representations A_3, B, we exploit operator

$$P_R = \frac{1}{|G|} \sum_{g \in G} \chi^{(R)*}(g) T(g) = \langle \chi^{(R)}(g) | T(g) \rangle_G .$$

It is a projector on invariant subspace corresponding to irreducible representations $R = E$, A_3 or B if the latter occurs precisely once in the decomposition, and $\chi^{(R)}(g)$ denotes the character of representations $R(g)$. If all the irreducible representations enter 0 or 1 time into the decomposition, the initial representation is called *simply reducible.*[2]

In our case, the irreducible representations A_3 and B meet exactly once. The easiest way is to build a projector on a subspace, transformed by the representation of B. For this, as seen from tables of irreducible characters, we need to know the explicit form of matrices of the initial representations T for only two elements e and σ_2:

$$P_B = \frac{1}{8}\left(2T(e) - 2T(\sigma_2)\right),$$

$$T(e) = \begin{pmatrix} 1\,0\,0\,0 \\ 0\,1\,0\,0 \\ 0\,0\,1\,0 \\ 0\,0\,0\,1 \end{pmatrix}, \quad T(\sigma_2) = \begin{pmatrix} 0\,0\,1\,0 \\ 0\,0\,0\,1 \\ 1\,0\,0\,0 \\ 0\,1\,0\,0 \end{pmatrix}.$$

Acting sequentially by projector P_B to basic vectors (the first and second basic vectors are sufficient), we get the subspace of two-dimensional representation:

$$b = (1,\ 0,\ -1,\ 0), \quad b' = (0,\ 1,\ 0,\ -1).$$

Substituting b or b' into motion equations

$$\begin{aligned} -\frac{m}{k}\omega^2 b_1 &= -2b_1 + b_2 + b_4, \\ -\frac{m}{k}\omega^2 b_2 &= b_1 - 2b_2 + b_3, \\ -\frac{m}{k}\omega^2 b_3 &= b_2 - 2b_3 + b_4, \\ -\frac{m}{k}\omega^2 b_4 &= b_1 + b_3 - 2b_4, \end{aligned}$$

we get expression $\omega_2 = \sqrt{2k/m}$ for frequency of double degenerate oscillation.

Vector b^a, which is transformed by representation A_3, is defined unambiguously by orthogonality conditions to vectors b^t, b, b':

$$b^a = (1,\ -1,\ 1,\ -1).$$

The normal frequency can be calculated from motion equations: $\omega_3 = 2\sqrt{k/m}$.

---

[2] If one irreducible representation occurs more than once, the operator P_R projects into a direct sum subspaces.

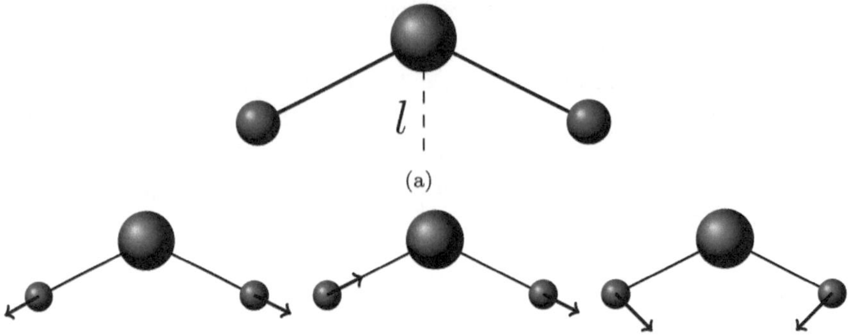

(b)

Figure 11.2 Molecule H_2O (a) and its normal oscillations (b). For visualization, we neglect a displacement of the oxygen atom.

11.3. Find the order of degeneracy for normal oscillations of the water molecule. The molecule has the shape of the equilateral triangle, Fig. 11.2(a).

Solution. In the case of H_2O molecule, the symmetry group $C_{2v} \approx \mathbf{D}_2$ consists of 4 elements: the identity element e, the rotation by 180° around axis l, the reflection ξ_1 in the plane of molecule, the reflection ξ_2 in the plane, perpendicular to the plane and passing through the axis l (see Fig. 11.2). The group is Abelian, and then each element forms its own conjugacy class. Then the character table of one-dimensional irreducible representations is:

	e	r	ξ_1	ξ_2
χ_E	1	1	1	1
χ_A	1	1	-1	-1
χ_B	1	-1	1	-1
χ_C	1	-1	-1	1

The trivial representation is denoted hereafter E. The irreducible representations are one-dimensional, then all the normal frequencies are non-degenerate. The initial representation Φ_i in space of dimension $3n = 9$ has the following values of character:

$$\chi_i(e) = 9, \quad \chi_i(r) = 1 + 2\cos\pi = -1, \quad \chi_i(\xi_1) = 3, \quad \chi_i(\xi_2) = 1.$$

Subtracting the character of zero modes (the translation and rotation of molecule as a whole), we get the character of vibrational representation of

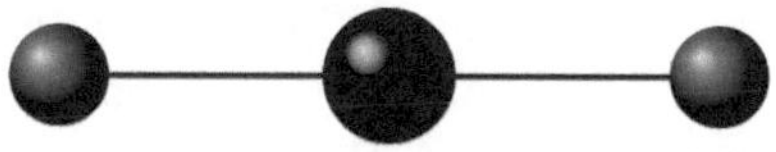

Figure 11.3 Linear molecule CO_2.

dimension $3n - 6 = 3$:

$$\chi_v(e) = 9 - 6 = 3, \quad \chi_v(r) = -1 - (-2) = 1, \quad \chi_v(\xi_1) = 3, \quad \chi_v(\xi_2) = 1.$$

Using the orthogonality of irreducible characters, we obtain the decomposition of Φ into a direct sum $\Phi = 2E \oplus B$. All three frequencies are simple. The character of ξ_1 is equal to unity within representations E and B. Therefore, all the normal oscillations occur in the plane of the molecule (Fig. 11.2(b)). With respect to axes x and y the normal oscillations are $s - s$, $s - a$, $a - s$ (from the left), where s denotes symmetric, a are antisymmetric oscillations.

11.4. Find the order of degeneracy for carbon dioxide molecule (Fig. 11.3).

Solution. Linear CO_2 molecule is a rotator with two rotational degrees of freedom, and then the number of vibrational ones is $3n - 5 = 4$. The symmetry group $D_{\infty v}$ is the direct product $C_i \otimes C_{\infty v}$ of the inversion group and group $C_{\infty v}$. The latter includes the continuous group **SO**(2) of rotations around line O-C-O and reflections in the plane passing through the line. It leads to double degeneration of the normal frequencies for vibrations violating the line: the vibrations in two orthogonal planes passing through the line have the same frequency. As for vibrations along the line, in its three-dimensional subspace the representation Φ of factor-group $D_{\infty v} / C_{\infty v} = C_i$ acts. The factor-group consists of two elements: the identity e and the inversion i.[3] From the character table of this group

	e	i
χ_E	1	1
χ_{E*}	1	-1

we get the decomposition of initial representation with character $\chi(e) = 3$, $\chi(i) = -1$:

$$\Phi = E \oplus 2E^*.$$

[3]The inversion is the rotation by 180° and reflection in the plane perpendicular to the rotation axis. It is the mirror-rotation of the second order $i = S_2$. In our case, $\chi(i) = -1$, not -3, since one-dimensional motion is considered.

The normal oscillations are symmetric (E) and antisymmetric (E^*). The other antisymmetric representation (E^*) corresponds to a zero mode, the translation of the whole molecule.

11.5. Find the degeneration multiplicity of ammonia molecule NH_3. The shape is a triangular pyramid with an equilateral triangle in the base, with the hydrogen atoms in its apexes.

Solution. The symmetry group is C_{3v}, which includes 6 elements split by 3 classes: the identity e, 2 rotations r, r^2 by $120°$ and $240°$ around the axis C_3 passing through the nitrogen atom, 3 reflections σ in planes passing through the nitrogen and hydrogen atoms. The character table is presented on page 334. The initial representation Φ, $\dim\Phi = 3n = 12$ has the character:

$$\chi_i(e) = 12, \quad \chi_i(r) = 1 + 2\cos\frac{2\pi}{3} = 0, \quad \chi(\sigma) = 2.$$

Subtracting the character of zero modes, we obtain the character of vibrational representation:

$$\chi_v(e) = 12 - 6 = 6, \quad \chi_v(r) = 0, \quad \chi_v(\sigma) = 2.$$

Then we decompose the vibrational representation:

$$\Phi = 2D_0 \oplus 2D_2.$$

Thus there are two simple normal frequencies and two double degenerated.

11.6. Find the multiplicity of ethane molecular oscillations, Fig. 11.4.

Solution. The symmetry group consists of 6 elements of subgroup $\mathbf{C}_{3v}$ and in addition 6 elements of the form $i\mathbf{C}_{3v} = \mathbf{C}_{3v}i$, where i is the inversion. The inversion belongs to the group center. The symmetry group is therefore the direct product of groups $\mathbf{C}_{3v}$ and $\mathbf{C}_2 = \{e, i\}$: $\mathbf{D}_{3h} = \mathbf{C}_{3v} \times \mathbf{C}_2$. Then the table is a product of their character tables:

	e	r	σ	i	ir	ip
χ_0	1	1	1	1	1	1
χ_1	1	1	-1	1	1	-1
χ_2	2	-1	0	2	-1	0
χ_3	1	1	1	-1	-1	-1
χ_4	1	1	-1	-1	-1	1
χ_5	2	-1	0	-2	1	0
χ_v	18	0	4	0	0	2

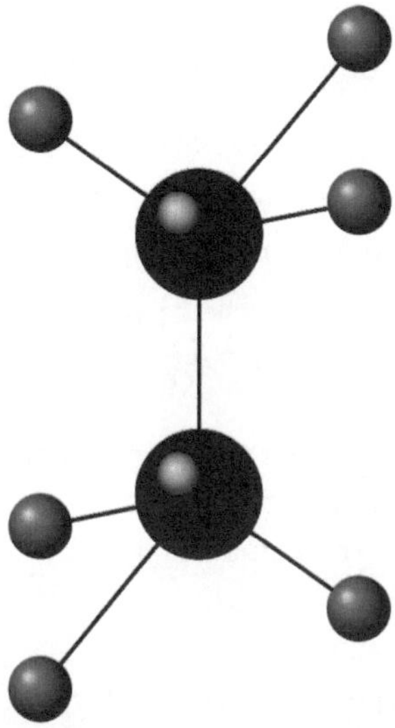

Figure 11.4 Molecule C_2H_6 having the center of inversion.

At the bottom, the character of vibrational representation is inserted. Its cells are filled according to Eq. (11.6). There are $n = 8$ atoms. For the identity element, the character is $\chi_v(e) = 3n - 6 = 18$. At rotation r, the value is zero, since the angle is 120°. There are 4 atoms in the mirror plane: two hydrogen and two carbon atoms, then $\chi_v(\sigma) = 4$. The inversion in three-dimensional space is a composition of rotation by 180° and reflection in the plane perpendicular to the axis: $i = r(\pi)\sigma$. Then the inversion can be considered as mirror rotation by 180°. Atoms are absent in the inversion center $N_S = 0$, then $\chi_v(i) = 0$. The next class involves mirror rotations by ±60°, since $ir^{\pm 1} = \sigma r(\pi) r(\pm 2\pi/3) = \sigma r(\mp \pi/3)$. The value is zero by the same reason: $N_S = 0$. The last conjugacy class $i\sigma$ includes the rotations with respect to the second-order axes, perpendicular to the $C - C$ segment in its center. Since $N_A = 0$, the value of character is $\chi_v(i\sigma) = -2(1 + 2\cos\pi) = 2$. Using orthogonality, we get $c_0 = 3$, $c_1 = 0$, $c_2 = 3$, $c_3 = 2$, $c_4 = 1$, $c_5 = 3$, then there are 12 normal frequencies, 6 among them are double degenerate.

11.7. Oscillations of a circular membrane with a fixed edge have the known angular dependence $\Phi(\varphi) = e^{im\varphi}$, where m is an integer azimuthal number, φ is an angular polar coordinate. Initial symmetry led to double degeneration: modes m and $-m$ $(m \neq 0)$ have equal frequencies. Three equal masses are placed on the membrane at equal distances from the center and between masses. Find the degeneration multiplicity of the oscillation due to the lower symmetry.

Solution. The initial symmetry was $G = \mathbf{SO}(2)$ falls to $H = \mathbf{C}_{3v}$. Let us introduce the initial functions in m-th subspace:

$$\psi(\varphi) = \begin{pmatrix} \mathrm{e}^{im\varphi} \\ \mathrm{e}^{-im\varphi} \end{pmatrix}$$

and build up the representation $T(g)$ of group H in this subspace. Identity element does not change function ψ then 2×2 unity matrix is its representative $T(1) = E$. The rotation r by 120° according to Eq. 10.1 acts as matrix $T(g)$:

$$r\psi(\varphi) = \psi(\varphi - 2\pi/3) = T(r)\psi(\varphi), \quad T(r) = \begin{pmatrix} \mathrm{e}^{-2im\pi/3} & 0 \\ 0 & \mathrm{e}^{+2im\pi/3} \end{pmatrix}.$$

If we choose the appropriate origin of coordinate φ, then the reflection σ with respect a height changes the sign $\varphi \to -\varphi$. Therefore

$$\sigma\psi(\varphi) = \psi(-\varphi) = T(\sigma)\psi(\varphi), \quad T(\sigma) = \begin{pmatrix} 0 & 1 \\ 1 & 0 \end{pmatrix}.$$

The representation is ready, and its character is $\chi(g) = \operatorname{tr} T(g)$:

	1	r	σ
$\chi(g)$	2	$2\cos(2m\pi/3)$	0

Then we decompose the representation and get

$$c_1 = c_2 = \frac{1}{3}(1 + 2\cos(2m\pi/3), \quad c_3 = 1 - c_1.$$

Thus, when 3 divides m, then $c_0 = c_1 = 1$, $c_2 = 0$, the degeneration is removed and we have two simple oscillations. While 3 does not divide m, then $c_0 = c_1 = 0$, $c_2 = 1$ and the oscillation remains double degenerate. The physics explains the result if we are imaging the azimuthal modes as angular standing waves. There are two waves with amplitudes $\sin m\varphi$ and $\cos m\varphi$. At $m = 3$, the first wave has nodes at mass locations ($\varphi = 0, \pm 2\pi/3$), while the second has the antinodes. Then the sine wave conserves the frequency, whereas the cosine wave increases its frequency.

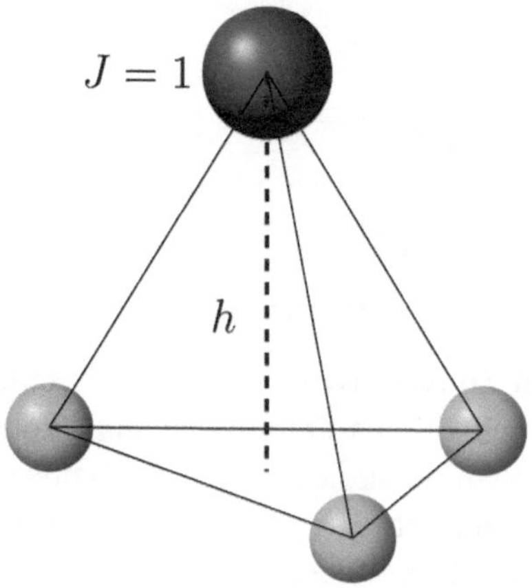

Figure 11.5 An atom in the field of three ions or atoms of other sort, placed in the apices of an equilateral triangle.

11.8. Consider an atom in the state with orbital momentum $J = 1$ in the tetrahedron's vertex. Its base has the shape of a regular triangle, consisting of identical ions or atoms of a different sort (see Fig. 11.5). Find the degeneracy multiplicity of the state, taking into account the influence of the external field.

Solution. The symmetry group F of the system consists of the identity element, two rotations by angles 120° and 240° around the height h (one class r, r^2) and three reflections with respect to planes passing through this height and one of the vertices of the triangle base (class σ, σr, σr^2). This group is isomorphic to group S_3, its character table has been calculated in problem **9.10**. The Hamiltonian is invariant with respect to the full group of rotations $\mathbf{O}(3)$, and the initial state has a degree of degeneration $2J + 1 = 3$. A three-dimensional representation of the group $\mathbf{O}(3)$ acts in the unperturbed space. The group F is a subgroup of $\mathbf{O}(3)$. The basis of the initial representation consists of spherical harmonics $Y_{J,m}(\theta, \varphi)$, where θ and φ are the angles in the spherical coordinate system with z-axis along the tetrahedron heights h:

$$U(r)Y_{J,m}(\theta,\varphi) = Y_{J,m}\left(\theta, \varphi - \frac{2\pi}{3}\right) = e^{-im\frac{2\pi}{3}}Y_{J,m}(\theta,\varphi), \quad (11.26)$$
$$U(\sigma)Y_{J,m}(\theta,\varphi) = Y_{J,m}(\theta,-\varphi) = Y_{J,-m}(\theta,\varphi).$$

Hence for $J = 1$

$$\chi(e) = 3, \quad \chi(r) = \sum_{m=-1}^{1} e^{im\frac{2\pi}{3}} = 0, \quad \chi(\sigma) = \mathrm{tr}\begin{pmatrix} 0 & 0 & 1 \\ 0 & 1 & 0 \\ 1 & 0 & 0 \end{pmatrix} = 1. \quad (11.27)$$

Using the orthogonality, we get

$$D = D_0 \oplus D_2, \quad (11.28)$$

where D_0 is the trivial, D_2 is the two-dimensional irreducible representation. Thus the initial level splits into a simple level and a double degenerated one. The wave function of a simple sublevel is a scalar being invariant with respect to any rotation. Consequently, the projection $m = 0$ corresponds to this sublevel (there are no φ-dependence):

$$\Psi_I = Y_{1,0}. \quad (11.29)$$

The wave functions of doublet level are orthogonal to singlet state (11.29), then they are linear combinations of spherical harmonics with $m = \pm 1$.

11.9. Consider an atom with angular momentum $J = 2$ in the center of an equilateral triangle, involving three identical of another sort. By how many sublevels does the initial state split?

Solution. Compared to the previous problem, an additional generatrix of the symmetry group F appears here, namely, the reflection $\tilde{\sigma}$ in the plane of the triangle. This element has the second order, $\tilde{\sigma}^2 = e$, and forms its own conjugacy class, since commutes with any group element. The number of elements doubles ($|F| = 12$), since for any element of the group has a pair, the element multiplied by $\tilde{\sigma}$. It follows that, along with classes of conjugate elements r and σ, appear classes r' and σ' as a result of multiplication by $\tilde{\sigma}$. Since $\tilde{\sigma}$ commutes with all elements of F, in any representation $T(\tilde{\sigma}) = \pm I$, where I is the identity matrix. The number of irreducible representations also doubles compared to the previous problem. Considering the character table as a matrix, we will get a table of characters as the direct product of tables of $\mathbf{C}_2$:

	1	$\tilde{\sigma}$
χ_1	1	1
χ_2	1	-1

and $\mathbf{D}_3$, considered as matrices.

The direct product is

	e	r	σ	$\tilde{\sigma}$	r'	σ'
χ_E	1	1	1	1	1	1
χ_A	1	1	-1	1	1	-1
χ_B	2	-1	0	2	-1	0
$\chi_{E'}$	1	1	1	-1	-1	-1
$\chi_{A'}$	1	1	-1	-1	-1	1
$\chi_{B'}$	2	-1	0	-2	1	0
χ_D	5	-1	1	1	1	1

The same product occurs in problem **11.6**. At the bottom, the character of initial representation D is written. Its components can be obtained if one chooses the basis of initial representation $Y_{J\,m}$ with $J = 2$ and $m = -2, -1, 0, 1, 2$ and use the transformation properties (11.26), and also

$$U(\tilde{\sigma})Y_{J\,m}(\theta,\varphi) = U(\tilde{\sigma})Y_{J\,m}(-\theta,\varphi) = (-1)^m Y_{J\,m}(\theta,\varphi), \qquad (11.30)$$
$$U(r') = U(\tilde{\sigma})U(r), \quad U(\sigma') = U(\tilde{\sigma})U(\sigma).$$

Note that the spherical harmonics with angular momentum 2 are transformed at rotations as the following functions of unity vector $\mathbf{n} = (n_x, n_y, n_z)$:

$$n_+^2, \quad n_+n_z, \quad n_z^2 - \frac{1}{3}, \quad n_-n_z, \quad n_-^2, \qquad (11.31)$$

where $n_\pm = n_x \pm in_y$ are its circular components.

Equation (9.4) gives

$$D = E \oplus B \oplus B'. \qquad (11.32)$$

Thus the initial state splits into three sublevels, one simple and two doublets.

11.10. Find the selection rules in parity of operator of the electric **d** and magnetic **m** dipole moments.

Solution. The inversion group $G = \{e, P\}$, $P^2 = e$ has only one-dimensional representations. Then a function of the final of initial state $\langle f|$, $|i\rangle$ can be only odd or even. They are transformed by one-dimensional representations $D_i(P)$, $D_f(P) = \pm 1$. Since the electric dipole moment changes its sign under the inversion while the magnetic does not change it, $D_\mathbf{d} = -1$, $D_\mathbf{m} = 1$. The direct product includes the trivial representation $D_f \otimes D_O \otimes D_i = 1$, only if $D_f D_i = -1$ for **d** or $D_f D_i = 1$ for **m**. Matrix element d_{fi} (m_{fi}) can be non-zero and transitions are not forbidden, when the initial and final states have the opposite (the identical) parities.

11.11. Find the selection rules for operator of electric dipole moment $\mathbf{d} = e\mathbf{r}$:

(a) in the full orbital momentum L;

(b) in the projection M of the orbital momentum.

Solution. 11.11(a) electric dipole operator is a vector, so is transformed by the three-dimensional irreducible representation $D^{(J)}$ of the group $\mathbf{SO}(3)$, with $J = 1$. Denote by L the orbital moment of the initial state, and by L' of final, then decomposition

$$D = D^{(L')*} \otimes D^{(1)} \otimes D^{(L)} = D^{(L')*} \left(D^{(L-1)} \oplus D^{(L)} \oplus D^{(L+1)}\right)$$

includes the trivial representation only at

$$L' = L,\ L \pm 1; \quad L \geq 1.$$

If $L = 0$, then $D^{(1)} \otimes D^{(0)} = D^{(1)}$, therefore $D = D^{(L')*} \otimes D^{(1)}$ contains a trivial representation only for $L' = 1$. Therefore, the matrix element between the two S states is zero. Note that in a hydrogen atom containing one electron, the transition to the state with $L' = L$ is forbidden by parity.

Solution. 11.11(b) Rotations around the quantization axis z form the Abelian group $\mathbf{SO}(2)$, all irreducible representations of which are one-dimensional: $D_M = \exp(\mathrm{i}M\varphi)$. Therefore, the product $D = D^*_{M'} \otimes D_m \otimes D_M$ contains trivial one if $M' = M + m$. Then the matrix element of the projection d_z is nonzero if $M' = M$. The matrix elements of circular components $d_\pm = d_x \pm \mathrm{i}d_y$ transformed by representations $D_{\pm 1}$ are nonzero when $M' = M \pm 1$.

11.12. Derive the selection rules for electrical dipole transition in a molecule with symmetry $\mathbf{D}_3$.

Solution. The electric dipole moment is a vector, then its character is $\chi(\varphi) = \operatorname{tr} D(\varphi) = 1 + 2\cos\varphi$. Substituting sequentially $\varphi = 0,\ 2\pi/3,\ \pi$, we get

$$\chi(1) = 3, \quad \chi(r) = 0, \quad \chi(p) = -1.$$

There are three irreducible representations of this group. Using the dipole character, we should decompose direct products $D_i \otimes D$, $i = 0, 1, 2$ over irreducibles. We get

$$D_0 \otimes D = D_1 \oplus D_2, \quad D_1 \otimes D = D_0 \oplus D_2, \quad D_2 \otimes D = D_0 \oplus D_1 \oplus D_2,$$

where D is the vector representation. If D_f meets in the decomposition of $D_i \otimes D$, then the transition $i \to f$ is allowed. Denoting the allowed transitions by a star, forbidden ones by zero, we obtain the following table of transitions:

	D_0	D_1	D_2
D_0	0	⋆	⋆
D_1	⋆	0	⋆
D_2	⋆	⋆	⋆

Thus the selection rules forbid transitions between equal one-dimensional representations.

11.13. Derive the selection rules for electrical dipole transition in a molecule with symmetry **T**.

Solution. The group of regular tetrahedron rotations includes rotations around the second and third-order axes, see the table on page **??**. The values of character are the same as in the previous problem:

$$\chi(1) = 3, \quad \chi(r) = \chi(r^2) = 0, \quad \chi(p) = -1.$$

Then we decompose products $D_i \otimes D$, $i = 0, 1, 2, 3$:

$$D_0 \otimes D = D_1 \otimes D = D_2 \otimes D = D_3, \quad D_3 \otimes D = D_0 \oplus D_1 \oplus D_2 \oplus 2D_3$$

and fill the i-th line in the table of transitions:

	D_0	D_1	D_2	D_3
D_0	0	0	0	⋆
D_1	0	0	0	⋆
D_2	0	0	0	⋆
D_3	⋆	⋆	⋆	⋆

Thus the dipole transitions are forbidden between one-dimensional representations.

11.14. Derive formulas for operators of the symmetrization and antisymmetrization:

$$\hat{S} = \frac{1}{p!}\sum_{\sigma\in P_p} \hat{\sigma}, \quad \hat{A} = \frac{1}{p!}\sum_{\sigma\in S_p} \hat{\sigma} \det\hat{\sigma},$$

where $\det\hat{\sigma} = \pm 1$ is the parity of permutation $\hat{\sigma}$.

Solution. Group S_p has an even trivial representation $[p]$ and odd one $[1^p]$, the representation of factor-group over alternating-sign subgroup is $C_2 = P_p/A_p$. Each symmetrization is the projector onto the invariant subspace with corresponding symmetry.

11.15. How many independent components does the symmetric third-rank tensor have?

Solution. Write down all the symmetry relations:

$$S_{112} = S_{211} = S_{121}, \quad S_{221} = S_{122} = S_{212},$$
$$S_{223} = S_{322} = S_{232}, \quad S_{332} = S_{233} = S_{323},$$
$$S_{113} = S_{311} = S_{131}, \quad S_{331} = S_{133} = S_{313},$$
$$S_{123} = S_{312} = S_{231} = S_{213} = S_{132} = S_{321}.$$

We have total 17 relations; then 10 independent components remain. Using the isomorphism, we could list all the corresponding 3-rd degree polynomials: $x^3, y^3, z^3, xy^2, x^2y, xz^2, x^2z, yz^2, y^2z, xyz$.

11.16. How many independent components does the antisymmetric third-rank tensor have?

Solution. Write down all the symmetry relations:

$$A_{111} = A_{222} = A_{333} = 0,$$
$$A_{112} = A_{211} = A_{121} = 0, \quad A_{221} = A_{122} = A_{212} = 0,$$
$$A_{223} = A_{322} = A_{232} = 0, \quad A_{332} = A_{233} = A_{323} = 0,$$
$$A_{113} = A_{311} = A_{131} = 0, \quad A_{331} = A_{133} = A_{313} = 0,$$
$$A_{123} = A_{312} = A_{231}, \quad A_{213} = A_{132} = A_{321}.$$

There are 26 relations; then it has only 1 component, like in problem **11.23**.

11.17. How many independent components does the partially symmetric third-rank tensor have?

Solution. The Jacobi identity of a partially symmetric tensor is

$$B_{ijk} + B_{kij} + B_{jki} = 0.$$

The relations between its components follow from the property

$$B_{111} = B_{222} = B_{333} = 0,$$
$$B_{112} + B_{211} + B_{121} = 0, \quad B_{221} + B_{122} + B_{212} = 0,$$
$$B_{223} + B_{322} + B_{232} = 0, \quad B_{332} + B_{233} + B_{323} = 0,$$
$$B_{113} + B_{311} + B_{131} = 0, \quad B_{331} + B_{133} + B_{313} = 0,$$
$$B_{123} + B_{312} + B_{231} = 0, \quad B_{213} + B_{132} + B_{321} = 0.$$

We have 11 relations, and then the tensor has 16 components.

11.18. How many independent components does second-rank tensor, invariant in $G = \mathbf{D_3}$, have?

Solution. Note that $\mathbf{D_3} < \mathbf{SO}(3)$. The character of rotation by angle φ is equal to $1 + 2\cos\varphi$. The trace of a tensor product is the product of traces. These two properties allow us to calculate the character of the triangle group. Let σ_1, σ_2, σ_3 be the conjugacy classes of rotations by $\varphi = 0, 120°, 180°$, see problem **9.10**. We obtain the character

	σ_1	σ_2	σ_3
χ	9	0	1

The trivial representation enters twice, then the tensor has two independent components.

11.19. How many independent components have a third-rank tensor, invariant with respect group $\mathbf{SO}(2)$? The same for group $\mathbf{SO}(3)$.

Solution. For a vector $\chi(\varphi) = 1 + 2\cos\varphi$, then

$$\mathbf{SO}(2) : c_0 = \overline{\chi^3(\varphi)} = 4;$$
$$\mathbf{SO}(3) : c_0 = \overline{\chi^3(\varphi)(1 - \cos\varphi)} = 1.$$

Thus 4 components are independent in $\mathbf{SO}(2)$ and only 1 in $\mathbf{SO}(3)$.

11.20. How many independent components does a third-rank tensor, invariant in the group $\mathbf{D_3}$, have? The same for the case of the symmetric tensor.

Solution. We decompose the character of tensor representation $\begin{array}{|c|c|c|}\hline 27 & 0 & -1 \\ \hline\end{array}$ over the table of irreducible representations of the triangle group. The first coefficient is $c_0 = 4$, and then there are 4 components. We find the character of the symmetric part by Eq. (11.35). We find the content of each cell using formula $\chi(\theta) = 1 + 2\cos\theta$, where $\theta = \alpha, 2\alpha, 3\alpha, \alpha = 0, 2\pi/3, \pi$:

$$\frac{1}{6}\left[\begin{array}{|c|c|c|}\hline 27 & 0 & -1 \\ \hline\end{array} + 3\begin{array}{|c|c|c|}\hline 3 & 0 & -1 \\ \hline\end{array} \cdot \begin{array}{|c|c|c|}\hline 3 & 0 & 3 \\ \hline\end{array} + 2\begin{array}{|c|c|c|}\hline 3 & 3 & -1 \\ \hline\end{array}\right] =$$

$\begin{array}{|c|c|c|}\hline 10 & 1 & -2 \\ \hline\end{array}$, hence $c_0 = 1$. Here the boxes, containing elements in cells, are multiplied as the lists:

$$\begin{array}{|c|c|c|}\hline a_1 & a_2 & a_3 \\ \hline\end{array} \cdot \begin{array}{|c|c|c|}\hline b_1 & b_2 & b_3 \\ \hline\end{array} = \begin{array}{|c|c|c|}\hline a_1b_1 & a_2b_2 & a_3b_3 \\ \hline\end{array}.$$

11.21. Find the number of independent components of the second-rank tensor, invariant in the group $G = \mathbf{SO}(2)$. How many components are in the symmetric and antisymmetric subspace?

Solution. The character of vector representation is $\chi(\varphi) = e^{-i\varphi} + 1 + e^{-i\varphi}$. For the tensor representation, the character is its square χ^2. We have to clarify how many times the trivial representation enters the tensor representation. For this purpose, we average the character over the angle φ: $c_0 = \overline{\chi^2} = \overline{(1 + 2\cos\varphi)^2} = 3$. Then the full number is 3. For the symmetric subspace we use Eq. (11.21):

$$\chi_s = \frac{1}{2}\left\{\chi^2(\varphi) + \chi(2\varphi)\right\} = \frac{1}{2}\left[(1 + 2\cos\varphi)^2 + (1 + 2\cos 2\varphi)\right],$$

then $c_0 = \overline{\chi_s} = 2$. The trivial representation enters twice. For the antisymmetric part, we explore Eq. (11.34) and get the same formula, but with "minus" sign in curly braces. Then the antisymmetric tensor has only 1 component. Indeed, the second-rank tensor in the group of rotations around axis **n** can be written as

$$T_{ij} = A\delta_{ij} + Bn_i n_j + Ce_{ijk}n_k.$$

Two former terms describe the 2-component of the symmetric tensor. The latter term is an antisymmetric part.

11.22. How many independent components does a symmetric third-rank tensor, invariant under the transformation of group $\mathbf{SO}(2)$, have? The same for group $\mathbf{SO}(3)$.

Solution. From Eq. (11.35), we find

$$\begin{aligned} c_0 &= \frac{1}{6}\left[\overline{(1 + 2\cos\varphi)^3} + 3\overline{(1 + 2\cos\varphi)(1 + 2\cos 2\varphi)} + 2\overline{(1 + 2\cos 3\varphi)}\right] \\ &= 2. \end{aligned}$$

Then in $\mathbf{SO}(2)$ there are 2 components:

$$T_{ijk} = A\left(\delta_{ij}n_k + \delta_{ki}n_j + \delta_{jk}n_i\right) + Bn_i n_j n_k.$$

In $\mathbf{SO}(3)$, according to Eq. (11.22), $D_s^{(3)} = D^{(3)} \oplus D^{(1)}$. If the trivial representation does not enter, then the tensor has 0 components.

11.23. How many independent components does an antisymmetric third-rank tensor, invariant under the transformation of group $\mathbf{SO}(3)$, have?

Solution. For the third-rank tensor, from Eq. (11.35), we get:

$$\chi_a(\varphi) = \frac{1}{6}\left[(1+2\cos\varphi)^3 - 3(1+2\cos\varphi)(1+2\cos 2\varphi) + 2(1+2\cos 3\varphi)\right],$$

then $c_0 = \overline{\chi_a(1-\cos\varphi)} = 1$. Indeed, the tensor is proportional to the antisymmetric identity tensor: $T_{ijk} = Ce_{ijk}$.

11.24. How many independent components does a symmetric tensor of rank 4, invariant in tetrahedron group $G = \mathbf{T}$, have?

Solution. From the character of the vector representation

$$\chi = \begin{array}{|c|c|c|c|}\hline 3 & -1 & 0 & 0 \\ \hline\end{array}$$

using Eq. (11.36), we find the symmetric part:

$$\chi_s = \frac{1}{24}\left[\begin{array}{|c|c|c|c|}\hline 3 & -1 & 0 & 0 \\ \hline\end{array}^{\,4} + 6\begin{array}{|c|c|c|c|}\hline 3 & -1 & 0 & 0 \\ \hline\end{array}^{\,2}\right.$$
$$\cdot\begin{array}{|c|c|c|c|}\hline 3 & 3 & 0 & 0 \\ \hline\end{array} + 8\begin{array}{|c|c|c|c|}\hline 3 & -1 & 0 & 0 \\ \hline\end{array}$$
$$\cdot\begin{array}{|c|c|c|c|}\hline 3 & -1 & 3 & 3 \\ \hline\end{array} + 3\begin{array}{|c|c|c|c|}\hline 3 & 3 & 0 & 0 \\ \hline\end{array}^{\,2}$$
$$\left.+6\begin{array}{|c|c|c|c|}\hline 3 & 3 & 0 & 0 \\ \hline\end{array}\right] = \begin{array}{|c|c|c|c|}\hline 15 & 3 & 0 & 0 \\ \hline\end{array}$$

At last, from the table of irreducible characters of the group of tetrahedron rotations, we average χ_s over the group:

$$c_0 = \frac{1}{12}(15+9) = 2.$$

One obtains the same answer using the uniform polynomials of 3 variables.

Exercises

11.25. Find the degeneracy order of molecule CH_3Cl vibrations.

11.26. Find the degeneracy order of molecule NH_3 vibrations.

11.27. Equal masses replace the vertices of a cube. Identical springs change their edges. Find the degeneracy order of the obtained "molecule."

11.28. Show the matrix elements of a scalar operator, between the states transformed by different irreducible representations, to be zero.

11.29. Find the selection rules for a vector operator under group $\mathbf{D}_3$.

11.30. Write all the symmetrization operators for a tensor of rank 4.

11.31. Find the symmetric and antisymmetric parts of the basis of the second-rank tensors $\mathbb{R}^N \otimes \mathbb{R}^N$.

11.32. Find all the invariant subspaces of the basis of the third-rank tensors $\mathbb{R}^N \otimes \mathbb{R}^N \otimes \mathbb{R}^N$.

11.33. Derive a formula for the character of second-rank tensor representation.

11.34. Derive formulas for the characters of fully and partially symmetric parts of the third-rank tensor.

11.35. Derive formulas for the characters of fully and partially symmetric parts of the fourth-rank tensor.

11.36. How many independent components have the second-rank tensor invariant with respect to group $G = \mathbf{SO}(3)$?

11.37. How many independent components does a third-rank tensor, invariant with respect to group $G = \mathbf{D}_4$, have?

11.38. How many independent components does a fourth-rank tensor, invariant with respect to group $G = \mathbf{D}_3$, have?

11.39. How many independent components does a fourth-rank tensor, invariant in group $G = \mathbf{SO}(3)$, have?

11.40. How many components does the partially symmetric third-rank tensor, invariant respect to group $G = \mathbf{SO}(2)$, have?

11.41. How many components does the third-rank tensor, invariant respect to group $G = \mathbf{O}$, have?

11.42. How many components are in the symmetric part of a third-rank tensor in four-dimensional space?

Answers

11.25. 3 single and 3 double.

11.26. 2 single and 2 double.

11.27 2 single, 2 double and 4 triple.

11.28. Hint. Use the orthogonality of irreducible characters.

11.29. The transitions between states, transformed by different one-dimensional representations, are forbidden.

11.30.

$$\hat{S} = \frac{1}{6}\left[\begin{pmatrix}1&2&3\\1&2&3\end{pmatrix} + \begin{pmatrix}1&2&3\\3&1&2\end{pmatrix} + \begin{pmatrix}1&2&3\\2&3&1\end{pmatrix}\right.$$
$$\left. + \begin{pmatrix}1&2&3\\2&1&3\end{pmatrix} + \begin{pmatrix}1&2&3\\1&3&2\end{pmatrix} + \begin{pmatrix}1&2&3\\3&2&1\end{pmatrix}\right],$$
$$\hat{A} = \frac{1}{6}\left[\begin{pmatrix}1&2&3\\1&2&3\end{pmatrix} + \begin{pmatrix}1&2&3\\3&1&2\end{pmatrix} + \begin{pmatrix}1&2&3\\2&3&1\end{pmatrix}\right.$$
$$\left. - \begin{pmatrix}1&2&3\\2&1&3\end{pmatrix} - \begin{pmatrix}1&2&3\\1&3&2\end{pmatrix} - \begin{pmatrix}1&2&3\\3&2&1\end{pmatrix}\right],$$
$$\hat{B} = \frac{1}{3}\left[2\begin{pmatrix}1&2&3\\1&2&3\end{pmatrix} - \begin{pmatrix}1&2&3\\3&1&2\end{pmatrix} - \begin{pmatrix}1&2&3\\2&3&1\end{pmatrix}\right]. \tag{11.33}$$

11.31.

$$\Psi^s_{ik} = \frac{1}{2}\left(|i\rangle \otimes |k\rangle + |k\rangle \otimes |i\rangle\right), \quad \Psi^a_{ik} = \frac{1}{2}\left(|i\rangle \otimes |k\rangle - |k\rangle \otimes |i\rangle\right).$$

11.32.

$$\Psi^s_{ijk} = \frac{1}{6}\Big(|i\rangle \otimes |j\rangle \otimes |k\rangle + |k\rangle \otimes |i\rangle \otimes |j\rangle + |j\rangle \otimes |k\rangle \otimes |i\rangle +$$
$$+|j\rangle \otimes |i\rangle \otimes |k\rangle + |i\rangle \otimes |k\rangle \otimes |j\rangle + |k\rangle \otimes |j\rangle \otimes |i\rangle\Big),$$
$$\Psi^a_{ijk} = \frac{1}{6}\Big(|i\rangle \otimes |j\rangle \otimes |k\rangle + |k\rangle \otimes |i\rangle \otimes |j\rangle + |j\rangle \otimes |k\rangle \otimes |i\rangle +$$
$$-|j\rangle \otimes |i\rangle \otimes |k\rangle - |i\rangle \otimes |k\rangle \otimes |j\rangle - |k\rangle \otimes |j\rangle \otimes |i\rangle\Big),$$
$$\Psi^b_{ijk} = \frac{1}{3}\Big(2|i\rangle \otimes |j\rangle \otimes |k\rangle - |k\rangle \otimes |i\rangle \otimes |j\rangle - |j\rangle \otimes |k\rangle \otimes |i\rangle\Big).$$

11.33.

$$\chi_a(g) = \frac{1}{2}\left(\chi^2(g) - \chi(g^2)\right). \qquad (11.34)$$

11.34.

$$\begin{aligned}
\chi_s(g) &= \frac{1}{6}\left[\chi^3(g) + 3\chi(g)\chi(g^2) + 2\chi(g^3)\right],\\
\chi_a(g) &= \frac{1}{6}\left[\chi^3(g) - 3\chi(g)\chi(g^2) + 2\chi(g^3)\right],\\
\chi_b(g) &= \frac{2}{3}\left[\chi^3(g) - \chi(g^3)\right].
\end{aligned} \qquad (11.35)$$

If we sum up all the lines, we get $\chi_s + \chi_a + \chi_b = \chi^3$, i.e., the character of tensor representation, of rank 3.

11.35.

$$\begin{aligned}
\chi_s(g) &= \frac{1}{24}\left[\chi^4(g) + 6\chi^2(g)\chi(g^2) + 8\chi(g)\chi(g^3) + 3\chi^2(g^2) + 6\chi(g^4)\right],\\
\chi_a(g) &= \frac{1}{24}\left[\chi^4(g) - 6\chi^2(g)\chi(g^2) + 8\chi(g)\chi(g^3) + 3\chi^2(g^2) - 6\chi(g^4)\right],\\
\chi_b(g) &= \frac{3}{8}\left[\chi^4(g) + 2\chi^2(g)\chi(g^2) - \chi^2(g^2) - 2\chi(g^4)\right],\\
\chi_c(g) &= \frac{1}{6}\left[\chi^4(g) - 4\chi(g)\chi(g^3) + 3\chi(g^4)\right],\\
\chi_d(g) &= \frac{3}{8}\left[\chi^4(g) - 2\chi^2(g)\chi(g^2) - \chi^2(g^2) + 2\chi(g^4)\right].
\end{aligned} \qquad (11.36)$$

11.36. 1.

11.37. 3.

11.38. 4.

11.39. The Clebsch–Gordan decomposition gives

$$\begin{aligned}
D_T = D^{(1)} \otimes D^{(1)} \otimes D^{(1)} \otimes D^{(1)} &= (D^{(0)} \oplus D^{(1)} \oplus D^{(2)}) \otimes (D^{(0)} \oplus D^{(1)} \oplus D^{(2)})\\
&= 3D^{(0)} \oplus 6D^{(1)} \oplus 6D^{(2)} + 3D^{(3)} \oplus D^{(4)}.
\end{aligned}$$

Then there are $c_0 = 3$ independent components.

Another way of calculation, the averaging over the group, gives the same results:

$$c_0 = \overline{(1 + 2\cos\varphi)^4(1 - \cos\varphi)} = 3.$$

The rank 4 tensor can be constructed from invariant tensors:

$$T_{ijkl} = a\delta_{ij}\delta_{kl} + b\delta_{il}\delta_{kj} + c\delta_{ik}\delta_{lj}.$$

11.40.

$$\chi_B(\varphi) = \frac{2}{3}\left[(1+2\cos\varphi)^3 - (1+2\cos 3\varphi)\right], \quad c_0 = \overline{\chi_B} = 4.$$

11.41 1.

11.42 Hint. Calculate the number of third-degree polynomials of 4 variables.

Appendix

Special Functions

Euler Gamma Function

$$\Gamma(x) = \int_0^\infty dz\, z^{x-1} e^{-z}, \tag{A.1}$$

$$\Gamma(x+1) = x\,\Gamma(x), \tag{A.2}$$

$$\Gamma(x)\Gamma(1-x) = \frac{\pi}{\sin \pi x}. \tag{A.3}$$

Hypergeometric Functions

The Gaussian hypergeometric function ${}_2F_1$

The differential equation for ${}_2F_1(a, b; c; x)$:

$$x(1-x)\,y'' + \Big[c - (a+b+1)x\Big]\,y' - ab\,y = 0. \tag{A.4}$$

Power series expansion near the point $x = 0$:

$${}_2F_1(a, b; c; x) = 1 + \frac{ab}{c}\frac{x}{1!} + \frac{a(a+1)b(b+1)}{c(c+1)}\frac{x^2}{2!} + \ldots \tag{A.5}$$

Euler transform:

$${}_2F_1(a, b; c; x) = (1-x)^{-b}\,{}_2F_1\left(c-a, b; c; \frac{x}{x-1}\right). \tag{A.6}$$

Integral representation:

$${}_2F_1(a, b; c; x) = \frac{\Gamma(c)}{\Gamma(a)\Gamma(b)} \int_0^1 dt\, \frac{t^{b-1}(1-t)^{c-b-1}}{(1-tx)^a}. \tag{A.7}$$

The confluent hypergeometric function ${}_1F_1$

The differential equation for ${}_1F_1(a;c;x)$:

$$x\,y'' + (c - x)\,y' - a\,y = 0. \tag{A.8}$$

Power series expansion near the point $x = 0$:

$${}_1F_1(a;c;x) = \lim_{b\to\infty} {}_2F_1(a,b;c;x/b) = 1 + \frac{a}{c}\frac{x}{1!} + \frac{a(a+1)}{c(c+1)}\frac{x^2}{2!} + \ldots \tag{A.9}$$

Second solution:

$$y = x^{1-c}{}_1F_1(a - c + 1; 2 - c; x). \tag{A.10}$$

Kummer transform:

$${}_1F_1(a;c;x) = \mathrm{e}^x\,{}_1F_1(c - a; c; -x). \tag{A.11}$$

Integral representation:

$${}_1F_1(a;c;x) = \frac{\Gamma(c)}{\Gamma(a)\Gamma(c-a)} \int_0^1 dt\, t^{a-1}(1-t)^{c-a-1}\mathrm{e}^{xt}, \tag{A.12}$$

$$\mathrm{Re}\, c > \mathrm{Re}\, a > 0\,.$$

Asymptotic behavior:

$$\begin{aligned} {}_1F_1(a;c;x) &\simeq \frac{\Gamma(c)}{\Gamma(a)}\mathrm{e}^x x^{a-c}, \qquad x \to +\infty, \\ {}_1F_1(a;c;x) &\simeq \frac{\Gamma(c)}{\Gamma(c-a)}(-x)^{-a},\ x \to -\infty. \end{aligned} \tag{A.13}$$

The Bessel Functions

Bessel functions J_ν and Neumann functions Y_ν

The differential equation for $J_\nu(x)$:

$$x^2\,y'' + x\,y' + (x^2 - \nu^2)\,y = 0. \tag{A.14}$$

Power series expansion near the point $x = 0$:

$$J_\nu(x) = \sum_{n=0}^{\infty} \frac{(-1)^n (x/2)^{2n+\nu}}{n!\Gamma(n+\nu+1)}. \tag{A.15}$$

Expression in terms of the hypergeometric function:

$$J_\nu(x) = \frac{(x/2)^\nu}{\Gamma(\nu+1)}\mathrm{e}^{-\mathrm{i}x}\,{}_1F_1\left(\nu + \frac{1}{2}; 2\nu + 1; 2\mathrm{i}x\right). \tag{A.16}$$

Recurrence relation:

$$\frac{2\nu}{x}J_\nu(x) = J_{\nu-1}(x) + J_{\nu+1}(x). \tag{A.17}$$

Formulas of differentiation:

$$2\frac{d}{dx}J_\nu(x) = J_{\nu-1}(x) - J_{\nu+1}(x),$$

$$\frac{d}{dx}\left(x^{\pm\nu}J_\nu(x)\right) = \pm x^{\pm\nu}J_{\nu\mp1}(x). \tag{A.18}$$

Integral representations:

$$J_\nu(x) = \frac{1}{2\pi \mathrm{i}}\int_{-\infty}^{(0+)} \frac{dz}{z^{\nu+1}} \exp\left(\frac{x}{2}\left(z - \frac{1}{z}\right)\right)$$

$$= \int_{-\pi}^{\pi} \frac{d\varphi}{2\pi} \mathrm{e}^{\mathrm{i}x\sin\varphi - \mathrm{i}\nu\varphi} - \frac{\sin\pi\nu}{\pi}\int_0^\infty dt\, \mathrm{e}^{-x\sinh t - \nu t}. \tag{A.19}$$

$$J_\nu(x) = \frac{2}{\sqrt{\pi}\,\Gamma(\nu+1/2)}\left(\frac{x}{2}\right)^\nu \int_0^{\pi/2} d\varphi \cos^{2\nu}(\varphi)\cos\left(x\sin(\varphi)\right)$$

$$= \frac{1}{\sqrt{\pi}\,\Gamma(\nu+1/2)}\left(\frac{x}{2}\right)^\nu \int_{-1}^{1} dt\, e^{\mathrm{i}x\,t}(1-t^2)^{\nu-1/2}. \tag{A.20}$$

In the first integral representation, the integration is performed along a path starting and ending at $-\infty$, going around the point $z = 0$ in the positive direction (Fig. A.1). In the Schläfli integral representation (A.19) for an integer ν, only the first term gives the nonzero contribution. In the Poisson integral (A.20) Re $\nu > -1/2$.
Second solution:

$$Y_\nu(x) = \frac{1}{\sin\pi\nu}\left[J_\nu(x)\cos\pi\nu - J_{-\nu}(x)\right]. \tag{A.21}$$

Asymptotic behavior:

$$J_\nu(x) \simeq \sqrt{\frac{2}{\pi x}}\cos\left(x - \frac{\nu\pi}{2} - \frac{\pi}{4}\right), \quad x \to +\infty. \tag{A.22}$$

The case of half-integer index:

$$J_{1/2}(x) = \sqrt{\frac{2}{\pi x}}\sin x, \quad J_{-1/2}(x) = \sqrt{\frac{2}{\pi x}}\cos x. \tag{A.23}$$

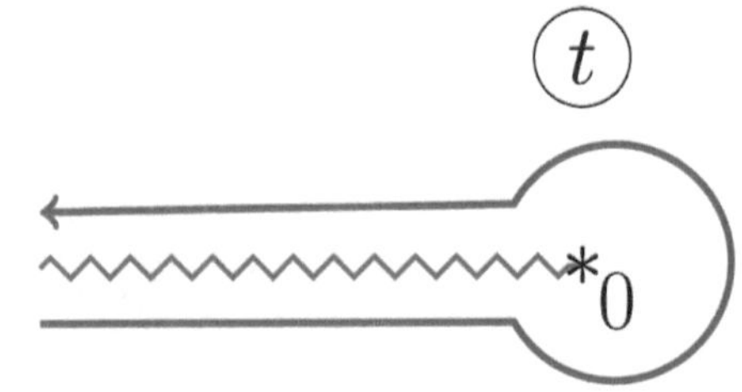

Figure A.1 Contour of integration going around the cut $t \in (-\infty, 0]$ in the positive direction.

Bessel functions J_n of integer order

$$J_{-n}(x) = (-1)^n J_n(x). \tag{A.24}$$

Generating function:

$$\exp\left(\frac{x}{2}\left(z - \frac{1}{z}\right)\right) = \sum_{n=-\infty}^{\infty} z^n J_n(x). \tag{A.25}$$

Orthogonality relations:

$$\int_0^1 dx\, x J_k(\gamma_n x) J_k(\gamma_m x) = \frac{\delta_{nm}}{2}\left(\frac{dJ_k(\gamma_m)}{d\gamma_m}\right)^2, \quad J_k(\gamma_m) = 0, \tag{A.26}$$

$$\int_0^1 dx\, x J_k(\lambda_n x) J_k(\lambda_m x) = \frac{\delta_{nm}}{2}\left(1 - \frac{k^2}{\lambda_m^2}\right) J_k^2(\lambda_m), \quad \frac{dJ_k(\lambda_m)}{d\lambda_m} = 0.$$

Modified Bessel function I_ν and MacDonald function K_ν

Differential equation for $I_\nu(x)$ and $K_\nu(x)$:

$$x^2 y'' + x y' - (x^2 + \nu^2) y = 0. \tag{A.27}$$

Power series expansion near the point $x = 0$:

$$I_\nu(x) = \sum_{n=0}^{\infty} \frac{(x/2)^{2n+\nu}}{n!\Gamma(n+\nu+1)}. \tag{A.28}$$

Expression in terms of ordinary Bessel functions:

$$I_\nu(x) = e^{-i\nu\pi/2} J_\nu(ix). \tag{A.29}$$

Expression for K_ν in terms of I_ν, $I_{-\nu}$:

$$K_\nu(x) = \frac{\pi\,[I_{-\nu}(x) - I_\nu(x)]}{2\,\sin\pi\nu}. \tag{A.30}$$

Integral representations:

$$I_\nu(x) = \frac{(x/2)^\nu}{\sqrt{\pi}\,\Gamma(\nu+1/2)} \int_{-1}^{1} dt\, e^{-xt}(1-t^2)^{\nu-1/2}, \quad \mathrm{Re}\,\nu > -1/2,$$

$$K_\nu(x) = \int_0^\infty dt\, e^{-x\cosh t}\cosh\nu t, \quad \mathrm{Re}\,x > 0, \tag{A.31}$$

$$K_\nu(2\sqrt{pq}) = \frac{1}{2}\left(\frac{p}{q}\right)^{\nu/2} \int_0^\infty x^{\nu-1} e^{-px-q/x}\,dx, \quad \mathrm{Re}\,p > 0, \mathrm{Re}\,q > 0.$$

Asymptotic behavior:

$$I_\nu(x) \simeq \sqrt{\frac{\pi x}{2}}\,e^x, \quad K_\nu(x) \simeq \sqrt{\frac{\pi}{2x}}\,e^{-x}, \quad x \to +\infty. \tag{A.32}$$

$$I_\nu(x) \simeq \frac{(x/2)^\nu}{\Gamma(\nu+1)}, \quad K_0(x) \simeq -\ln x, \quad x \to +0; \tag{A.33}$$

$$K_\nu(x) \simeq \frac{\Gamma(\nu)}{2}\left(\frac{x}{2}\right)^{-\nu}, \quad x \to +0,\ \nu \neq 0. \tag{A.34}$$

Orthogonal Polynomials

Legendre polynomials P_l and associated Legendre functions P_l^m

Differential equation for $P_l(x)$:

$$(1-x^2)\,y'' - 2x\,y' + l(l+1)\,y = 0. \tag{A.35}$$

Differential equation for $P_l^m(x)$:

$$(1-x^2)\,y'' - 2x\,y' + \left(l(l+1) - \frac{m^2}{1-x^2}\right) y = 0. \tag{A.36}$$

Rodrigues' formulas:

$$P_l(x) \equiv P_l^0(x) = \frac{1}{2^l l!}\frac{d^l}{dx^l}(x^2-1)^l, \tag{A.37}$$

$$P_l^m(x) = (1-x^2)^{m/2}\frac{d^m}{dx^m}P_l(x).$$

First three polynomials:

$$P_0(x) = 1, \quad P_1(x) = x, \quad P_2(x) = \frac{3x^2 - 1}{2}. \tag{A.38}$$

Orthogonality relation:

$$\int_{-1}^{1} dx\, P_l^m(x) P_{l'}^m(x) = \frac{2}{2l+1} \frac{(l+m)!}{(l-m)!} \delta_{ll'}. \tag{A.39}$$

Recurrence relation:

$$x(2l+1)P_l(x) = (l+1)P_{l+1}(x) + l P_{l-1}(x). \tag{A.40}$$

Formulas of differentiation:

$$\begin{aligned} (2l+1)P_l(x) &= \frac{d}{dx} P_{l+1}(x) - \frac{d}{dx} P_{l-1}(x), \\ l P_l(x) &= x \frac{d}{dx} P_l(x) - \frac{d}{dx} P_{l-1}(x). \end{aligned} \tag{A.41}$$

Generating functions:

$$\frac{1}{\sqrt{1 - 2xr + r^2}} = \begin{cases} \sum_{l=0}^{\infty} r^l P_l(x), & r < 1; \\ \sum_{l=0}^{\infty} \frac{1}{r^{l+1}} P_l(x), & r > 1; \end{cases} \quad -1 < x < 1. \tag{A.42}$$

Integral representations:

$$P_l(x) = \frac{1}{2\pi i} \int^{(0+)} \frac{dz\, z^{-l-1}}{\sqrt{1 - 2xz + z^2}}, \tag{A.43}$$

$$P_l(\cos\theta) = \frac{1}{\pi} \int_0^{\pi} d\varphi \, (\cos\theta + \mathrm{i}\, \sin\theta\, \cos\varphi)^l . \tag{A.44}$$

In formula (A.43), the integration is performed along a closed contour around point $t = 0$ in the positive direction.
Asymptotic behavior:

$$P_l(\cos\theta) \simeq \sqrt{\frac{2}{\pi l}} \frac{\sin\left[\left(l + \frac{1}{2}\right)\theta + \frac{\pi}{4}\right]}{\sqrt{\sin\theta}}, \quad l|\sin\theta| \gg 1. \tag{A.45}$$

Spherical harmonics Y_{lm}

$$Y_{lm}(\theta,\varphi) = C_{lm} e^{im\varphi} P_l^{|m|}(\cos\theta). \tag{A.46}$$

Differential equations for Y_{lm}:

$$\Delta_\Omega Y_{lm} = -l(l+1)Y_{lm}, \quad i\frac{d}{d\varphi}Y_{lm} = -mY_{lm}, \tag{A.47}$$

where Δ_Ω is the angular part of the three-dimensional Laplace operator in spherical coordinates.
Orthogonality relation:

$$\int \sin\theta\, d\theta\, d\varphi\, Y_{lm}(\theta,\varphi)Y_{l'm'}(\theta,\varphi) = \delta_{ll'}\delta_{mm'}. \tag{A.48}$$

Completeness relation:

$$\sum_{l=0}^{\infty}\sum_{m=-l}^{l} Y_{lm}(\mathbf{n})Y_{lm}^*(\mathbf{n}') = \delta(\mathbf{n}-\mathbf{n}'). \tag{A.49}$$

Hermite polynomials H_n

Differential equation for $H_n(x)$:

$$y'' - 2x\,y' + 2n\,y = 0. \tag{A.50}$$

Rodrigues' formula:

$$H_n(x) = (-1)^n e^{x^2}\frac{d^n}{dx^n}e^{-x^2}. \tag{A.51}$$

First three polynomials:

$$H_0(x) = 1, \quad H_1(x) = 2x, \quad H_2(x) = 4x^2 - 2. \tag{A.52}$$

Orthogonality relation:

$$\int_{-\infty}^{\infty} dx\, e^{-x^2} H_m(x)H_n(x) = \sqrt{\pi}\,2^n n!\,\delta_{mn}. \tag{A.53}$$

Completeness relation:

$$\frac{e^{-(x^2+x'^2)/2}}{\sqrt{\pi}}\sum_{n=0}^{\infty}\frac{H_n(x)H_n(x')}{2^n n!} = \delta(x-x'). \tag{A.54}$$

Recurrence relation:

$$H_{n+1}(x) - 2xH_n(x) + 2nH_{n-1}(x) = 0. \tag{A.55}$$

Formula of differentiation:

$$\frac{d}{dx}H_n(x) = 2nH_{n-1}(x). \tag{A.56}$$

Generating function:

$$\exp\left(2xz - z^2\right) = \sum_{n=0}^{\infty} \frac{z^n}{n!} H_n(x). \tag{A.57}$$

Integral representation:

$$\begin{aligned} H_n(x) &= \frac{2^{n+1}e^{x^2}}{\sqrt{\pi}} \int_0^{\infty} dz\, z^n e^{-z^2} \cos\left(2xz - \frac{n\pi}{2}\right) \\ &= \frac{2^n}{\sqrt{\pi}} \int_{-\infty}^{\infty} (x+it)^n\, e^{-t^2}\, dt. \end{aligned} \tag{A.58}$$

Laguerre polynomials L_n^{ν}

Differential equation for $L_n^{\nu}(x)$:

$$x\,y'' + (\nu + 1 - x)\,y' + n\,y = 0. \tag{A.59}$$

Rodrigues' formula:[4]

$$L_n^{\nu}(x) = \frac{x^{-\nu}e^x}{n!}\frac{d^n}{dx^n} e^{-x} x^{n+\nu}. \tag{A.60}$$

First three polynomials:

$$\begin{aligned} L_0^{\nu}(x) = 1, \quad L_1^{\nu}(x) = \nu + 1 - x, \\ L_2^{\nu}(x) = \tfrac{(\nu+1)(\nu+2)}{2} - (\nu+2)x + \tfrac{x^2}{2}. \end{aligned} \tag{A.61}$$

Orthogonality relation:

$$\int_0^{\infty} dx\, e^{-x} x^{\nu} L_m^{\nu}(x) L_n^{\nu}(x) = \frac{\Gamma(n+\nu+1)}{n!}\,\delta_{mn}. \tag{A.62}$$

[4] In the book (Landau and Lifshitz, 1977), the generalized Laguerre polynomials are defined differently.

Completeness relation:

$$(xx')^{\nu/2}\mathrm{e}^{-(x+x')/2}\sum_{n=0}^{\infty}\frac{n!L_n^{\nu}(x)L_n^{\nu}(x')}{\Gamma(n+\nu+1)}=\delta(x-x'). \tag{A.63}$$

Recurrence relation:

$$(n+1)L_{n+1}^{\nu}(x)-(2n+\nu+1-x)L_n^{\nu}(x)+(n+\nu)L_{n-1}^{\nu}(x)=0. \tag{A.64}$$

Formulas of differentiation:

$$x\frac{d}{dx}L_n^{\nu}(x)=nL_n^{\nu}(x)-(n+\nu)L_{n-1}^{\nu}(x), \tag{A.65}$$

$$\frac{d}{dx}L_n^{\nu}(x)=-L_{n-1}^{\nu+1}(x).$$

Generating function:

$$(1-z)^{-\nu-1}\exp\left(\frac{xz}{z-1}\right)=\sum_{n=0}^{\infty}z^nL_n^{\nu}(x). \tag{A.66}$$

Integral representation:

$$L_n^{\nu}(x)=\frac{1}{2\pi\mathrm{i}}\int^{(0+)}\left(1+\frac{x}{t}\right)^n\mathrm{e}^{-t}\left(1+\frac{t}{x}\right)^{\nu}\frac{dt}{t}$$

$$=\frac{(-1)^n}{2\pi\mathrm{i}}\int^{(0+)}\frac{(1-t)^{n+\nu}}{t^n}\mathrm{e}^{tx}\frac{dt}{t}. \tag{A.67}$$

Here the integration is performed along a closed contour (Fig. A.1) in the positive direction around $t=0$.

Tables

Irreducible Characters of Permutation Groups

Conjugacy classes are market by curly braces $\{1^{\tau_1}\dots n^{\tau_n}\}$. Number $i=1,\dots,n$ means the length of cycle, τ_i is the number of cycles with given length in the class, $\sum_{k=1}^{n}k\tau_k=n$, h denotes the order of class $\{1^{\tau_1}\dots n^{\tau_n}\}$, given by combinatorics formula:

$$h=n!\left[\prod_{k=1}^{n}\tau_k!k^{\tau_k}\right]^{-1}.$$

The irreducible representations are marked by square brackets: $[1^{\tau_1}\dots n^{\tau_n}]$

S_2	$\{1^2\}$	$\{2\}$
h	1	1
[2]	1	1
$[1^2]$	1	-1

S_3	$\{1^3\}$	$\{12\}$	$\{3\}$
h	1	3	2
[3]	1	1	1
[12]	2	0	-1
$[1^3]$	1	-1	1

S_4	$\{1^4\}$	$\{1^2 2\}$	$\{13\}$	$\{2^2\}$	$\{4\}$
h	1	6	8	3	6
[4]	1	1	1	1	1
[13]	3	1	0	-1	-1
$[2^2]$	2	0	-1	2	0
$[1^2 2]$	3	-1	0	-1	1
$[1^4]$	1	-1	1	1	-1

S_5	$\{1^5\}$	$\{1^3 2\}$	$\{1^2 3\}$	$\{14\}$	$\{12^2\}$	$\{23\}$	$\{5\}$
h	1	10	20	30	15	20	24
[5]	1	1	1	1	1	1	1
[14]	4	2	1	0	0	-1	-1
[23]	5	1	-1	-1	1	1	0
$[1^2 3]$	6	0	0	0	-2	0	1
$[12^2]$	5	-1	-1	1	1	-1	0
$[1^3 2]$	4	-2	1	0	0	1	-1
$[1^5]$	1	-1	1	-1	1	-1	1

Irreducible Characters of Point Groups

In the first line of each table the conjugacy classes σ_i are listed. E is unity, C_n is the axis of n-th order. The the number of elements in the class $h_i = |\sigma_i|$ is indicated in front of axis C_n name. If there are several axes of the same order in the group, then the second and the third are marked by primes, e.g., C_2, C_2', C_2'' etc., $\varepsilon = e^{2\pi i/3}$.

$\mathbf{C}_2$	E	C_2
χ_0	1	1
χ_1	1	-1

$\mathbf{D}_2$	E	C_2	C_2'	C_2''
χ_0	1	1	1	1
χ_1	1	-1	1	-1
χ_2	1	1	-1	-1
χ_3	1	-1	-1	1

$\mathbf{D}_3$	E	$2C_3$	$3C_2$
χ_0	1	1	1
χ_1	1	1	−1
χ_2	2	−1	0

$\mathbf{D}_4$	E	C_4^2	$2C_4$	$2C_2$	$2C_2'$
χ_0	1	1	1	1	1
χ_1	1	1	1	−1	−1
χ_2	1	1	−1	1	−1
χ_3	1	1	−1	−1	1
χ_4	2	−2	0	0	0

$\mathbf{T}$	E	$3C_2$	$4C_3$	$4C_3^2$
χ_0	1	1	1	1
χ_1	1	1	ε	ε^2
χ_2	1	1	ε^2	ε
χ_3	3	−1	0	0

$\mathbf{O}$	E	$8C_3$	$3C_4^2$	$6C_2$	$6C_4$
χ_0	1	1	1	1	1
χ_1	1	1	1	−1	−1
χ_2	2	−1	2	0	0
χ_3	3	0	−1	−1	1
χ_4	3	0	−1	1	−1

$\mathbf{Y}$	E	$15C_2$	$20C_3$	$12C_5$	$12C_5^2$
χ_0	1	1	1	1	1
χ_1	3	−1	0	$\frac{1+\sqrt{5}}{2}$	$\frac{1-\sqrt{5}}{2}$
χ_2	3	−1	0	$\frac{1-\sqrt{5}}{2}$	$\frac{1+\sqrt{5}}{2}$
χ_3	4	0	1	−1	−1
χ_4	5	1	−1	0	0

Bibliography

Ablowitz, M. J., Fokas, A. S., and Fokas, A. S. (2003). *Complex Variables: Introduction and Applications* (Cambridge University).

Ablowitz, M. J., and Segur, H. (1981). *Solitons and the Inverse Scattering Transform* (SIAM, Philadelphia).

Abramowitz, M., and Stegun, I. A. (1965). *Handbook of Mathematical Functions with Formulas, Graphs, and Mathematical Tables* (U.S. Govt. Print. Off., Washington).

Ahlfors, L. V. (1953). *Complex Analysis: An Introduction to the Theory of Analytic Functions of One Complex Variable*, Vol. 177 (McGraw-Hill, New York, London).

Arfken, G. B., and Weber, H. J. (1999). *Mathematical Methods for Physicists* (Elsevier, California—Massachusetts—London).

Arnold, V. I. (1983). *Geometrical Methods in the Theory of Ordinary Differential Equations* (Springer-Verlag, New York).

Arnold, V. I. (1992). *Ordinary Differential Equations: With 272 Figures* (Springer-Verlag, Berlin—New York).

Barut, A. O., and Raczka, R. (1977). *Theory of Group Representations and Applications* (Polish Scientific, Warszawa).

Bateman, H., and Erdélyi, A. (1953-1955). *Higher Transcendental Functions* (McGraw-Hill, New York).

Bender, C. M., Orszag, S., and Orszag, S. A. (1999). *Advanced Mathematical Methods for Scientists and Engineers I: Asymptotic Methods and Perturbation Theory*, Vol. 1 (Springer-Verlag, Berlin—New York—Heidelberg).

Bogoliubov, N. N., and Mitropolsky, Y. A. (1961). *Asymptotic Methods in the Theory of Non-Linear Oscillations* (Gordon and Breach Science Publishers, New York).

Brown, J. W., and Churchill, R. V. (2009). *Complex Variables and Applications* (McGraw-Hill).

Cole, J. D. (1968). *Perturbation Methods in Applied Mathematics* (Blaisdell Pub. Co., Waltham, Massachusetts).

Courant, R., and Hilbert, D. (1989). *Methods of Mathematical Physics* (Wiley, New York).

Dirac, P. A. M. (1967). *The Principles of Quantum Mechanics* (Clarendon Press, Oxford).

Dubrovin, B. A., Fomenko, A. T., and Novikov, S. P. (1992). *Modern Geometry—Methods and Applications* (Springer-Verlag, New York).

Duffy, D. G. (2015). *Green's Functions with Applications* (Chapman and Hall/CRC, New York).

Erdélyi, A. (1952). *Asymptotic Expansions* (Dover, New York).

Fedoriuk, M. V. (1977). *Method of Steepest Descent (in Russian)* (Nauka, Moscow).

Hall, M. (1999). *The Theory of Groups* (AMS Chelsea, Providence, R.I.).

Hamermesh, M. (1989). *Group Theory and Its Application to Physical Problems* (Dover, New York).

Heading, J. (1962). *An Introduction to Phase-Integral Methods* (Wiley, London).

Heine, V. (1977). *Group Theory in Quantum Mechanics* (Pergamon Press, Oxford—Toronto).

Jahnke, E., Emde, F., and Lösch, F. (1960). *Tables of Higher Functions* (McGraw-Hill, New York).

Kabwal, R. P. (1971). *Linear Integral Equations* (Academic Press, London).

Kamke, E. (1969). *Differentialgleichungen* (Akademische Verlagsgesellschaft, Leipzig).

Kolmogorov, A. N., and Fomin, S. V. (1963). *Elements of the Theory of Functions and Functional Analysis* (Graylock Press, Rochester, New York).

Landau, L. D., and Lifshitz, E. M. (1977). *Quantum Mechanics: Non-Relativistic Theory* (Pergamon Press, New York).

Landau, L. D., and Lifshitz, E. M. (1987). *Fluid Mechanics* (Pergamon Press, New York).

Mathews, J., and Walker, R. L. (1970). *Mathematical Methods of Physics* (W.A. Benjamin, New York).

Messiah, A. (1962). *Quantum Mechanics* (NHPC, Amsterdam).

Mikhlin, S. G. (1960). *Linear Integral Equations* (Hindustan Pub. Corp., Delhi).

Mikhlin, S. G. (1967). *Linear Equations of Mathematical Physics* (Holt, Rinehart and Winston, New York).

Morse, P. M., and Feshbach, H. (1953). *Methods of Theoretical Physics* (McGraw-Hill, New York).

Nayfeh, A. H. (1985). *Problems in Perturbation* (Wiley, New York).

Newell, A. C. (1985). *Solitons in Mathematics and Physics* (SIAM, Philadelphia).

Novikov, S., Manakov, S. V., Pitaevskii, L. P., and Zakharov, V. E. (1984). *Theory of Solitons: The Inverse Scattering Method* (Springer Science & Business Media, New York).

Olver, F. W. J. (1974). *Introduction to Asymptotics and Special Functions* (Academic Press, New York).

Ovler, F. W. J., Lozier, D. W., Boisvert, R. F., and Clark, C. W. (2010). *NIST Handbook of Mathematical Functions* (Cambridge University, New York).

Perelomov, A. M., and Zel'dovich, Y. B. (1998). *Quantum Mechanics: Selected Topics* (World Scientific, Singapore; River Edge, NJ).

Petrashen, M. I. (1969). *Applications of Group Theory in Quantum Mechanics* (MIT Press, Cambridge, Massachusetts).

Pipkin, A. C. (1991). *A Course on Integral Equations* (Springer-Verlag, New York).

Polyanin, A. D., and Manzhirov, A. V. (2008). *Handbook of Integral Equations* (Chapman & Hall, London).

Polyanin, A. D., and Zaitsev, V. F. (2012). *Handbook of Nonlinear Partial Differential Equations* (Chapman & Hall, London—New York).

Richtmyer, R. D. (1978). *Principles of Advanced Mathematical Physics* (Springer-Verlag, New York).

Roach, G. F. (1992). *Green's Functions*, 2nd ed. (Cambridge University Press, Cambridge).

Sanders, J. A., Verhulst, F., and Murdock, J. (2007). *Averaging Methods in Nonlinear Dynamical Systems*, Vol. 59 (Springer, New York).

Smirnov, V. I. (1964). *A Course of Higher Mathematics, Volumes 1-5* (Pergamon, Oxford).

Sobolev, S. L. (1964). *Partial Differential Equations of Mathematical Physics* (Pergamon Press, New York).

Sommerfeld, A. (1949). *Partial Differential Equations in Physics* (Academic Press, New York).

Stakgold, I., and Holst, M. J. (2011). *Green's Functions and Boundary Value Problems* (John Wiley & Sons, New Jersey).

Tricomi, F. (1957). *Integral Equations* (Interscience Publishers, New York).

Tychonov, A. N., and Samarski, A. A. (1964). *Partial Differential Equations of Mathematical Physics* (Holden-Day, San Francisco).

Vladimirov, V. S. (1971). *Equations of Mathematical Physics* (M. Dekker, New York).

Whitham, G. B. (1974). *Linear and Nonlinear Waves* (Wiley, New York).

Wigner, E. P. (1959). *Group Theory and Its Application to the Quantum Mechanics of Atomic Spectra* (Academic Press, New York).

Zel'dovich, Y. B., and Raizer, Y. P. (1966). *Physics of Shock Waves and High-Temperature Hydrodynamic Phenomena* (Academic Press, New York).

Zhelobenko, D. (2006). *Principal Structures and Methods of Representation Theory* (AMS, Providence, R.I.).

Index